AF413449

Essentials of Quantum Mechanics and Relativity

Other Related Titles from World Scientific

Application-Driven Quantum and Statistical Physics: A Short Course for Future Scientists and Engineers
Volume 3: Transitions
by Jean-Michel Gillet
ISBN: 978-1-78634-788-6
ISBN: 978-1-78634-801-2 (pbk)

The Basic Physics of Quantum Theory
by Basil S Davis
ISBN: 978-981-121-939-9
ISBN: 978-981-121-995-5 (pbk)

General Relativity: A First Examination
Second Edition
by Marvin Blecher
ISBN: 978-981-122-043-2
ISBN: 978-981-122-108-8 (pbk)

Loop Quantum Gravity for Everyone
by Rodolfo Gambini and Jorge Pullin
ISBN: 978-981-121-195-9

Essentials of Quantum Mechanics and Relativity

Shangwu Qian

Peking University, China

NEW JERSEY · LONDON · SINGAPORE · BEIJING · SHANGHAI · HONG KONG · TAIPEI · CHENNAI · TOKYO

Published by

World Scientific Publishing Co. Pte. Ltd.
5 Toh Tuck Link, Singapore 596224
USA office: 27 Warren Street, Suite 401-402, Hackensack, NJ 07601
UK office: 57 Shelton Street, Covent Garden, London WC2H 9HE

Library of Congress Control Number: 2020947395

British Library Cataloguing-in-Publication Data
A catalogue record for this book is available from the British Library.

Essentials of Quantum Mechanics and Relativity
Qian Shangwu
This edition is published by World Scientific Publishing Company Pte Ltd by arrangement with Science Press, Beijing, China.
All rights reserved. No reproduction and distribution without permission.

ISBN 978-981-122-118-7 (hardcover)
ISBN 978-981-122-119-4 (ebook for institutions)
ISBN 978-981-122-120-0 (ebook for individuals)

For any available supplementary material, please visit
https://www.worldscientific.com/worldscibooks/10.1142/11854#t=suppl

Desk Editor: Ng Kah Fee

Typeset by Stallion Press
Email: enquiries@stallionpress.com

Preface

This book abides by the less-is-more principle and adopts an approach that is easy to understand. It introduces in a concise manner the essentials of quantum mechanics and relativity, and along the way some related basic mathematical knowledge.

This book contains two independent parts: the first is the essentials of quantum mechanics. It is my lecture notes of the course 4310.001, given in the 1990 fall semester (14 weeks) for senior students of Physics Department of the University of North Texas (UNT). The second part is the essentials of relativity, which is my lecture notes of the course 4900, given in the 1991 spring semester (14 weeks) for graduate students of Physics Department of UNT. Both lectures received the highest student evaluations from the students of Physics Department of UNT.

I am greatly indebted to Mr. Zhu Zhaoxuan (诸兆轩) for the arrangement and preparation of the manuscripts based on my lecture notes of the course 4310.001. Thanks very much for the financial support coming from the School of Physics, Peking University.

This book is written mainly for the readers in the training and provides necessary elementary knowledge of the two pillars of modern physics: quantum mechanics and relativity.

Qian Shangwu
School of Physics, Peking University

Contents

Relativity   117

PART 1
Quantum Mechanics

Chapter 1

Preliminaries

1.1 Introduction

Firstly we want to answer three basic questions: 1) What is quantum mechanics (QM)? 2) Why do we need QM? 3) How do QM explain atomic phenomena? That is to say, we want to answer the 3 W's (what, why, how) about QM.

1.1.1 *What is QM?*

In short, QM is the mechanics of atomic phenomena, i.e. it is a fundamental theory of microscopic phenomena, which can *describe*, *correlate* and *predict* the behavior of a vast range of physical systems, from elementary particles, through nuclei, atoms and radiation, to molecules and condensed matters. Newtonian mechanics can be considered as its limiting case for macroscopic phenomena.

It is well known that every *viable* theory should contain four necessary elements: *self-consistency*; *completeness*; *agreement with past experiments*; and *power of explanation and prediction of new phenomena*. Both Newtonian mechanics and QM satisfy the above requirement, hence they are both viable theories.

QM was invented as the culmination of the effort to understand atomic and radiation phenomena. The entire edifice of QM is constructed in the mid-twenties of 20th century. Though the achievement of QM is immense, i.e. it is very successful in explaining microscopic phenomena, its very foundation is still in debate, and there are many controversies up till now. In short, the very foundation of QM is still very difficult to understand. A very famous physicist in the last century, R.P. Feynman, in his book *The Character of Physical Law* (1965) has a famous comment about this situation:

3

"I think I can safely say that nobody understands QM." Surely, this is the situation for the very foundation up till now, but the fundamental principles and essential methods of QM are not difficult to grasp. In this chapter we shall prepare some preliminaries for understanding the essentials of QM and applying it to explain microscopic phenomena.

1.1.2   *Why do we need QM?*

It has been over 300 years since Newton laid the foundations of his mechanics, and more than a hundred years since Maxwell formulated the equations of electromagnetic theory, i.e. the Maxwell equations. Since the spectacular and uniform success of Newtonian mechanics and Maxwell equations, at the end of 19th century, it was generally believed that they were the complete and final expressions of the basic laws of physics, only the details, the minor refinements and more exact solutions remained to be worked out. Some physicists even said that there was nothing left to be discovered by the next generations. But the situation soon changed since the very beginning of 20th century. Physics at the turn of the century was in a state of turmoil, there were so many experimental facts which could not be explained by classical physics, i.e. Newtonian mechanics and Maxwell equations. There was a Pandora's Box of experimental observations, which, on the grounds of classical theory, were totally inexplicable.

There are two kinds of experiments in the Pandora's Box which would compel a major theoretical change in physics. One is the wave–particle duality, i.e. wave and particle natures are considered complementary aspects of matter and radiation; they are equally essential for a full description of matter and radiation. The other is the discreteness of observables of physical quantities, such as energy, momentum, angular momentum etc. We shall talk about these experiments later. In order to explain these experimental facts, QM was invented in the mid-twenties. It can successfully explain these effects and, furthermore, it can predict a lot of new phenomena which are confirmed by experiments afterwards. Therefore, QM is a very powerful theory for treating microscopic phenomena and predicting new experimental results and we need it. Now, we can even say without QM there would be no physics.

For natural science, we know the principle: "only experiment can be the criterion of truth". By the vast number of experiments, we can definitely say without any doubt that QM is correct. Perhaps you would ask such questions as "Will QM survive?" and "Will QM be modified as

Newtonian mechanics has been?" The answers should be "Yes". The history of science suggests that we should not accept any theory as necessarily final, however beautiful and however reasonable it may seem. Actually we can only say that all scientific theories are relative truths, not the absolute truth. Theories are developed such that they can approach closer and closer the absolute truth but never arrive at the destination. The absolute truth can be imagined as an infinite series, all existing theories only contain finite terms of the series. The modifications of existing viable theories can only increase the included number of terms of the series, i.e. increase the precision of approximation of the absolute truth.

1.1.3 *How do QM explain atomic phenomena?*

The core of QM comprises fundamental laws or fundamental assumptions. This is the general characteristic for all mechanics, e.g. for classical mechanics (CM), the fundamental laws are Newton's three laws; for thermodynamics, they are the zeroth law, first law, second law and third law of thermodynamics; for electrodynamics, they are the Maxwell equations and Lorentz force law. Usually there are various formulations for fundamental laws, just as the case that there are various languages for the same statement. For CM, besides Newton's formulation, there are Lagrangian formulation, Hamiltonian formulation and some other formulations. For QM, there are also a few different formulations. In this book, we mainly use the so-called Schrödinger's formulation; in this formulation, the fundamental law is the Schrödinger equation, which is

$$i\hbar\frac{\partial\psi}{\partial t} = \hat{H}\psi, \tag{1.1}$$

where $\hbar = \frac{h}{2\pi}$, $h = 6.62\times10^{-34}$ joule-seconds is the famous Planck constant, $\hat{H}$ is the so-called Hamiltonian operator, ψ is the so-called wavefunction. The microscopic system is completely described by the wavefunction; the fundamental problem in QM is to find the wavefunction from Schrödinger equation as a function of time under certain given conditions. Whereas in CM, the macroscopic system is completely described by a set of generalized coordinates $(q_1, q_2, \ldots, q_n)$; the fundamental problem in CM is to find $(q_1, q_2, \ldots, q_n)$ as a function of time under certain given conditions.

When we have found the wavefunction ψ, then we can give the answer about the probability to find this system in certain state. For one particle, it gives the probability $\psi\psi^*\mathrm{d}\tau$ of finding this particle in certain volume element $\mathrm{d}\tau$. It should be emphasized now that QM can only give the probability rather than the exact value, i.e. QM can only solve the problem statistically, not deterministically.

For a firm understanding of QM, we shall give a brief review about the elements of CM in the next section, and give a historical review about the early experiments and theories of microscopic phenomena in Sec. 1.3.

1.2   Review of some basic concepts of CM

1.2.1   *Generalized coordinates*

A particle constrained to move on a given curve has one degree of freedom, since only one variable s (distance from a chosen fixed point O) is needed to uniquely specify the location of the particle in space. For a particle constrained to move on a plane, we need two variables (x, y) or (r, θ) to uniquely specify its location; we say that it has two degrees of freedom (DOF). Evidently, if a particle is constrained to a three-dimensional space, its DOF is three. In short, the number of variables needed to locate a definite system is called the DOF of this system. A rigid body is fixed by 6 variables, three for the center of mass, two for the orientation of the axis through the center of mass and one for the angle of rotation about this axis, hence the DOF of rigid body is 6. Now we shall give the definition of generalized coordinates of a given system. The independent coordinates which serve to uniquely determine the exact location of a given system are called generalized coordinates. The number of generalized coordinates is equal to the number of DOF. Hence, for a system with n DOF, the corresponding generalized coordinates will be $q_1, q_2, \ldots, q_n$, and the time derivatives $\dot{q}_1, \dot{q}_2, \ldots, \dot{q}_n$ are called the generalized velocities, where $\dot{q}_i = \frac{\mathrm{d}}{\mathrm{d}t} q_i$.

1.2.2   *The Lagrange equations*

For a conservative system with kinetic energy T, there exists a potential function $V(q_1, q_2, \ldots, q_n)$, then the Lagrangian of this system is defined as

$$L = T - V = L(q_1, q_2, \ldots, q_n; \dot{q}_1, \dot{q}_2, \ldots, \dot{q}_n), \tag{1.2}$$

and the so-called Lagrange equations are

$$\frac{\mathrm{d}}{\mathrm{d}t}\left(\frac{\partial L}{\partial \dot{q}_i}\right) - \frac{\partial L}{\partial q_i} = 0. \tag{1.3}$$

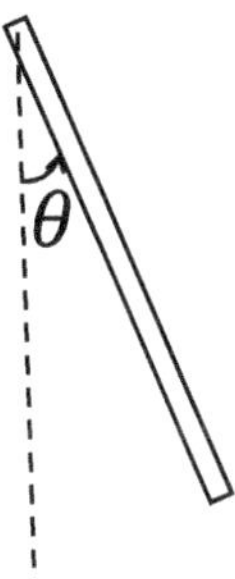

Fig. 1.1   Compound pendulum.

This is the Lagrange formulation of CM.

Example 1.1. Let us consider the compound pendulum. A uniform rod of length $2a$ and mass m is free to rotate about a horizontal axis through one end, as shown in Fig. 1.1.

In this case, the DOF is one, the generalized coordinate is θ, the kinetic energy T is

$$T = \frac{1}{2}I\dot{\theta}^2 = \frac{2}{3}ma^2\dot{\theta}^2, \tag{1.4}$$

where $I = \frac{4}{3}ma^2$ is the moment of inertia of the compound pendulum. The potential energy is

$$V = -mga\cos\theta. \tag{1.5}$$

Hence, we have

$$L - \frac{2}{3}ma^2\dot{\theta}^2 + mga\cos\theta. \tag{1.6}$$

Then the Lagrange equation is

$$\frac{4}{3}a\ddot{\theta} + g\sin\theta = 0. \tag{1.7}$$

For small oscillation $\sin\theta \approx \theta$, we have

$$\ddot{\theta} + \frac{3g}{4a}\theta = \ddot{\theta} + \omega^2\theta = 0. \tag{1.8}$$

The period is

$$T = \frac{2\pi}{\omega} = 2\pi\sqrt{\frac{4a}{3g}}. \tag{1.9}$$

1.2.3 *Hamilton's equations*

Firstly we generalize the concept of momentum. The Lagrangian of a particle moving in a conservative field is

$$L = \frac{m}{2}(\dot{x}^2 + \dot{y}^2 + \dot{z}^2) - V(x, y, z). \tag{1.10}$$

Since the three components of momentum for the rectangular coordinates (x, y, z) are respectively

$$p_x = m\dot{x} = \frac{\partial L}{\partial \dot{x}}, \quad p_y = m\dot{y} = \frac{\partial L}{\partial \dot{y}}, \quad p_z = m\dot{z} = \frac{\partial L}{\partial \dot{z}}, \tag{1.11}$$

we introduce the generalized momentum p_i for generalized coordinate q_i as

$$p_i = \frac{\partial L}{\partial \dot{q}_i}. \tag{1.12}$$

Example 1.2. Consider a particle moving along a circle in a horizontal plane with radius r. In this case the kinetic energy $T = \frac{1}{2}mv^2 = \frac{1}{2}m(r\dot{\theta})^2 = \frac{1}{2}mr^2\dot{\theta}^2$, potential energy $V = 0$, hence $L = T = \frac{1}{2}mr^2\dot{\theta}^2$, then the generalized momentum conjugate to the rotating angle θ is

$$p_\theta = \frac{\partial L}{\partial \dot{\theta}} = mr^2\dot{\theta} = mvr, \tag{1.13}$$

i.e. the angular momentum.

Example 1.3. For the compound pendulum discussed above, the generalized momentum conjugate to θ is

$$p_\theta = \frac{\partial L}{\partial \dot{\theta}} = I\dot{\theta} = \frac{4}{3}ma^2\dot{\theta}. \tag{1.14}$$

The Hamiltonian function is defined as

$$H = \sum_i p_i\dot{q}_i - L = H(q_1, q_2, \ldots, q_n;\ p_1, p_2, \ldots, p_n) = H(q_i, p_i). \tag{1.15}$$

Only in a conservative field, Hamiltonian equals total energy, i.e. $H = T + V$. It must be emphasized that Lagrangian is a function of generalized coordinates and generalized velocities, whereas Hamiltonian is a function of generalized coordinates and generalized momenta.

The so-called Hamilton's equations are

$$\frac{\partial H}{\partial q_i} = -\dot{p}_i, \qquad \frac{\partial H}{\partial p_i} = \dot{q}_i. \tag{1.16}$$

This is the Hamilton formulation of CM.

Example 1.4. Consider a particle with mass m moving under gravity, then

$$L = \frac{m}{2}(\dot{x}^2 + \dot{y}^2 + \dot{z}^2) - mgz, \tag{1.17}$$

$$p_x = \frac{\partial L}{\partial \dot{x}} = m\dot{x}, \quad p_y = \frac{\partial L}{\partial \dot{y}} = m\dot{y}, \quad p_z = \frac{\partial L}{\partial \dot{z}} = m\dot{z} \tag{1.18}$$

and

$$H = T + V = \frac{1}{2m}(p_x^2 + p_y^2 + p_z^2) + mgz. \tag{1.19}$$

In the conservative field, the total energy $E = T + V = H(q_i, p_i)$, hence from Hamilton's equations we have

$$\frac{dE}{dt} = \frac{dH}{dt} = \sum_j \left(\frac{\partial H}{\partial q_j}\dot{q}_j + \frac{\partial H}{\partial p_j}\dot{p}_j \right) = \sum_j \left(\frac{\partial H}{\partial q_j}\frac{\partial H}{\partial p_j} - \frac{\partial H}{\partial p_j}\frac{\partial H}{\partial q_j} \right) = 0, \tag{1.20}$$

and therefore, total energy is conserved.

1.2.4 *Poisson brackets*

If there are two functions $F(q_i, p_i)$ and $G(q_i, p_i)$, then we can define the so-called Poison bracket $\{\cdot, \cdot\}$ such that

$$\{F, G\} = \sum_j \left(\frac{\partial F}{\partial q_j}\frac{\partial G}{\partial p_j} - \frac{\partial F}{\partial p_j}\frac{\partial G}{\partial q_j} \right). \tag{1.21}$$

Obviously we have

$$\{F, G\} = -\{G, F\}, \tag{1.22}$$

$$\{F, F\} = 0, \tag{1.23}$$

and

$$\{q_i, q_j\} = 0, \tag{1.24}$$

$$\{p_i, p_j\} = 0, \tag{1.25}$$

$$\{q_i, p_j\} = \delta_{ij}, \tag{1.26}$$

where δ_{ij} is the Kronecker delta symbol: it equals zero when $i \neq j$, equals one when $i = j$.

From the definition of Poisson bracket, we have

$$\frac{\mathrm{d}F}{\mathrm{d}t} = \sum_j \left(\frac{\partial F}{\partial q_j} \dot{q}_j + \frac{\partial F}{\partial p_j} \dot{p}_j \right) = \{F, H\}. \tag{1.27}$$

From the above equation we easily obtain

$$\dot{p}_i = \{p_i, H\} = -\frac{\partial H}{\partial q_i}, \tag{1.28}$$

$$\dot{q}_i = \{q_i, H\} = \frac{\partial H}{\partial p_i}, \tag{1.29}$$

i.e. the Hamilton's equations. Obviously, if $\{F, H\} = 0$, then F is a constant of the motion.

1.3 Historical review: Experiments and theories

We have already mentioned that around the end of the 19th century, it was generally accepted that classical theories are the complete and final theories of nature, i.e. they are the ultimate description of nature, hence theoretically there are nothing left to be discovered by the next generation. From the physical point of view, *nature* is composed of *matter* and *radiation*. At that time physicists acknowledge that matter is described by Newtonian mechanics, while radiation is described by Maxwell's equations, and that Newtonian mechanics and Maxwell's equations are two pillars of classical theories, which are unshakable. But exceeding all expectations, at the turn of the century new experimental facts about structure of matter, structure of radiation, interaction between radiation and matter have sprung up like mushrooms, which force physicists to give up the above-mentioned classical picture of nature, and have no choice but to accept some fundamentally different new ideas. The foundations of physics were radically re-examined, and the first quarter of the 20th century was to see the development of the theories of relativity and quantum mechanics. This section attempts to trace the history of the development of quantum mechanics, and to show how the obvious failures of classical theory were overcome by quite a few physicists in the frontiers of physics. These failures and new ideas comprise the genesis of QM.

Firstly let us list the important dates related to the development of QM in Table 1.1.

Table 1.1   The development of QM.

Date	Physicist	New idea
1900	Planck, Max	Quantum hypothesis
1905	Einstein, Albert	Photon hypothesis
1913	Bohr, Niels	Quantum theory of spectra
1922	Compton, Arthur	Collision of photon and free electron
1924	de Broglie, Louis	Matter wave
1925	Schrödinger, Erwin	Wave equation
1926	Heisenberg, Werner	Uncertainty principle
1927	Davisson, Clinton and Germer, Lester	Diffraction pattern of electrons
1927	Born, Max	Interpretation of the wavefunction

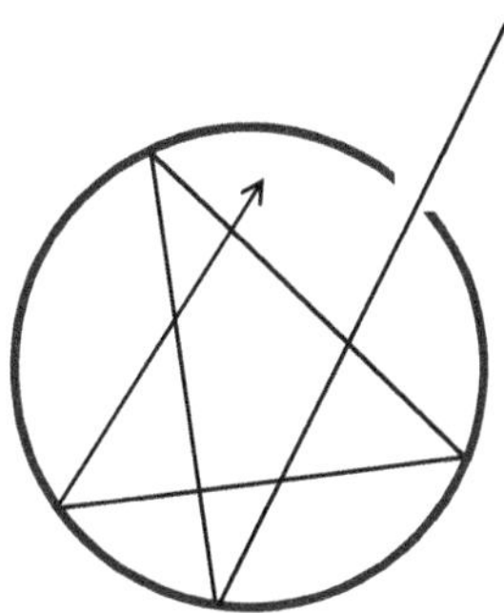

Fig. 1.2   The cavity contains radiant energy which is in thermal equilibrium with the cavity wall. The window acts as a perfect radiator and a perfect absorber at the same time.

1.3.1   *The work of Planck: Black body radiation*

A body which can absorb all incident electromagnetic radiations is called a black body. A true black body does not exist, but the radiation emitted from a small opening of a hollow enclosure is equivalent to black body radiation, since all incident electromagnetic radiations on the small opening will be almost completely absorbed by the wall of the hollow enclosure after many reflections, see Fig. 1.2.

Before the year 1900, physicists already obtained the experimental results about black body radiation which is shown in Fig. 1.3 by solid lines, and tried to explain these results by means of classical theory, but all such attempts are proved to be in vain.

The black body radiation is described by a physical quantity — intensity. The intensity $I(\lambda)$ is a function of wavelength λ. The physical meaning

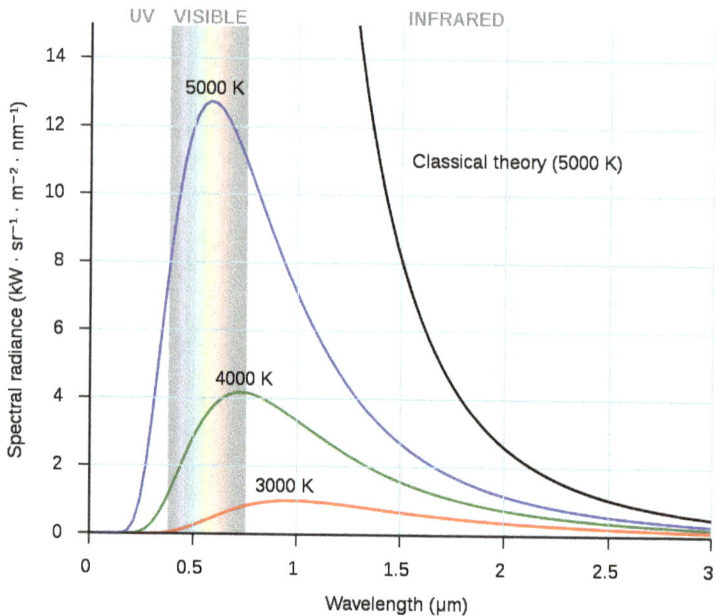

Fig. 1.3   The observed black body radiation spectra and the prediction from classical theory (Rayleigh–Jeans law). Notice the discrepancy at short wavelengths.

of $I(\lambda)$ is as follows: $I(\lambda)\,d\lambda$ is the energy density of the black body radiation in the wavelength interval $(\lambda, \lambda + d\lambda)$. Hence the total energy density U of the black body radiation is given by

$$U = \int_0^\infty I(\lambda)\,d\lambda. \tag{1.30}$$

In the classical theory, light waves can exchange any amount of energy with the surroundings. In this framework, Lord Rayleigh and Sir James Jeans derived the very important result, which is called the Rayleigh–Jeans law. Under this law, $I(\lambda) \to \infty$ when $\lambda \to 0$; only in the limit of long wavelength, their result approaches the experimental values. This behavior at short wavelength is known as the "ultraviolet catastrophe" for classical physics, which is impermissible.

In 1900, Max Planck discovered that in order to explain the experimental results, and eliminate the ultraviolet catastrophe brought by the classical theory, he must suggest an extraordinary new idea about the emission and absorption of electromagnetic radiation. His suggestion is as follows: *the electromagnetic radiation of frequency ν can exchange energy with the* *surroundings only in units of* $h\nu$: $h\nu$, $2h\nu$, ... That is to say, the emission and absorption of electromagnetic radiation cannot be continuous, i.e. they must be discrete, and the smallest energy unit $h\nu$ is called an energy quantum. The universal constant $h = 6.62 \times 10^{-34}$ Js is called the Planck's constant. Then the $I(\lambda)$ deduced from Planck's assumption fits the experimental result very well.

1.3.2 *The work of Einstein: Photoelectric effect*

Though Planck had suggested that oscillators in the wall of the hollow enclosure absorb or emit radiation in energy bundles but he still persisted in that the radiation in free space obeys the continuous laws of Maxwell. Einstein pointed out that these two ideas are not compatible, he suggested that electromagnetic radiation itself only exists in discrete energy bundles, no matter in the process of absorption and emission or in propagation in free space. Later these energy bundles are called photons.

The first major success of this idea was Einstein's explanation of the photoelectric effect, i.e. electrons were released from metals when they were irradiated with light. Experiments tell us that this effect can be observed only when the frequency of light ν is greater than a certain threshold frequency ν_0, i.e. when $\nu > \nu_0$, there is photoelectric effect; when $\nu < \nu_0$, there is no photoelectric effect. That is to say, no amount of radiation could release the electrons from the metal, no matter how large the intensity of the radiation may be. According to Maxwell equations, the energy density of electromagnetic waves is proportional to the intensity of the waves, then increasing the intensity of light should result in increasing the number of electrons released from the metal. Why the frequency ν plays such an important role in the photoelectric effect? Classical theory has no answer for this question.

In 1905 Einstein solved this puzzle very simply by using only the photon hypothesis, which says that light is composed of localized bundles of electromagnetic energy called photons, at frequency ν, and the energy of a photon is $h\nu$. The electrons in metal are bound to the metal, so we must do some work W to extract the electron. Such work is called the work function. By the photon hypothesis, when a photon strikes the metal surface, it will interact with an electron and eject it from the metal. Then from the relation of energy conservation we have

$$h\nu - W = E_k, \tag{1.31}$$

where E_k is the kinetic energy of the ejected electron. The work function W can be written as $W = h\nu_0$, then we have

$$h(\nu - \nu_0) = E_k \geq 0. \tag{1.32}$$

This relation clearly explains why radiation below a critical frequency ν_0 cannot release the electron. It also supports the hypothesis about the particle nature of light waves.

Later, in 1916 Robert Millikan used just this relation to measure the Planck constant h. It is perhaps more well known that in 1909 Millikan applied the famous oil drop experiment to measure the charge of an electron.

1.3.3 *The work of Bohr: Hydrogen atom*

In 1913, Niels Bohr combined the nuclear atom concept of Rutherford with Planck's quantum hypothesis to give a theory which accounted for the observed spectrum of hydrogen. This theory was called Bohr's old quantum theory of the hydrogen atom. In this theory there are two basic postulates.

(1) *Existence of stationary states*: Hydrogen exists in discrete energy states, which are characterized by discrete values of the angular momentum p_θ:

$$\oint p_\theta d\theta = nh, \qquad n = 1, 2, 3, \ldots \tag{1.33}$$

The symbol $\oint$ means integration over one complete orbit about the nucleus. From this quantum condition we have the discrete angular momenta, and hence the discrete energies of these stationary states. In these states the hydrogen atom is stationary and does not radiate electromagnetic waves.

(2) *Frequency condition*: When an atom undergoes a transition from state n with energy E_n to state m with energy E_m, a photon with energy $h\nu$ will be emitted according to the following rule:

$$h\nu = E_n - E_m. \tag{1.34}$$

In order to explain the spectrum of hydrogen atom, Bohr only considers that the stationary states correspond to circular orbits. In this case $p_\theta = mvr$, and the quantum condition becomes $p_\theta \times 2\pi = nh$, or $p_\theta = mvr = n\frac{h}{2\pi} = n\hbar$. The radius of the circular orbit r can be obtained easily from Newton's law $F = ma = m\frac{v^2}{r}$, and Coulomb's law $F = \frac{e^2}{4\pi\varepsilon_0 r^2}$

($\varepsilon_0 = 8.854 \times 10^{-12}$ coulomb/volt·meter is the permittivity of free space, e is the charge of electron).

$$m\frac{v^2}{r} = \frac{e^2}{4\pi\varepsilon_0 r^2} \tag{1.35}$$

$$\Longrightarrow \frac{p_\theta^2}{mr^3} = \frac{n^2\hbar^2}{mr^3} = \frac{e^2}{4\pi\varepsilon_0 r^2}. \tag{1.36}$$

For the nth orbit the radius is

$$r_n = \frac{4\pi\varepsilon_0 n^2\hbar^2}{me^2} = n^2 a_0, \tag{1.37}$$

where $a_0 = \frac{4\pi\varepsilon_0\hbar^2}{me^2} = 5.29 \times 10^{-11}$ m.

The total energy of the electron is

$$E = E_k + V = \frac{1}{2}mv^2 - \frac{e^2}{4\pi\varepsilon_0 r} = \frac{1}{2}\frac{e^2}{4\pi\varepsilon_0 r} - \frac{e^2}{4\pi\varepsilon_0 r} = -\frac{e^2}{8\pi\varepsilon_0 r}. \tag{1.38}$$

For the nth orbit

$$E_n = -\frac{e^2}{8\pi\varepsilon_0 r_n} = -\frac{R_y}{n^2}, \tag{1.39}$$

where $R_y = \frac{me^4}{2(4\pi\varepsilon_0\hbar)^2} = 2.18 \times 10^{-11}$ erg $= 13.6$ eV is the Rydberg energy.

Then, by using directly the frequency condition, Bohr obtained $\nu = \frac{R_y}{h}\left(\frac{1}{m^2} - \frac{1}{n^2}\right)$. The case $n = 1, m > 1$ corresponds to the Lyman series, whereas the case $n = 2, m > 2$ corresponds to the Balmer series. Thus Bohr successfully explains the spectra of hydrogen by his two basic postulates of old quantum theory.

However, there are some shortcomings in Bohr's theory.

(1) Mixture of classical and quantum concepts. Bohr's theory is a hybrid theory combining principles taken from classical theory with new postulates that break sharply from it. Classical physics demanded that the accelerating electrons must emit radiation, but Bohr postulated that this did not happen for electrons in a stationary orbit.

(2) Inability for calculating intensities of spectral lines.

(3) Inability for generalizing to deal with atomic systems containing two or more electrons.

Even though there are shortcomings, Bohr's theory is successful in explaining the hydrogen spectrum, and it is helpful to us in the search for a new theory on microscopic phenomena.

1.3.4   *The work of Compton: Compton effect and the dual nature of light*

1.3.4.1   *Compton effect*

The particle nature of light is decisively demonstrated by the Compton effect. In 1922, Compton discovered that the frequency of light could be changed by scattering from free electrons. This effect could be easily explained by the collision process of photon and free electron.

From special relativity we know that the mass m of a moving particle with velocity v equals $m = m_0/\sqrt{1 - \frac{v^2}{c^2}}$, where c is the speed of light in vacuum, m_0 is the rest mass. The total energy $E = mc^2$. Since the momentum of the particle equals $p = mv$, hence from $m = m_0/\sqrt{1 - \frac{v^2}{c^2}}$ and $E = mc^2$ we readily obtain $E^2 = c^2p^2 + m_0^2c^4$. For photon, the rest mass m_0 equals zero, hence $p = \frac{E}{c} = \frac{h\nu}{c}$. Before the collision the energy of the photon is $h\nu$, the momentum is $\frac{h\nu}{c}$. Simple calculation by the conservation of momentum and conservation of energy in the collision process shows that the frequency of the scattered light is changed from ν to ν', where the value of ν' depends on the scattering angle, which agrees with the experiment. See Fig. 1.4.

Before collision: Electron has momentum 0 and energy m_ec^2; photon has momentum $p_0 = \frac{h\nu_0}{c}$ in x direction and energy $E_0 = h\nu_0$.

After collision: Electron has momentum $\boldsymbol{p}_e$ (x component $p_e\cos\theta$, y component $-p_e\sin\theta$), energy $E_e = \sqrt{p_e^2c^2 + m_e^2c^4}$; photon has momentum $\boldsymbol{p}_1$ (x component $p_1\cos\theta_1$, y component $p_1\sin\theta_1$), energy $E_1 = h\nu_1$.

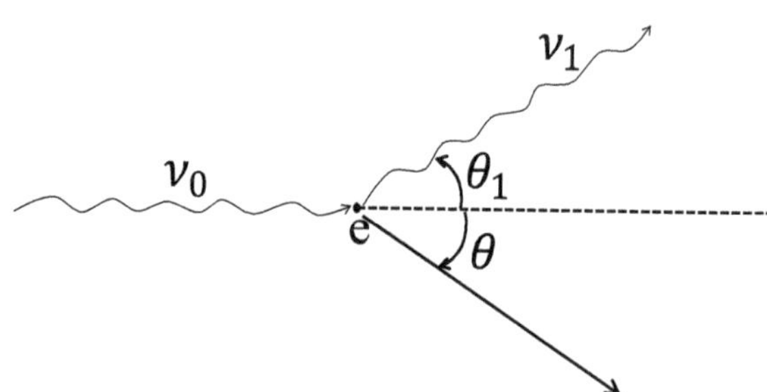

Fig. 1.4   Scattering of a photon from a free electron. The frequency of the scattered light is changed.

From the conservation of momentum,

$$\begin{cases} p_0 = p_e \cos\theta + p_1 \cos\theta_1 \\ 0 = -p_e \sin\theta + p_1 \sin\theta_1 \end{cases} \tag{1.40}$$

$$\implies p_e^2 = p_0^2 - 2p_0 p_1 \cos\theta_1 + p_1^2. \tag{1.41}$$

From the conservation of energy,

$$E_0 + m_e c^2 = (p_e^2 c^2 + m_e^2 c^4)^{\frac{1}{2}} + E_1 \tag{1.42}$$

$$\implies p_e^2 = p_0^2 + p_1^2 - 2p_0 p_1 + 2m_e c(p_0 - p_1). \tag{1.43}$$

Combine (1.41) and (1.43), we get

$$\frac{1}{p_1} - \frac{1}{p_0} = \frac{1 - \cos\theta_1}{m_e c} \tag{1.44}$$

$$\implies \lambda_1 - \lambda_0 = \frac{h}{m_e c}(1 - \cos\theta_1). \tag{1.45}$$

This result agrees with experiments.

Therefore, from the photoelectric effect and the Compton effect light shows its particle property, i.e. light is composed of photons with energy $E = h\nu$ and momentum $p = \frac{h\nu}{c}$.

1.3.4.2 *Dual nature of light*

It is well known that light has three typical wave phenomena: interference, diffraction and polarization. Light is shown to be a kind of electromagnetic wave. After the work of Einstein and Compton, we know that light also has the properties which can only belong to particles. Such appearance of light is called the dual nature or wave–particle duality. The characteristic wave properties, frequency ν and wavelength λ, are related to the particle properties, energy E and momentum p by the relations $E = h\nu = \hbar\omega$ (the angular frequency $\omega = 2\pi\nu$) and $p = \frac{h\nu}{c} = \frac{h}{\lambda} = \hbar k$ (the wave number $k = \frac{2\pi}{\lambda}$).

Now let's review some fundamental concepts of wave optics.

(1) *Expression of a plane wave*

$$E = A\cos\omega\left(t - \frac{r}{v}\right) = A\cos(\omega t - kr). \tag{1.46}$$

We can write it in the complex form

$$E = Ae^{i(kr-\omega t)}. \tag{1.47}$$

We must remember that, for wave optics, complex representation is used only for convenience, while the actual representation is always the real part.

(2) *Principle of superposition*: $E = E_1 + E_2$

If $E_1 = Ae^{i(kr_1-\omega t)}$, $E_2 = Ae^{i(kr_2-\omega t)}$, then by the principle of superposition,

$$E = E_1 + E_2. \tag{1.48}$$

$$I^2 = |E|^2 = EE^* = (E_1 + E_2)(E_1 + E_2)^*$$
$$= E_1 E_1^* + E_2 E_2^* + E_1 E_2^* + E_2 E_1^*$$
$$= A^2 + A^2 + A^2 e^{ik(r_1-r_2)} + A^2 e^{-ik(r_1-r_2)}.$$

Clearly, we see that $|E|^2 \neq |E_1|^2 + |E_2|^2$, i.e. there is interference.

If $k(r_1 - r_2) = 2n\pi$, $r_1 - r_2 = n\lambda$, then $I = 4A^2$ (max).

If $k(r_1 - r_2) = (2n + 1)\pi$, $r_1 - r_2 = (2n + 1)\dfrac{\lambda}{2}$, then $I = 0$ (min).

(3) *Double-slit experiment*

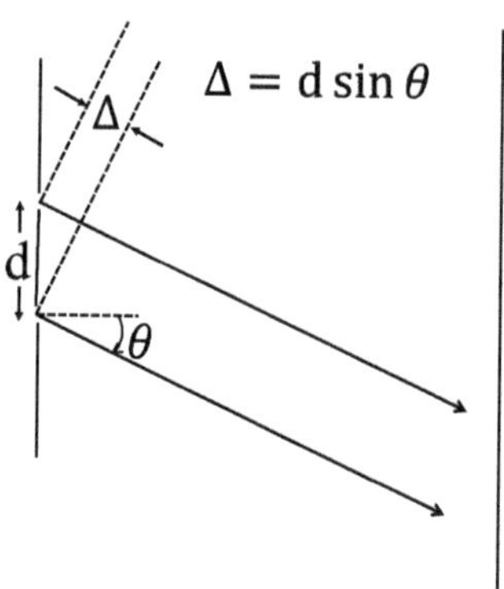

$$\Delta = n\lambda \qquad \text{constructive interference}$$
$$\Delta = (2n + 1)\frac{\lambda}{2} \qquad \text{destructive interference}$$
$$n = 0, 1, 2, \ldots$$

(4) *Diffraction grating*

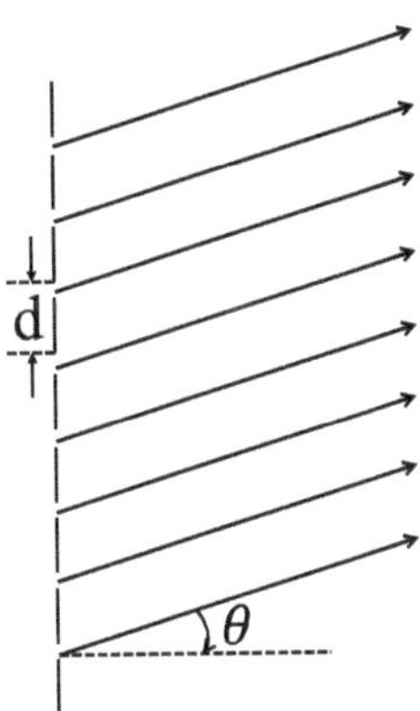

Positions of the spectral lines are determined by equation

$$d \sin \theta = n\lambda, \qquad (1.49)$$

where d is the spacing of the grating and n is the order of the maxima.

(5) *Single-slit diffraction pattern*

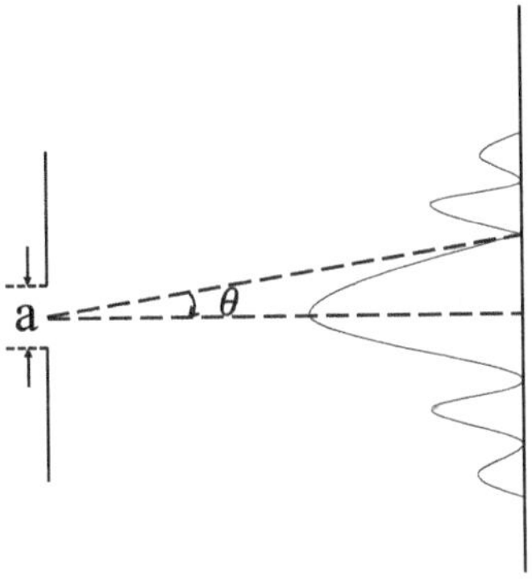

The first diffraction minimum occurs at

$$a \sin \theta = \lambda. \qquad (1.50)$$

In summary, when two non-interacting beams of particles combine in the same region of space, intensities add; when waves interact, amplitudes add, while the intensity is proportional to the time average of the absolute square of the resultant amplitude.

1.3.5 *The work of de Broglie: Matter waves*

It is well known that in 1897 English physicist J.J. Thomson succeeded in deflecting cathode rays by an electrostatic field and proved conclusively that the rays consist of material particles with a negative charge. For the next quarter century no doubt was expressed as to the particle nature of the electron. In 1924, when the dual nature of light is popularly accepted, de Broglie considered the question: "Does matter also has the dual nature?" He solved this conundrum, the answer is that both radiation and matter have the dual nature, i.e. the "wave–particle duality". De Broglie suggested that for any material particle with energy E and momentum p, there is a corresponding matter wave with wavelength $\lambda = \frac{h}{p}$. This is the de Broglie hypothesis, also called the de Broglie relation.

When de Broglie introduced the relation $\lambda = \frac{h}{p}$, there was no experimental motivation. The motivation lay to a large degree in the mystery surrounding Bohr's first postulate on the existence of stationary states, see Sec. 1.3.5.1. Later, in 1927, Davisson and Germer experimentally showed the wave nature of electrons by diffraction on crystal (Fig. 1.5), and verified the de Broglie relation.

In this experiment, electrons are diffracted by a crystal. The idea is similar to the diffraction grating of optics. The electron beam takes on the role of the light and the crystal the diffraction grating. When the path difference of adjacent layers of atoms equals to integer multiple of the wavelength of electron $\lambda = \frac{h}{p}$, there will be a diffraction peak.

1.3.5.1 *Bohr's first postulate and the de Broglie hypothesis*

For circular orbits $p_\theta = rp$, from Eq. (1.33) we obtain $2\pi rp = nh$. When the circumference $2\pi r$ equals integer multiple of wavelength, i.e. $2\pi r = n\lambda$, then we obtain $n\lambda p = nh$, i.e. $p = \frac{h}{\lambda}$, which is just the de Broglie relation.

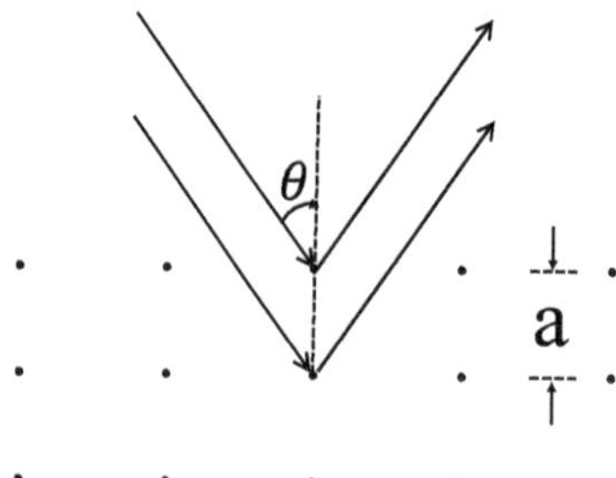

Fig. 1.5   Constructive interference occurs when $\Delta = 2a\cos\theta = n\lambda$.

Hence we obtain the conclusion that Bohr's stationary state corresponds to the standing wave state of the matter wave of electron.

In summary, de Broglie suggested that matter like radiation, also has a dual nature, i.e. the wave–particle duality. The relations between matter wave and particle are $E = h\nu$, $p = \frac{h}{\lambda}$, which have been verified by experiments.

1.3.5.2 *The phase velocity of plane wave and group velocity of wave packet*

It is well known that a simple harmonic plane wave is represented by

$$A \cos \omega \left(t - \frac{r}{v} \right) = A \cos(\omega t - kr) = A \cos(kr - \omega t), \tag{1.51}$$

where $k = \frac{2\pi}{\lambda}$ is the magnitude of the wave vector $\boldsymbol{k}$. The vector representation of de Broglie relation is $\boldsymbol{p} = \hbar \boldsymbol{k}$.

Since $A \cos(\omega t - kr)$ is the real part of the complex function $A \exp i(\omega t - kr)$ or $A \exp i(kr - \omega t)$, we also use this complex representation to represent one-dimensional simple harmonic plane wave. But we must remember that the complex representation is used only for convenience, the actual wave is always real not complex; it is the real or imaginary part of the complex function.

According to the Einstein relation $E = h\nu = \hbar\omega$ and de Broglie relation $p = \frac{h}{\lambda} = \hbar k$, for a particle with definite energy E and definite momentum p, the related de Broglie wave is represented by $\Psi = A \exp \frac{i}{\hbar}(pr - Et)$. However, if we calculate the phase velocity of the plane wave,

$$v_p = \nu\lambda - \frac{\omega}{k} = \frac{E}{p} = \frac{\frac{p^2}{2m}}{p} = \frac{p}{2m} = \frac{v_{cl}}{2}, \tag{1.52}$$

it does not equal the classical velocity v_{cl} of a particle.

The reason is that a particle, lying in a particular region in space, corresponds to a wave packet rather than a pure plane wave which spreads universally in the whole space. And the particle velocity corresponds to the group velocity of the wave packet,

$$v_g = \frac{\partial \omega}{\partial k} = \frac{\partial E}{\partial p} = \frac{p}{m} = v_{cl}. \tag{1.53}$$

1.3.6 *The work of Schrödinger: Wave equation*

According to de Broglie's hypothesis, the wavefunction corresponding to a free particle with energy E and momentum p is $\Psi = A \exp \frac{i}{\hbar}(pr - Et)$. Then, two questions arose: the first question is "How to find the wavefunction in general?"; the second question is "What is the physical meaning of the wavefunction?" In 1925, Schrödinger suggested the equation determining the wavefunction, the so-called Schrödinger's equation, which is the fundamental law of QM.

First of all, we must point out that fundamental equations in any branch of physics cannot be derived. They are first principles, they can only be suggested as axioms. Their viability can only be verified by the agreement of their predictions with experimental facts. For example, Newton's three laws for classical mechanics; Maxwell's equations for electrodynamics; the zeroth, first, second, and third laws of thermodynamics; all these laws are fundamental laws, which cannot be derived and are suggested as first principles for their corresponding branches of physics. Here, we only point out that generalization is one method for suggestion of the fundamental laws in a certain branch of physics.

Now we shall suggest a procedure by which Schrödinger's wave equation can be constructed; of course this is not a derivation. For one-dimensional plane wave propagating in the x direction, we have $\Psi = A \exp \frac{i}{\hbar}(px - Et)$, then $\frac{\partial \Psi}{\partial t} = -\frac{iE}{\hbar}\Psi \longrightarrow E\Psi = i\hbar\frac{\partial}{\partial t}\Psi$. Now if we consider E as an operator $\hat{E}$, then we have $\hat{E} \longleftrightarrow i\hbar\frac{\partial}{\partial t}$. Similarly we obtain $\frac{\partial \Psi}{\partial x} = \frac{ip}{\hbar}\Psi \longrightarrow p\Psi = \frac{\hbar}{i}\frac{\partial}{\partial x}\Psi$. If we again consider p as an operator $\hat{p}$, then we have $\hat{p} \longleftrightarrow \frac{\hbar}{i}\frac{\partial}{\partial x}$. In the three-dimensional case $\hat{\mathbf{p}} \longleftrightarrow \frac{\hbar}{i}\nabla$, where $\nabla = \mathbf{i}\frac{\partial}{\partial x} + \mathbf{j}\frac{\partial}{\partial y} + \mathbf{k}\frac{\partial}{\partial z}$ is the so-called Del operator. The two relations $\hat{E} \longleftrightarrow i\hbar\frac{\partial}{\partial t}$ and $\hat{\mathbf{p}} \longleftrightarrow \frac{\hbar}{i}\nabla$ are obtained in the particular case of plane matter wave. For a particle of mass m moving in a conservative potential field $V(\mathbf{r}, t)$, Hamiltonian $H = \frac{1}{2m}\mathbf{p}^2 + V(\mathbf{r}, t)$ equals the total energy E, if we consider H as operator we must have $\hat{H}\Psi = E\Psi$. If we generalize the two relations $\hat{E} \longleftrightarrow i\hbar\frac{\partial}{\partial t}$ and $\hat{\mathbf{p}} \longleftrightarrow \frac{\hbar}{i}\nabla$ obtained in the particular case of plane matter wave (i.e. free particle) to this case, then we obtain

$$i\hbar\frac{\partial}{\partial t}\Psi = \left[\frac{1}{2m}\mathbf{p}^2 + V(\mathbf{r}, t)\right]\Psi = -\frac{\hbar^2}{2m}\nabla^2\Psi + V(\mathbf{r}, t)\Psi. \qquad (1.54)$$

This is the Schrödinger wave equation which describes the motion of a particle moving in a potential field. Once more we must point out that this equation is not obtained by derivation but by suggestion, actually by generalizing two relations true only in the particular case to a more general case. The final equation thus obtained must then be judged, not on the merits of its genesis, but by its agreement with the experimental facts.

A further generalization to any system should lead to

$$ i\hbar\frac{\partial}{\partial t}\Psi = \hat{H}\Psi. \tag{1.55} $$

This is the so-called time-dependent Schrödinger wave equation. This means the wavefunction of a system is completely determined by the Hamiltonian of the system.

The above process of generalization for suggestion of the fundamental laws of QM can be simply expressed as follows:

$$ \Psi = Ae^{i(kx-\omega t)} = Ae^{\frac{i}{\hbar}(px-Et)} $$

$$ \longrightarrow \quad E \to -i\hbar\frac{\partial}{\partial t}, \quad p \to \frac{\hbar}{i}\frac{\partial}{\partial x} $$

$$ \longrightarrow \quad i\hbar\frac{\partial}{\partial t}\Psi = -\frac{\hbar^2}{2m}\nabla^2\Psi + V\Psi $$

$$ \longrightarrow \quad i\hbar\frac{\partial\Psi}{\partial t} = \hat{H}\Psi $$

In some particular cases, $V(\mathbf{r},t)$ is not a function of t, i.e. V is a function of $\mathbf{r}$ only, then we can use the method of separation of variables and let $\Psi(\mathbf{r},t) = \psi(\mathbf{r})\phi(t)$. From the above Schrödinger equation we obtain

$$ \frac{i\hbar}{\phi}\frac{d\phi}{dt} = \frac{1}{\psi}\left[-\frac{\hbar^2}{2m}\nabla^2 + V(\mathbf{r})\right]\psi. \tag{1.56} $$

Since the left side depends only on t and the right side depends only on $\mathbf{r}$, both sides must equal the same separation constant E, hence we obtain

$$ \left[-\frac{\hbar^2}{2m}\nabla^2 + V(\mathbf{r})\right]\psi(\mathbf{r}) = E\psi(\mathbf{r}), $$

$$ i\hbar\frac{d\phi}{dt} = E\phi. $$

The first equation $\hat{H}\psi(\mathbf{r}) = E\psi(\mathbf{r})$ is called the time-independent Schrödinger equation; $\psi(\mathbf{r})$ is the amplitude function, E is the energy eigenvalue. In this case the solution of the wave equation is

$$ \Psi(\mathbf{r},t) = \psi(\mathbf{r})\exp\left(-\frac{i}{\hbar}Et\right). \tag{1.57} $$

This solution is called a stationary state. Clearly, the position dependence of $\Psi(\boldsymbol{r},t)$ is all included in $\psi(\boldsymbol{r})$, while the time dependence of $\Psi(\boldsymbol{r},t)$ is just a pure imaginary phase $e^{-\frac{i}{\hbar}Et}$.

Here we should note that, though we can draw an analogy between wavefunction of a quantum state and optical wave from the perspective of their wave nature, the complex form of wavefunction is not just a convenient paraphrase of the real physical quantity — wavefunction is by nature complex.

The relation $\hat{p} \longleftrightarrow \frac{\hbar}{i}\frac{\partial}{\partial x}$ is actually a central concept in QM, i.e. every physical observable corresponds to an operator. Operators are not commutative in general, so we can define the *commutator*:

$$[\hat{A}, \hat{B}] \equiv \hat{A}\hat{B} - \hat{B}\hat{A}. \tag{1.58}$$

If $\hat{A}$ and $\hat{B}$ have their classical counterparts, an important property of their commutator is that the commutator in QM can be connected with the Poisson bracket in CM by

$$[\hat{A}, \hat{B}] \equiv i\hbar\{A, B\}. \tag{1.59}$$

For example,

$$\begin{aligned}
[\hat{x}, \hat{p}]f(x) &= \left[x\frac{\hbar}{i}\frac{\mathrm{d}}{\mathrm{d}x} - \frac{\hbar}{i}\frac{\mathrm{d}}{\mathrm{d}x}x\right]f(x) \\
&= x\frac{\hbar}{i}\frac{\mathrm{d}}{\mathrm{d}x}f(x) - \left[x\frac{\hbar}{i}\frac{\mathrm{d}}{\mathrm{d}x}f(x) + \frac{\hbar}{i}f(x)\right] \\
&= i\hbar f(x),
\end{aligned} \tag{1.60}$$

so $[\hat{x}, \hat{p}] = i\hbar$, while the Poisson bracket is $\{x, p\} = 1$.

As regard to differential equations,

$$\hat{A}\psi = a\psi, \tag{1.61}$$

only for certain values of a, these equations have solutions. These values a_n are called eigenvalues, and the corresponding solution ψ_n are called eigenfunctions. If $\hat{A}$ corresponds to a physical operator, then the quantum states correspond to eigenfunctions can be called eigenstates of $\hat{A}$.

Mathematically, solving a time-independent Schrödinger equation $\hat{H}\psi = E\psi$ is an eigenvalue problem, which will give a series of eigenvalues $E_1, E_2, E_3, \ldots, E_n, \ldots$ and corresponding eigenfunctions $\psi_1, \psi_2, \psi_3, \ldots, \psi_n, \ldots$

The totality of all eigenvalues for an operator is called the spectrum of the operator. Hence the totality of all eigenvalues E_n of $\hat{H}$ is called the energy spectrum.

1.3.7 *The work of Born: Probability interpretation*

The Schrödinger equation gives the answer to the first question "How to find the wavefunction in general?" Then it is ineluctable to find an answer to the second question "What is the physical meaning of the wavefunction?" This answer, proposed by Born, is the probability interpretation of wavefunction. The crucial point is how to reconcile the wave and particle nature in one wavefunction. The particle nature means that, when we observe an electron, we detect a whole electron at one point in space rather than an extended distribution of charge. Thus what the wavefunction describes is not a physical quantity like the electric field in electromagnetic waves or displacement in mechanical waves, but the probability distribution of the particle, i.e. the wave is a probability wave. So the position we find the electron has a probability distribution but it is a whole electron wherever we find it. Knowing the probabilistic nature of the wave, the probability distribution is just the modular square of the wavefunction $|\psi(x)|^2$, and thus the wavefunction is also called the probability amplitude. The probability of finding the particle in a volume element $d\tau$ is $|\psi(\boldsymbol{r})|^2\,d\tau$.

According to the probability interpretation, $C\Psi$ (C is a constant number) and Ψ describe the same state, because

$$\frac{|C\Psi(\boldsymbol{r_1},t)|^2}{|C\Psi(\boldsymbol{r_2},t)|^2} = \frac{|\Psi(\boldsymbol{r_1},t)|^2}{|\Psi(\boldsymbol{r_2},t)|^2}. \tag{1.62}$$

For definiteness, we normalize it in accordance with

$$\int \Psi\Psi^*\,d\tau = 1. \tag{1.63}$$

Even for a normalized wavefunction, there still exists a "phase uncertainty". If ψ is normalized, $e^{i\alpha}\psi$ is also normalized. ψ and $e^{i\alpha}\psi$ describe the same state.

This probabilistic physical interpretation puts some natural restriction on wavefunctions. Wavefunctions should be single-valued, continuous and square-integrable; these are the standard conditions of wavefunctions.

For a stationary state solution, the probability of finding the particle in a volume element $d\tau$ is

$$\Psi(\boldsymbol{r},t)\Psi^*(\boldsymbol{r},t)\,d\tau = \psi(\boldsymbol{r})e^{-\frac{i}{\hbar}Et}\psi^*(\boldsymbol{r})e^{\frac{i}{\hbar}Et}\,d\tau$$

$$= \psi(\boldsymbol{r})\psi^*(\boldsymbol{r})\,d\tau, \tag{1.64}$$

which is independent of time, hence the name "stationary".

Given the probability distribution, the expectation value of $f(x)$ is naturally

$$\langle f(x) \rangle = \int \psi^*(x) f(x) \psi(x)\, \mathrm{d}x. \tag{1.65}$$

After we observe the position of a particle, the probability distribution turns into a definite location, just like the dice lies still on the table, without any probabilistic feature. Here emerges a question: since a state is described by a wavefunction, what is the wavefunction that describes the state after the measurement of the particle's position? Such a wavefunction should depict a distribution which concentrates extremely sharply at one point, for example x_0. This function is the δ function introduced by Dirac. The function $\delta(x - x_0)$ can be intuitively understood as

$$\delta(x - x_0) = \begin{cases} 0, & \text{if } x \neq x_0 \\ \infty, & \text{if } x = x_0 \end{cases} \tag{1.66}$$

while

$$\int_{-\infty}^{+\infty} \delta(x - x_0)\, \mathrm{d}x = 1. \tag{1.67}$$

The δ function should be understood in the sense of integration,

$$\int_{-\infty}^{+\infty} \delta(x - x_0) f(x)\, \mathrm{d}x = f(x_0). \tag{1.68}$$

In fact, δ function is very useful in quantum mechanics. Here we note some of its features that will be utilized soon:

$$\delta(ax) = \frac{\delta(x)}{|a|}, \tag{1.69}$$

$$\delta(x - x_0) = \frac{1}{2\pi} \int_{-\infty}^{+\infty} \mathrm{d}k\, e^{ik(x-x_0)}. \tag{1.70}$$

Generally, in QM, when we measure a physical quantity, the wavefunction will collapse to an eigenfucntion of the operator that corresponds to the physical observable we measured.

Eigenfunctions of an operator (representing certain physical quantity) with different eigenvalues are orthogonal (which will be explained later). For discrete eigenvalues,

$$\int \psi_m^* \psi_n\, \mathrm{d}\tau = \delta_{mn}. \tag{1.71}$$

For continuous eigenvalues,

$$\int \psi_k^* \psi_{k'} \, \mathrm{d}\tau = \delta(k - k').$$

(1.72)

For eigenfunction ψ_p of momentum operator $\hat{p}$ with eigenvalue p,

$$\hat{p}\psi_p = p\psi_p$$

$$\implies \frac{\hbar}{i} \frac{\mathrm{d}}{\mathrm{d}x} \psi_p = p\psi_p$$

$$\implies \psi_p = A\mathrm{e}^{\frac{i}{\hbar}px},$$

(1.73)

so ψ_p is a plane wave. A plane wave has a definite wavelength λ. In a momentum eigenstate, momentum has a definite value $p = \frac{\hbar}{\lambda}$. In this case, the probability density $\rho = \psi_p^* \psi_p = |A|^2$. Everywhere ρ is the same, this means the position is totally undefined. So a plane wave represents a particle whose position is completely uncertain.

To normalize ψ_p, we firstly realize

$$\int_{-\infty}^{+\infty} \psi_p^* \psi_p \, \mathrm{d}x = \int_{-\infty}^{+\infty} |A|^2 \, \mathrm{d}x = \infty.$$

(1.74)

Now we can use the δ function to normalize ψ_p, namely

$$\int_{-\infty}^{+\infty} \psi_p^* \psi_{p'} \, \mathrm{d}x = \delta(p - p').$$

(1.75)

Since

$$\int_{-\infty}^{+\infty} \psi_p^* \psi_{p'} \, \mathrm{d}x = |A|^2 \int_{-\infty}^{+\infty} \mathrm{e}^{\frac{i}{\hbar}(p'-p)x} \, \mathrm{d}x = |A|^2 2\pi\hbar\delta(p - p'),$$

(1.76)

so we set $A = \frac{1}{\sqrt{2\pi\hbar}}$ and the normalized ψ_p is

$$\psi_p = \frac{1}{\sqrt{2\pi\hbar}}\mathrm{e}^{\frac{i}{\hbar}px}.$$

(1.77)

The eigenfunction of position operator $\hat{x}$ is a δ function,

$$x\psi_{x'} = x'\psi_{x'}, \quad \psi_{x'} = \delta(x - x').$$

(1.78)

And the orthogonal relation is

$$\int_{-\infty}^{+\infty} \psi_{x'} \psi_{x''} \, \mathrm{d}x = \int_{-\infty}^{+\infty} \delta(x - x')\delta(x - x'') \, \mathrm{d}x = \delta(x' - x'').$$

(1.79)

Now we know how to calculate the probability distribution of position from the wavefunction. However, the wavefunction contains all information of a quantum state, so we can calculate the probability distribution of momentum from the wavefunction as well. As a preparation, we first introduce the Fourier transform,

$$f(x) = \frac{1}{\sqrt{2\pi}} \int_{-\infty}^{+\infty} dk\, e^{ikx} g(k), \tag{1.80}$$

$$g(k) = \frac{1}{\sqrt{2\pi}} \int_{-\infty}^{+\infty} dx\, e^{-ikx} f(x). \tag{1.81}$$

We have

$$\int_{-\infty}^{+\infty} dk\, g^*(k)g(k) = \int_{-\infty}^{+\infty} dx\, f^*(x)f(x). \tag{1.82}$$

Proof:

$$\int_{-\infty}^{+\infty} g^*(k)g(k)\, dk = \frac{1}{2\pi} \int_{-\infty}^{+\infty} \int_{-\infty}^{+\infty} \int_{-\infty}^{+\infty} f^*(x)f(x')e^{ik(x-x')}\, dx\, dx'\, dk$$

$$= \int_{-\infty}^{+\infty} \int_{-\infty}^{+\infty} f^*(x)f(x')\delta(x - x')\, dx\, dx'$$

$$= \int_{-\infty}^{+\infty} f^*(x) \int_{-\infty}^{+\infty} f(x')\delta(x - x')\, dx'\, dx$$

$$= \int_{-\infty}^{+\infty} f^*(x)f(x)\, dx.$$

Fourier transform is a tool of spectrum analysis. Wavefunction $\psi(x)$ can be seen as a superposition of plane waves. Conduct Fourier transform on $\psi(x)$ to get $\phi(p)$ as

$$\psi(x) = \frac{1}{\sqrt{2\pi\hbar}} \int_{-\infty}^{+\infty} \phi(p)e^{ipx/\hbar}\, dp, \tag{1.83}$$

$$\phi(p) = \frac{1}{\sqrt{2\pi\hbar}} \int_{-\infty}^{+\infty} \psi(x)e^{-ipx/\hbar}\, dx. \tag{1.84}$$

Then $\phi(p)$ is called the wavefunction in p representation, while $\psi(x)$ is the wavefunction in x representation. And $|\phi(p)|^2$ is just the probability density of momentum p. The role of $\phi(p)$ in p representation is equivalent to the role of $\psi(x)$ in x representation. Therefore $\phi(p)$ contains all information of the quantum state as well.

As Fourier transform, we have

$$\int_{-\infty}^{+\infty} \psi^*(x)\psi(x)\, dx = 1 = \int_{-\infty}^{+\infty} \phi^*(p)\phi(p)\, dp, \tag{1.85}$$

which is in accordance with the physical meaning of $\phi^*(p)\phi(p)$ as the probability density of p.

The expectation value of momentum is

$$\bar{p} = \int p\phi^*(p)\phi(p)\,\mathrm{d}p$$

$$= \frac{1}{2\pi\hbar}\iiint p\psi^*(x)\mathrm{e}^{\frac{i}{\hbar}px}\psi(x')\mathrm{e}^{-\frac{i}{\hbar}px'}\,\mathrm{d}x\,\mathrm{d}x'\,\mathrm{d}p$$

$$= \frac{1}{2\pi\hbar}\iiint \psi^*(x)\psi(x')p\,\mathrm{e}^{\frac{i}{\hbar}p(x-x')}\,\mathrm{d}x\,\mathrm{d}x'\,\mathrm{d}p$$

$$= \frac{1}{2\pi\hbar}\frac{-\hbar}{i}\iiint \psi^*(x)\psi(x')\frac{\mathrm{d}}{\mathrm{d}x'}\mathrm{e}^{\frac{i}{\hbar}p(x-x')}\,\mathrm{d}x\,\mathrm{d}x'\,\mathrm{d}p$$

$$= \frac{1}{2\pi\hbar}\frac{-\hbar}{i}\iint \psi^*(x)\left[\int \psi(x')\frac{\mathrm{d}}{\mathrm{d}x'}\mathrm{e}^{\frac{i}{\hbar}p(x-x')}\,\mathrm{d}x'\right]\mathrm{d}x\,\mathrm{d}p$$

$$= \frac{1}{2\pi\hbar}\frac{-\hbar}{i}\iint \psi^*(x)\left[\psi(x')\mathrm{e}^{\frac{i}{\hbar}p(x-x')}\Big|_{-\infty}^{+\infty} - \int \mathrm{e}^{\frac{i}{\hbar}p(x-x')}\frac{\mathrm{d}}{\mathrm{d}x'}\psi(x')\,\mathrm{d}x'\right]\mathrm{d}x\,\mathrm{d}p$$

$$= \frac{1}{2\pi\hbar}\iint \psi^*(x)\left[\frac{\hbar}{i}\frac{\mathrm{d}}{\mathrm{d}x'}\psi(x')\right]\mathrm{d}x'\,\mathrm{d}x\int \mathrm{e}^{\frac{i}{\hbar}p(x-x')}\,\mathrm{d}p$$

$$= \iint \psi^*(x)\frac{\hbar}{i}\frac{\mathrm{d}}{\mathrm{d}x'}\psi(x')\delta(x-x')\,\mathrm{d}x'\,\mathrm{d}x$$

$$= \int \psi^*(x)\frac{\hbar}{i}\frac{\mathrm{d}}{\mathrm{d}x}\psi(x)\,\mathrm{d}x$$

$$= \int \psi^*(x)\hat{p}\psi(x)\,\mathrm{d}x.$$

Then we can calculate the expectation value of momentum p in x representation using

$$\bar{p} = \int \psi^*(x)\hat{p}\psi(x)\,\mathrm{d}x. \tag{1.86}$$

In general, we have

$$\langle f(x,p)\rangle = \int \psi^*(\boldsymbol{r})f(x,\hat{p})\psi(\boldsymbol{r})\,\mathrm{d}\tau. \tag{1.87}$$

From probability interpretation, we can define the probability current density $\boldsymbol{j}$. An analogy can be made between probability and charge: probability density ρ is like charge density while probability current density $\boldsymbol{j}$ is like current density. We have known that probability density ρ is related to wavefuntion as $\rho = \psi^*\psi$. In the following we shall deduce how to express probability current density $\boldsymbol{j}$ by wavefunction.

From the conservation of probability, i.e. there is no source or drain of probability, we know that the quantity of charge flow out of the given volume V through the boundary surface S per unit time $\oint \boldsymbol{j} \cdot \mathrm{d}\boldsymbol{s}$ equals the decrease of charge in the volume V per unit time $-\frac{\mathrm{d}}{\mathrm{d}t} \int \rho \, \mathrm{d}V$. This implies that

$$\int \nabla \cdot \boldsymbol{j} \, \mathrm{d}V = -\int \frac{\partial \rho}{\partial t} \, \mathrm{d}V. \tag{1.88}$$

Since the volume is arbitrary, we should have

$$\nabla \cdot \boldsymbol{j} + \frac{\partial \rho}{\partial t} = 0. \tag{1.89}$$

The Schrödinger equation is

$$i\hbar \frac{\partial \psi}{\partial t} = -\frac{\hbar^2}{2m} \nabla^2 \psi + V\psi. \tag{1.90}$$

Take the complex conjugate of Eq. (1.90),

$$-i\hbar \frac{\partial \psi^*}{\partial t} = -\frac{\hbar^2}{2m} \nabla^2 \psi^* + V\psi^*. \tag{1.91}$$

Subtract $\psi \times$ Eq. (1.91) from $\psi^* \times$ Eq. (1.90),

$$i\hbar \frac{\partial(\psi^* \psi)}{\partial t} = -\frac{\hbar^2}{2m} [\psi^* \nabla^2 \psi - \psi \nabla^2 \psi^*] \tag{1.92}$$

$$\implies \frac{\partial \rho}{\partial t} = -\nabla \cdot \boldsymbol{j} = \frac{i\hbar}{2m} [\psi^* \nabla^2 \psi - \psi \nabla^2 \psi^*] \tag{1.93}$$

$$\implies \boldsymbol{j} = -\frac{i\hbar}{2m} [\psi^* \nabla \psi - (\nabla \psi^*)\psi]. \tag{1.94}$$

In the last step we used the formula $\nabla \cdot (a\boldsymbol{V}) = a\nabla \cdot \boldsymbol{V} + \boldsymbol{V} \cdot \nabla a$, which can be proved as

$$\nabla \cdot (a\boldsymbol{V}) = \frac{\partial}{\partial x}(aV_x) + \frac{\partial}{\partial y}(aV_y) + \frac{\partial}{\partial z}(aV_z)$$

$$= a\left(\frac{\partial V_x}{\partial x} + \frac{\partial V_y}{\partial y} + \frac{\partial V_z}{\partial z}\right) + V_x \frac{\partial a}{\partial x} + V_y \frac{\partial a}{\partial y} + V_z \frac{\partial a}{\partial z}$$

$$= a\nabla \cdot \boldsymbol{V} + \boldsymbol{V} \cdot \nabla a.$$

For a stationary state,

$$\nabla \cdot \boldsymbol{j} = -\frac{i\hbar}{2m}[\psi^*\nabla^2\psi - \psi\nabla^2\psi^*] = -\frac{\partial\rho}{\partial t} = 0. \qquad (1.95)$$

For a plane wave $\psi = Ae^{ikx}$,

$$\begin{aligned}
\boldsymbol{j} &= -\frac{i\hbar}{2m}[A^*e^{-ikx}A(ik)e^{ikx}\boldsymbol{e}_x - (-ik)A^*e^{-ikx}Ae^{ikx}\boldsymbol{e}_x] \\
&= -\frac{i\hbar}{2M}2ikAA^*\boldsymbol{e}_x = \frac{\hbar k}{m}AA^*\boldsymbol{e}_x \\
&= \frac{p}{m}\rho\boldsymbol{e}_x = \rho v\boldsymbol{e}_x \\
&= \rho\boldsymbol{v}, \qquad\qquad\qquad\qquad\qquad\qquad\qquad\qquad (1.96)
\end{aligned}$$

which is reminiscent of a formula in electromagnetism,

$$\boldsymbol{j} = \rho\boldsymbol{v}, \qquad (1.97)$$

where j is the current density and ρ is the charge density.

1.3.8 *The work of Heisenberg: Uncertainty principle*

The uncertainty principle is a manifestation of the wave–particle duality. The statement of this principle is as follows: It is impossible to know precisely a particle's exact position and its momentum simultaneously, and this statement can be quantitatively expressed by

$$\Delta x\Delta p_x \sim \hbar, \qquad (1.98)$$

where Δx is the variation of x, and Δp_x is the variation of p_x. If x is definite, i.e. $\Delta x = 0$, then $\Delta p_x \to \infty$, i.e. p_x is totally indefinite; similarly, if $\Delta p_x = 0$, then $\Delta x \to \infty$.

This conclusion is indifferent to the particular wavefunction. However, in order to test this principle, we shall give some examples as follows.

Example 1.5. Consider an eigenstate of momentum with eigenvalue p_0, $\psi_{p_0}(x) = \frac{1}{\sqrt{2\pi\hbar}}e^{ip_0x/\hbar}$. Since this state is an eigenstate of momentum, its momentum is definite, $p = p_0$, $\Delta p = 0$. It is easy to see that $|\psi_{p_0}|^2$ is independent of x, which means that the probability distribution of x is identical everywhere, i.e. x is completely uncertain.

Example 1.6. Consider an eigenstate of position with eigenvalue x_0, $\psi_{x_0}(x) = \delta(x - x_0)$. The particle is definitely located at $x = x_0$, $\Delta x = 0$.

To find the probability distribution of momentum, we calculate the Fourier transform of the wavefunction (transform the wavefunction from x representation to p representation).

$$\psi_{x_0}(p) = \frac{1}{\sqrt{2\pi\hbar}} \int_{-\infty}^{+\infty} \psi_{x_0}(x)e^{-ipx/\hbar}\mathrm{d}x = \frac{1}{\sqrt{2\pi\hbar}}e^{-ipx_0/\hbar}. \tag{1.99}$$

Then

$$|\psi_{x_0}(p)|^2 = \frac{1}{2\pi\hbar} = \text{const}, \tag{1.100}$$

so the momentum is completely uncertain.

Example 1.7. Consider the Gaussian wave packet, $\psi(x) = e^{-\frac{1}{2}\alpha^2 x^2}$. The wave packet can be decomposed to a set of plane waves with different wave number k and different intensity, so the wave number k has a distribution as well. In this case, we look upon $\psi(x) = e^{-\frac{1}{2}\alpha^2 x^2}$ as a pure mathematical function and try to find the relation between the extended range of x and k. The result can be easily turned to the relation between x and p if we set $k = p/\hbar$ and regard $\psi(x)$ as a wavefunction in x representation.

$$|\psi(x)|^2 = e^{-\alpha^2 x^2} \implies \Delta x \sim \frac{1}{\alpha}. \tag{1.101}$$

$$\phi(k) = \frac{1}{\sqrt{2\pi}} \int_{-\infty}^{+\infty} \psi(x)e^{-ikx}\,\mathrm{d}x = \frac{1}{\alpha}e^{-\frac{k^2}{2\alpha^2}}$$

$$\implies \Delta k \sim \alpha. \tag{1.102}$$

So we have $\Delta x \cdot \Delta k \sim 1$ and $\Delta x \cdot \Delta p \sim \hbar$ if $k = p/\hbar$.

Thus the uncertainty principle seems to be just a natural mathematical consequence if a quantum state is described by a wavefunction and we accept the probability interpretation. However, the uncertainty principle has a very profound physics meaning. (If you refer to Table 1.1, you will find that the uncertainty principle was proposed before Born's interpretation of wavefunction.)

If an identical experiment involving an electron is performed many times, and in each run of the experiment the position x of the electron is measured, then although the experimental setup is identical in each run, measurement of the position of the electron does not always give the same result. The average of x is

$$\langle x \rangle = \sum_k x_k P_k. \tag{1.103}$$

In QM,

$$\langle f(x) \rangle = \int f(x)\psi^*(x)\psi(x)\, dx. \tag{1.104}$$

The variance of x is defined as

$$\langle (\Delta x)^2 \rangle = \langle (x - \langle x \rangle)^2 \rangle, \tag{1.105}$$

which measures the deviation from the mean. The standard deviation is the square root of variance.

If Δx is small compared to $\langle x \rangle$, one is more certain to find the value $x = \langle x \rangle$ in any given run; if Δx is large, it is not certain what the measurement of x will yield. For this reason Δx is also called the uncertainty in x.

The variance can be expressed by *moments* of x,

$$\langle (\Delta x)^2 \rangle = \langle (x - \langle x \rangle)^2 \rangle = \langle x^2 \rangle - 2\langle x \rangle^2 + \langle x \rangle^2 = \langle x^2 \rangle - \langle x \rangle^2. \tag{1.106}$$

Chapter 2

Postulates and One-Dimensional Problems

2.1 Postulates of quantum mechanics

We have learned some elementary knowledge of quantum mechanics (QM), now it is the time to systematize what we have already learned, i.e. it is the time to formulate into a system.

Just like classical mechanics (CM) is based on Newton's three laws, QM is based on five postulates. These postulates cannot be proved directly but can be verified by experimental results, i.e. the validity of these postulates is confirmed by the agreement of the theory with experiment.

2.1.1 *Postulate 1: Quantum state*

Any state of a system with n degrees of freedom is described as completely as possible by a wavefunction $\Psi(q_1, q_2, \ldots, q_n, t)$ which depends upon the coordinates q_i and the time t.

Recall that the wavefunctions Ψ and $C\Psi$ (C is a constant number) represent the same state. Also the wavefunction Ψ should be single-valued, continuous and finite. These three properties are the standard conditions of a wavefunction. Hereafter we shall introduce two definitions.

Definition 2.1 (Linear operator). If an operator $\hat{A}$ satisfies

$$\hat{A}(\Psi_1 + \Psi_2) = \hat{A}\Psi_1 + \hat{A}\Psi_2, \tag{2.1}$$

$$\hat{A}(a\Psi) = a\hat{A}\Psi, \tag{2.2}$$

then $\hat{A}$ is a linear operator.

To give an intuitive impression of this, we give an example of a non-linear operator. A square operator satisfies

$$\hat{S}\psi = \psi^2, \tag{2.3}$$

then

$$\hat{S}(\psi_1 + \psi_2) = (\psi_1 + \psi_2)^2 \neq \hat{S}\psi_1 + \hat{S}\psi_2, \tag{2.4}$$

so $\hat{S}$ is a non-linear operator.

Now we introduce a new notation for quantum states.

Notation 2.1 (Dirac notation). Dirac notation is a notation of quantum state without referring to a particular representation. We note wavefunction $\psi(x)$ as $|\psi\rangle$, and its complex conjugate $\psi^*(x)$ as $\langle\psi|$.

The meaning of $\langle\psi|\phi\rangle$ and its properties are as follows (a is a number):

$$\langle\psi|\phi\rangle = \int_{-\infty}^{+\infty} \psi^*(x)\phi(x)\,\mathrm{d}x \tag{2.5}$$

$$\langle\psi|a\phi\rangle = a\langle\psi|\phi\rangle \tag{2.6}$$

$$\langle a\psi|\phi\rangle = a^*\langle\psi|\phi\rangle \tag{2.7}$$

$$\langle\psi|\phi\rangle^* = \langle\phi|\psi\rangle \tag{2.8}$$

$$\langle\psi + \phi| = \langle\psi| + \langle\phi| \tag{2.9}$$

$$\langle\psi_1 + \psi_2|\phi_1 + \phi_2\rangle = \langle\psi_1|\phi_1\rangle + \langle\psi_1|\phi_2\rangle + \langle\psi_2|\phi_1\rangle + \langle\psi_2|\phi_2\rangle \tag{2.10}$$

The relation between a quantum state and a representation is similar to the relation between a vector and a coordinate system. A vector $\boldsymbol{a}$, for example in 3-dimensional space, can be represented by its components, namely its projections on the coordinate frames, as (a_x, a_y, a_z). A component is the inner product between vector $\boldsymbol{a}$ and basis vector of the coordinate frame, e.g. $a_x = \boldsymbol{a} \cdot \boldsymbol{e}_x$. The vector itself is indifferent to the coordinate system. However, different coordinate systems will give different a_x, a_y, a_z.

Similarly, a quantum state $|\psi\rangle$, indifferent to the particular representation, can be represented by wavefunction $\psi(x)$ in x representation or $\phi(p)$ in p representation. Using Dirac notation means that we refer to the quantum state $|\psi\rangle$, i.e. we refer to the vector instead of components of the vector.

The wavefunction $\psi(x)$, like (a_x, a_y, a_z), is the set of projections of the quantum state $|\psi\rangle$ onto the axes, i.e. wavefunction $\psi(x)$ is the set of inner products between $|\psi\rangle$ and the eigenfunctions which span the representation.

If we denote $\delta(x - x')$, the eigenfunction of operator $\hat{x}$ with eigenvalue x', as $|x'\rangle$ in Dirac notation, then

$$\psi(x') = \langle x'|\psi\rangle = \int_{-\infty}^{+\infty} \delta(x - x')\psi(x)\,\mathrm{d}x. \tag{2.11}$$

Dropping the $'$, we get

$$\psi(x) = \langle x|\psi\rangle. \tag{2.12}$$

The difference is that, in x representation or p representation, which is spanned by eigenfunctions with continuous eigenvalues, infinite components assemble into a function, like $\psi(x)$ or $\phi(p)$, instead of a finite set of numbers, like (a_x, a_y, a_z).

Definition 2.2 (Hermitian adjoint and Hermitian operator).
The Hermitian adjoint of an operator $\hat{A}$ is denoted by $\hat{A}^\dagger$, which has the property

$$\langle \psi_1|\hat{A}\psi_2\rangle = \langle \hat{A}^\dagger\psi_1|\psi_2\rangle. \tag{2.13}$$

If $\hat{A}^\dagger = \hat{A}$, then $\hat{A}$ is called a Hermitian operator.

Example 2.1. Examples of Hermition operator.

(1) Clearly, for a number a, its Hermitian adjoint $a^\dagger = a^*$.
(2) The Hermitian adjoint of $\frac{\mathrm{d}}{\mathrm{d}x}$ is $(\frac{\mathrm{d}}{\mathrm{d}x})^\dagger = -\frac{\mathrm{d}}{\mathrm{d}x}$, because (using the standard conditions of wavefunction)

$$\left\langle \psi_1 \left| \frac{\mathrm{d}}{\mathrm{d}x}\psi_2 \right. \right\rangle = \int_{-\infty}^{+\infty} \mathrm{d}x\, \psi_1^* \frac{\mathrm{d}\psi_2}{\mathrm{d}x}$$

$$= \psi_1^*\psi_2 \Big|_{-\infty}^{\infty} - \int_{-\infty}^{+\infty} \mathrm{d}x\, \frac{\mathrm{d}\psi_1^*}{\mathrm{d}x}\psi_2$$

$$= \int_{-\infty}^{+\infty} \mathrm{d}x \left(-\frac{\mathrm{d}\psi_1^*}{\mathrm{d}x} \right) \psi_2$$

$$= \left\langle -\frac{\mathrm{d}}{\mathrm{d}x}\psi_1 \left| \psi_2 \right. \right\rangle.$$

(3) The Hermitian adjoint of $\hat{A}\hat{B}$ is $(\hat{A}\hat{B})^\dagger = \hat{B}^\dagger\hat{A}^\dagger$, because

$$\langle \psi_1|\hat{A}\hat{B}\psi_2\rangle = \langle \hat{A}^\dagger\psi_1|\hat{B}\psi_2\rangle = \langle \hat{B}^\dagger\hat{A}^\dagger\psi_1|\psi_2\rangle.$$

(4) Since

$$\hat{p}^\dagger = \left(\frac{\hbar}{i}\frac{\partial}{\partial x}\right)^\dagger = \left(-\frac{\partial}{\partial x}\right)\left(-\frac{\hbar}{i}\right) = \frac{\hbar}{i}\frac{\partial}{\partial x} = \hat{p}, \qquad (2.14)$$

hence the momentum operator $\hat{p}$ is a Hermitian operator.

A linear combination of Hermitian operators is itself a Hermitian operator,

$$(\hat{A} + \hat{B})^\dagger = \hat{A}^\dagger + \hat{B}^\dagger = \hat{A} + \hat{B} \longrightarrow \hat{A} + \hat{B} \text{ is Hermitian.} \qquad (2.15)$$

And any power of a Hermitian operator is also a Hermitian operator,

$$(\hat{A}^n)^\dagger = (\hat{A}^\dagger)^n = \hat{A}^n \longrightarrow \hat{A}^n \text{is Hermitian.} \qquad (2.16)$$

2.1.2   *Postulate 2: Physical observable*

With every physical observable there is associated a linear operator and this linear operator is Hermitian.

In general, two different operators $\hat{A}$ and $\hat{B}$ do not commute, i.e. $\hat{A}\hat{B} \neq \hat{B}\hat{A}$; it means $\hat{A}\hat{B}\psi \neq \hat{B}\hat{A}\psi$.

Definition 2.3 (Commutator). The commutator of two operators $\hat{A}$ and $\hat{B}$ is defined as

$$[\hat{A}, \hat{B}] = \hat{A}\hat{B} - \hat{B}\hat{A}. \qquad (2.17)$$

The commutator has the following properties:

$$[\hat{A}, \hat{B}] = -[\hat{B}, \hat{A}],$$

$$[\hat{A}, (\hat{B} + \hat{C})] = [\hat{A}, \hat{B}] + [\hat{A}, \hat{C}],$$

$$[\hat{A}, \hat{B}\hat{C}] = [\hat{A}, \hat{B}]\hat{C} + \hat{B}[\hat{A}, \hat{C}],$$

$$[\hat{A}, [\hat{B}, \hat{C}]] + [\hat{B}, [\hat{C}, \hat{A}]] + [\hat{C}, [\hat{A}, \hat{B}]] = 0,$$

$$[\hat{A}, \hat{B}]^\dagger = [\hat{B}^\dagger, \hat{A}^\dagger].$$

Remark 2.1. In the classical limit $h \to 0$, the commutator $[\hat{A}, \hat{B}]$ is related to the Poisson bracket:

$$\lim_{h \to 0} \frac{[\hat{A}, \hat{B}]}{i\hbar} = \{A, B\}. \qquad (2.18)$$

The result of the operation on wavefunction Ψ by the commutator depends on the "representation" used. In the Schrödinger representation, q_i, t are represented by q_i and t respectively, and p_i is represented by $\frac{\hbar}{i}\frac{\partial}{\partial q_i}$.

$$[q_i, p_j]\Psi = \left[q_i, \frac{\hbar}{i}\frac{\partial}{\partial q_j}\right]\Psi = i\hbar\delta_{ij}\Psi, \tag{2.19}$$

hence we have

$$[q_i, p_j] = i\hbar\delta_{ij}. \tag{2.20}$$

As another example, we calculate $[\hat{p}_x^2, f(x)]$.

$$\begin{aligned}
[\hat{p}_x^2, f(x)]\psi &= \hat{p}_x^2\left(f(x)\psi\right) - f(x)\hat{p}_x^2\psi \\
&= \hat{p}_x\frac{\hbar}{i}(f'\psi + f\psi') - f(x)(-\hbar^2)\psi'' \\
&= -\hbar^2(f''\psi + f'\psi' + f'\psi' + f\psi'') + \hbar^2 f\psi'' \\
&= -\hbar^2(f''\psi + 2f'\psi') \\
&= (-\hbar^2 f'' - 2i\hbar f'\hat{p}_x)\psi,
\end{aligned}$$

hence

$$[\hat{p}_x^2, f(x)] = -\hbar^2 f'' - 2i\hbar f'\hat{p}_x. \tag{2.21}$$

In particular,

$$i[\hat{p}_x^2, x] = 2\hbar\hat{p}_x, \tag{2.22}$$

so $i[\hat{p}_x^2, x]$ is a Hermitian operator.

2.1.3 *Postulate 3: Time evolution of quantum state*

The state wavefunction Ψ must satisfy Schrödinger's time-dependent equation

$$\hat{H}\psi = i\hbar\frac{\partial\Psi}{\partial t}. \tag{2.23}$$

If the state function is known at some initial time, the Schrödinger equation determines Ψ at any other time.

In CM, Hamilton's equations determine the time variation of a state $(\dot{p}_i = \{p_i, H\} = -\frac{\partial H}{\partial q_i}, \ \dot{q}_i = \{q_i, H\} = \frac{\partial H}{\partial p_i})$. Similarly, the Hamiltonian operator for a quantum system determines the time variation of a state.

Definition 2.4 (Eigenvalues and eigenfunctions). For an operator $\hat{A}$, if we have

$$\hat{A}\psi = a\psi, \tag{2.24}$$

then ψ is an eigenfunction of $\hat{A}$ and a is the corresponding eigenvalue. The totality of all eigenvalues for an operator is called the spectrum of the operator.

Definition 2.5 (Orthogonality). Functions ψ and ϕ are orthogonal to each other if

$$\int \psi^* \phi \, d\tau = 0 \tag{2.25}$$

or

$$\langle \psi | \phi \rangle = 0 \tag{2.26}$$

in Dirac notation.

An important property of Hermitian operator is that its eigenfunctions with different eigenvalues are orthogonal to each other. This will be proved later in Theorem 2.2.

If there is no degeneration in the eigenfunctions of a Hermitian operator, i.e. different eigenfunctions $|\Psi_n\rangle$ have different eigenvalues, then

$$\langle \Psi_n | \Psi_m \rangle = 0, \quad n \neq m. \tag{2.27}$$

If in addition the functions are normalized to unity, then

$$\langle \Psi_n | \Psi_m \rangle = \delta_{nm}, \tag{2.28}$$

and the functions are called orthogonal normalized.

A well known example of orthogonal functions is the set

$$\Phi_n(x) = \frac{1}{\sqrt{\pi}} \sin n\pi x, \tag{2.29}$$

the members of which are orthogonal over the range $0 \leq x \leq 2\pi$, i.e.

$$\int_0^{2\pi} \Psi_n \Psi_m \, dx = \delta_{nm}. \tag{2.30}$$

Definition 2.6 (Vector space). In Cartesian 3-space a vector $\boldsymbol{v}$ is a set of 3 numbers, called components (v_x, v_y, v_z). Any vector in this space can be expanded in terms of the 3 unit vectors $\boldsymbol{e}_x$, $\boldsymbol{e}_y$, $\boldsymbol{e}_z$. Under such conditions one terms the triad $\boldsymbol{e}_x$, $\boldsymbol{e}_y$, $\boldsymbol{e}_z$ a basis.

$$\boldsymbol{v} = v_x \boldsymbol{e}_x + v_y \boldsymbol{e}_y + v_z \boldsymbol{e}_z. \tag{2.31}$$

The vectors $\boldsymbol{e}_x$, $\boldsymbol{e}_y$, $\boldsymbol{e}_z$ are said to span the vector space.

The inner ("dot") product of two vectors ($\boldsymbol{v}$ and $\boldsymbol{u}$) in the space is defined as

$$\boldsymbol{v} \cdot \boldsymbol{u} = v_x u_x + v_y u_y + v_z u_z. \tag{2.32}$$

The length of the vector $\boldsymbol{v}$ is $\sqrt{\boldsymbol{v} \cdot \boldsymbol{v}}$.

Definition 2.7 (Hilbert space). A Hilbert space is closely akin to a vector space. Mathematicians call it an infinite-dimensional vector space. Its elements are functions instead of 3-dimensional vectors. The similarity is so close that the functions are sometimes called vectors. Elements of this space have length and one can form an inner product between any two elements.

Hilbert space is linear, i.e. for a Hilbert space $\mathcal{H}$,

$$\text{if} \quad \psi \in \mathcal{H}, \quad \text{then} \quad a\psi \in \mathcal{H},$$
$$\text{if} \quad \psi \in \mathcal{H}, \ \phi \in \mathcal{H}, \quad \text{then} \quad \psi + \phi \in \mathcal{H}.$$

It also has an inner product:

$$\langle \phi | \psi \rangle = \int_a^b \phi^* \psi \, \mathrm{d}x, \tag{2.33}$$

when the functions in the space are defined in the interval $a \le x \le b$. Any element of $\mathcal{H}$ has a norm ("length") that is related to the inner product as follows:

$$(\text{norm of } \phi)^2 \equiv \|\phi\|^2 = \langle \phi | \phi \rangle. \tag{2.34}$$

The reason we complex-conjugate the first "vector" in the inner product is to ensure that the "length" of a vector ϕ is real.

Just like the triad e_x, e_y, e_z in Cartesian 3-space, a Hilbert space $\mathcal{H}$ has a set of basis vectors $\{\psi_n\}$ which are orthogonal and normalized, i.e.

$$\langle \psi_n | \psi_{n'} \rangle = \delta_{nn'}. \tag{2.35}$$

However, this basis is composed of infinite number of vectors.

Like v can be expressed in terms of e_x, e_y, e_z in Eq. (2.31), a vector $|\psi\rangle$ in $\mathcal{H}$ can be expressed as

$$|\psi\rangle = \sum_n c_n |\psi_n\rangle. \tag{2.36}$$

The components can be found by taking the inner product with the basis vectors.

$$\langle \psi_{n'} | \psi \rangle = \sum_n c_n \langle \psi_{n'} | \psi_n \rangle = \sum_n c_n \delta_{n'n} = c_{n'}. \tag{2.37}$$

All the vectors $|\psi_n\rangle$ together are said to span the Hilbert space.

All $|\psi_n\rangle$ are linearly independent, which means

$$\sum_n c_n |\psi_n\rangle = 0 \quad \text{if and only if all } c_n = 0. \tag{2.38}$$

This means that none of these functions can be expressed in terms of the others.

The correspondence between Cartesian 3-space and Hilbert space is listed as follows:

$$\text{vector } v \longrightarrow \text{function or state vector } |\psi\rangle$$

$$\{e_x, e_y, e_z\}, \; e_i \cdot e_j = \delta_{ij} \longrightarrow \{|\psi_n\rangle\}, \; \langle \psi_n | \psi_m \rangle = \delta_{nm}$$

$$v = v_x e_x + v_y e_y + v_z e_z \longrightarrow |\psi\rangle = \sum_i c_i |\psi_i\rangle$$

$$v_x = e_x \cdot v, \; v_y = e_y \cdot v, \; v_z = e_z \cdot v \longrightarrow c_i = \langle \psi_i | \psi \rangle.$$

We can write $|\psi\rangle$ as

$$|\psi\rangle = \sum_n \langle \psi_n | \psi \rangle |\psi_n\rangle = \sum_n |\psi_n\rangle \langle \psi_n | \psi \rangle = \left(\sum_n |\psi_n\rangle \langle \psi_n| \right) |\psi\rangle. \tag{2.39}$$

Since $|\psi\rangle$ is arbitrary, we get the "closure relation", namely

$$\sum_n |\psi_n\rangle \langle \psi_n| = 1. \tag{2.40}$$

In summary, a Hilbert space $\mathcal{H}$ has the following properties:

(1) The space is linear:

$$\text{if} \quad \psi \in \mathcal{H}, \quad \text{then} \quad a\psi \in \mathcal{H}, \quad a \text{ is const,}$$

$$\text{if} \quad \psi \in \mathcal{H}, \ \phi \in \mathcal{H}, \quad \text{then} \quad \psi + \phi \in \mathcal{H}.$$

(2) There is an inner product: $\langle \phi | \psi \rangle$.
(3) Any element of $\mathcal{H}$ has a norm ("length"):

$$(\text{norm of } \psi)^2 \equiv \|\psi\|^2 = \langle \psi | \psi \rangle.$$

Theorem 2.1 (Real eigenvalues). *The eigenvalues of a Hermitian operator are real.*

Proof. Suppose $\hat{A}$ is a Hermitian operator with eigenvalues a_n and corresponding eigenfunctions $|\psi_n\rangle$, we have

$$\langle \psi_n | \hat{A} \psi_n \rangle = a_n \langle \psi_n | \psi_n \rangle = a_n. \tag{2.41}$$

On the other hand, because $\hat{A}$ is Hermitian,

$$\langle \psi_n | \hat{A} \psi_n \rangle = \langle \hat{A} \psi_n | \psi_n \rangle = a_n^* \langle \psi_n | \psi_n \rangle = a_n^*. \tag{2.42}$$

Hence from the above two equations we have

$$a_n = a_n^*. \tag{2.43}$$

$\square$

Theorem 2.2 (Orthogonal eigenfunctions). *Any two eigenfunctions of a Hermitian operator with different eigenvalues are orthogonal.*

Proof. Similar to the previous proof,

$$\langle \psi_l | \hat{A} \psi_n \rangle = a_n \langle \psi_l | \psi_n \rangle, \tag{2.44}$$

and

$$\langle \psi_l | \hat{A} \psi_n \rangle = \langle \hat{A} \psi_l | \psi_n \rangle = a_l^* \langle \psi_l | \psi_n \rangle = a_l \langle \psi_l | \psi_n \rangle. \tag{2.45}$$

Hence from the above two equations we have

$$(a_n - a_l)\langle \psi_l | \psi_n \rangle = 0 \xrightarrow{a_n \neq a_l} \langle \psi_l | \psi_n \rangle = 0. \tag{2.46}$$

$\square$

2.1.4   *Postulate 4: Measurement*

Suppose the wavefunction $\Psi(q_1, \ldots, q_n, t)$ describes a state which is an eigenstate of the operator $\hat{A}$ representing observable A, i.e.

$$\hat{A}\Psi = a\Psi. \tag{2.47}$$

Then in this state a measurement of the observable A will yield precisely the constant real value a.

Example 2.2. Examples of definite measurements.

(1) If the system is in an eigenstate of the Hamiltonian, its energy is fixed.

$$\hat{H}\psi_n = E_n\psi_n. \tag{2.48}$$

(2) Plane waves have definite momenta. They are eigenstates of the momentum operator.

$$\hat{p}\psi_{p_0} = p_0\psi_{p_0}, \quad \psi_{p_0} = \frac{1}{\sqrt{2\pi\hbar}}e^{\frac{i}{\hbar}p_0 x}. \tag{2.49}$$

(3) The eigenstates of the position operator are the δ functions.

$$x\psi_{x_0} = x_0\psi_{x_0}, \quad \psi_{x_0} = \delta(x - x_0). \tag{2.50}$$

2.1.5   *Postulate 5: Expectation value*

The expectation value of a physical observable A for a state $|\Psi\rangle$ is given by its corresponding operator $\hat{A}$.

$$\langle A \rangle = \int \Psi^* \hat{A}\Psi \, d\tau = \langle \Psi | \hat{A} | \Psi \rangle. \tag{2.51}$$

If $\hat{A}$ is a Hermitian operator, i.e. $\hat{A} = \hat{A}^\dagger$, then we have

$$\langle A \rangle = \langle \Psi | \hat{A}\Psi \rangle = \langle \hat{A}\Psi | \Psi \rangle. \tag{2.52}$$

By the definition of Dirac notation,

$$\langle \Psi | \hat{A}\Psi \rangle = \langle \hat{A}\Psi | \Psi \rangle^*, \tag{2.53}$$

hence we have

$$\langle A \rangle = \langle A \rangle^*. \tag{2.54}$$

So for Hermitian operators, their expectation value is real.

If the eigenfunctions of $\hat{A}$, $\{|\psi_i\rangle\}$, span the Hilbert space, then any vector $|\psi\rangle$ can be expanded as

$$|\psi\rangle = \sum_i c_i|\psi_i\rangle, \quad c_i = \langle\psi_i|\psi\rangle. \tag{2.55}$$

The expectation value is

$$\langle\psi|\hat{A}|\psi\rangle = \sum_i \langle\psi|c_i\hat{A}|\psi_i\rangle$$

$$= \sum_i c_i a_i \langle\psi|\psi_i\rangle$$

$$= \sum_i c_i a_i \langle\psi_i|\psi\rangle^*$$

$$= \sum_i c_i c_i^* a_i$$

$$= \sum_i |c_i|^2 a_i$$

$$= \sum_i P_i a_i.$$

By postulate 4, in state $|\psi_i\rangle$ the measured value of the dynamical observable is a_i, where $\hat{A}|\psi_i\rangle = a_i|\psi_i\rangle$. The measured value of A in state $|\psi\rangle$ should be the expectation value,

$$\langle\psi|\hat{A}|\psi\rangle = \sum_i P_i a_i. \tag{2.56}$$

So $P_i = |c_i|^2$ is the probability of observing the eigenvalue a_i. Note that

$$\int \psi^*\psi \, d\tau = \sum_i |c_i|^2 = 1, \tag{2.57}$$

which is expected since the sum of probabilities of all results should equals to 1.

In the next section, we will give a few examples of finding stationary states, which means solving $\hat{H}\psi = E\psi$ in a specific potential $V(\mathbf{r})$, within which some important basic knowledge will be introduced.

2.2 Preliminary applications on one-dimensional problems

2.2.1 One-dimensional free particle

To find stationary states, we firstly write down again the eigenvalue problem

$$\hat{H}\psi = E\psi. \tag{2.58}$$

"Free particle" means that the potential $V(\mathbf{r})$ is universally zero, which leaves only kinetic energy in the Hamiltonian. In one dimension, the representation $-i\hbar\nabla$ of momentum operator $\hat{p}$ reduces to $-i\hbar\frac{\partial}{\partial x}$. Therefore, Eq. 2.58) turns into an ordinary differential equation

$$-\frac{\hbar^2}{2m}\frac{\mathrm{d}^2}{\mathrm{d}x^2}\psi(x) = E\psi(x) \tag{2.59}$$

or

$$\frac{\mathrm{d}^2}{\mathrm{d}x^2}\psi(x) + \frac{2mE}{\hbar^2}\psi(x) = 0. \tag{2.60}$$

The solution is

$$\psi(x) = A\exp(ikx) + B\exp(-ikx) \tag{2.61}$$

and every eigenfunction $\psi(x)$ (or $\psi_k(x)$ when we put the index explicitly) has a corresponding energy eigenvalue E (or E_k), which can be solved easily by substituting the solution (2.61) into the differential equation (2.59) or (2.60). The result is

$$E = \frac{\hbar^2 k^2}{2m}, \tag{2.62}$$

which means that the energy spectrum of free particle is continuous and any $E > 0$ is permissible.

Combining $\psi(x)$ and time-dependent part $\exp(-\frac{i}{\hbar}Et)$ gives the wavefunction $\Psi(x, t)$, namely

$$\Psi(x, t) = Ae^{i(kx-\omega t)} + Be^{-i(kx+\omega t)}, \qquad \omega = \frac{E}{\hbar} = \frac{\hbar k}{2m}, \tag{2.63}$$

which is clearly a plane-wave solution. The first term in Eq. (2.63) represents the plane wave propagating in the positive x direction, whereas the second term represents the plane wave propagating in negative x direction.

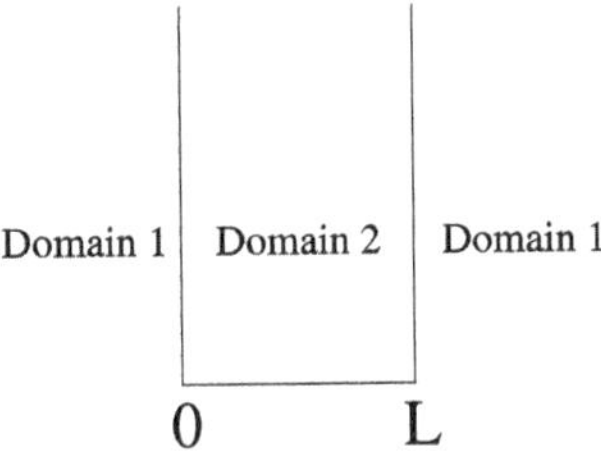

Fig. 2.1   Linear box potential.

2.2.2 *Linear box*

In a one-dimensional linear box, the potential is given by

$$V(x) = \begin{cases} \infty, & x \leq 0, \ x \geq L \quad \text{(domain 1)} \\ 0, & 0 < x < L \quad \text{(domain 2)}. \end{cases} \tag{2.64}$$

This is illustrated in Fig. 2.1.

In domain 1, the Hamiltonian is

$$\hat{H}_1 = \frac{\hat{p}^2}{2m} + \infty = \infty \tag{2.65}$$

and the eigenvalue equation is

$$\hat{H}_1 \psi = E\psi. \tag{2.66}$$

Since E and ψ are finite, right-hand side of Eq. (2.66) is finite, then the left-hand side should also be finite, which leads to

$$\psi = 0. \tag{2.67}$$

So, there is zero probability that the particle is found in domain 1.

In domain 2, the Hamiltonian is

$$\hat{H}_2 = \frac{\hat{p}^2}{2m} = \frac{-\hbar^2}{2m} \frac{\mathrm{d}^2}{\mathrm{d}x^2} \tag{2.68}$$

and the eigenvalue equation is

$$-\frac{\hbar^2}{2m} \frac{\mathrm{d}^2}{\mathrm{d}x^2} \psi_n = E\psi_n, \tag{2.69}$$

which can be rewritten as

$$\frac{\mathrm{d}^2}{\mathrm{d}x^2} \psi_n + k_n^2 \psi_n = 0 \tag{2.70}$$

where

$$k_n^2 = \frac{2mE_n}{\hbar^2}.$$

(2.71)

Since ψ in domain 1 is universally zero, continuity of ψ at $x = 0$ and $x = L$ leads to

$$\psi_n(0) = \psi_n(L) = 0,$$

(2.72)

which is the boundary condition of differential equation (2.70).

The general solution of Eq. (2.70) is

$$\psi_n = A \sin k_n x + B \cos k_n x.$$

(2.73)

Using the boundary condition (2.72), we have

$$\psi_n(0) = 0 \longrightarrow B = 0$$
$$\psi_n(L) = 0 \longrightarrow A \sin k_n L = 0$$
$$\longrightarrow k_n L = n\pi, \quad n = 0, 1, 2, \ldots$$

and the wavefunction is

$$\psi_n(x) = A \sin \frac{n\pi x}{L}.$$

(2.74)

Since $k_n L = n\pi$, we have $L = n\frac{\lambda_n}{2}$, i.e. in stationary states there is an integer number of half wavelengths between the two walls. Hence we see that the quantum number n appears naturally in QM.

To determine A, we use the normalization condition

$$\int \psi\psi^* \, d\tau = 1,$$

(2.75)

which gives

$$\int_0^L \psi_n^2 \, dx = 1 = A^2 \int_0^L \sin^2\left(\frac{n\pi x}{L}\right) dx$$
$$= \frac{A^2 L}{n\pi} \int_0^{n\pi} \sin^2 \theta \, d\theta$$
$$= \frac{A^2 L}{2}$$
$$\Longrightarrow A = \sqrt{\frac{2}{L}}$$

(2.76)

So the stationary-state wavefunction in domain 2 is

$$\psi_n = \sqrt{\frac{2}{L}} \sin\left(\frac{n\pi x}{L}\right), \tag{2.77}$$

and the corresponding eigenvalue of energy is

$$E_n = n^2 E_1, \qquad E_1 = \frac{\hbar^2 k_1^2}{2m} = \frac{\hbar^2 \pi^2}{2mL^2}. \tag{2.78}$$

It can be easily confirmed that the eigenfunctions are orthogonal.

$$\int_0^L \psi_n \psi_{n'}\, \mathrm{d}x = 0, \qquad n \neq n'. \tag{2.79}$$

2.2.3 *Potential barrier of infinite width*

In this case, the potential is given by

$$V(x) = \begin{cases} 0, & x < 0 \\ V_0, & x > 0. \end{cases} \tag{2.80}$$

This is illustrated in Fig. 2.2.

In domain 1, the time-independent Schrödinger equation reads

$$-\frac{\hbar^2}{2m}\frac{\mathrm{d}^2}{\mathrm{d}x^2}\psi = E\psi, \tag{2.81}$$

and the solution is

$$\psi = Ae^{ik_1 x} + Be^{-ik_1 x}, \qquad E = \frac{\hbar^2 k_1^2}{2m}. \tag{2.82}$$

In domain 2, we have to consider two cases. When $E > V_0$, the time-independent Schrödinger equation reads

$$-\frac{\hbar^2}{2m}\frac{\mathrm{d}^2}{\mathrm{d}x^2}\psi + V_0\psi = E\psi \tag{2.83}$$

$$\implies \frac{\mathrm{d}^2\psi}{\mathrm{d}x^2} + \frac{2m(E - V_0)}{\hbar^2}\psi = 0 \tag{2.84}$$

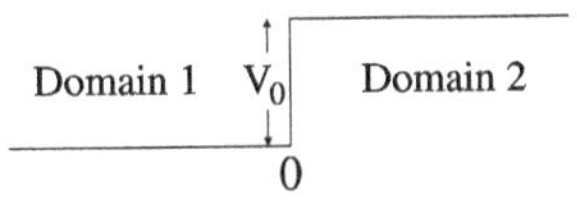

Fig. 2.2 Potential barrier of infinite width.

or

$$\frac{\mathrm{d}^2\psi}{\mathrm{d}x^2} + k_2^2\psi = 0, \quad k_2^2 = \frac{2m(E - V_0)}{\hbar^2}. \tag{2.85}$$

The solution is

$$\psi(x) = Ce^{ik_2 x} + De^{-ik_2 x}. \tag{2.86}$$

Both wavefunctions (2.82) and (2.86) can be seen as a combination of two plane waves with opposite direction. Then this stationary-state problem can be seen as a problem of particle transmission from domain 1 to domain 2. Classically, when $E > V_0$, all particles would be transmitted through the barrier. In QM, this is no longer true and $\left|\frac{B}{A}\right|^2$ gives the intensity of the reflection wave relative to that of the incident wave. For $x > 0$, no particles can flow to the left, so D must be zero.

For a finite barrier, ψ and $\frac{\mathrm{d}\psi}{\mathrm{d}x}$ are both continuous. At $x = 0$,

$$\begin{cases} A + B = C \\ (A - B)k_1 = Ck_2 \end{cases} \implies \begin{cases} \dfrac{B}{A} = \dfrac{k_1 - k_2}{k_1 + k_2} = \dfrac{\sqrt{E} - \sqrt{E - V_0}}{\sqrt{E} + \sqrt{E - V_0}} \\ \dfrac{C}{A} = \dfrac{2k_1}{k_1 + k_2} = \dfrac{2\sqrt{E}}{\sqrt{E} + \sqrt{E - V_0}} \end{cases} \tag{2.87}$$

In the limit of $E \to \infty$, $B = 0$ and $C = A$, then all particles are transmitted and QM is in agreement with CM.

When $E < V_0$, the time-independent Schrödinger equation is

$$\frac{\mathrm{d}^2\psi}{\mathrm{d}x^2} - \frac{2m(V_0 - E)}{\hbar^2}\psi = 0 \tag{2.88}$$

or

$$\frac{\mathrm{d}^2\psi}{\mathrm{d}x^2} - k_2'^2\psi = 0, \quad k_2'^2 = \frac{2m(V_0 - E)}{\hbar^2}. \tag{2.89}$$

The solution is

$$\psi = Ce^{-k_2' x} + De^{k_2' x}. \tag{2.90}$$

The finiteness of ψ requires $D = 0$, otherwise $\psi \to \infty$ when $x \to \infty$.

Again, because of the continuity of ψ and $\frac{\mathrm{d}\psi}{\mathrm{d}x}$ at $x = 0$,

$$\begin{cases} A + B = C \\ i(A - B)k_1 = -Ck_2' \end{cases} \implies \begin{cases} \dfrac{B}{A} = \dfrac{k_1 - ik_2'}{k_1 + ik_2'} \\ \dfrac{C}{A} = \dfrac{2k_1}{k_1 + ik_2'} \end{cases} \tag{2.91}$$

The reflectance $\left|\frac{B}{A}\right|^2 = 1$, which means that no particle is transmitted to domain 2, in accordance with the exponentially decaying wavefunction (2.90) (with $D = 0$) in domain 2.

$$|C|^2 = |A|^2 \frac{4k_1^2}{k_1^2 + k_2^2} = \frac{4E}{V_0}|A|^2, \tag{2.92}$$

so

$$|\psi(x)|^2 = |A|^2 \frac{4E}{V_0} \exp\left[-2\sqrt{2m(V_0 - E)}x/\hbar\right], \quad x > 0. \tag{2.93}$$

When $V_0 \to \infty$, $k_2' \to \infty$, Eq. (2.91) gives $B = -A$ and $C = 0$, then the wavefunction is

$$\begin{cases} \psi(x) = 2iA \sin k_1 x, & x < 0 \\ \psi(x) = 0, & x \geq 0. \end{cases} \tag{2.94}$$

2.2.4 *Tunneling effect*

Consider a finite potential barrier with finite width.

$$V(x) = \begin{cases} 0, & x < 0 \\ V_0, & 0 \leq x \leq a \\ 0, & x > a. \end{cases} \tag{2.95}$$

This is illustrated in Fig. 2.3. When $E < V_0$, classically none of the particles would be transmitted, but this is not the case in QM.

The time-independent Schrödinger equation is

$$\begin{cases} -\dfrac{\hbar^2}{2m}\dfrac{d^2\psi}{dx^2} = E\psi, & x < 0,\ x > a \\[2ex] -\dfrac{\hbar^2}{2m}\dfrac{d^2\psi}{dx^2} + V_0\psi = E\psi, & 0 \leq x \leq a. \end{cases} \tag{2.96}$$

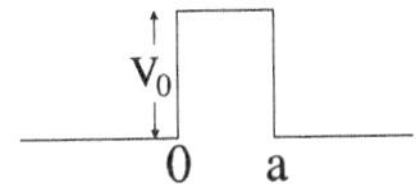

Fig. 2.3   Potential barrier with finite width.

When $E < V_0$, the solutions are

$$\begin{cases} \psi(x) = Ae^{ikx} + Be^{-ikx}, & x < 0 \\ \psi(x) = Ce^{\alpha x} + De^{-\alpha x}, & 0 \le x \le a \\ \psi(x) = Fe^{ikx}, & x > a \end{cases} \tag{2.97}$$

with $k = \dfrac{\sqrt{2mE}}{\hbar}$ and $\alpha = \dfrac{\sqrt{2m(V_0 - E)}}{\hbar}$.

Since ψ and $\frac{\mathrm{d}\psi}{\mathrm{d}x}$ are continuous at $x = 0$ and $x = a$,

$$\begin{aligned} \text{at} \quad x = 0, \quad & \begin{cases} A + B = C + D \\ ik(A - B) = \alpha(C - D) \end{cases} \\ \text{at} \quad x = a, \quad & \begin{cases} Ce^{\alpha a} + De^{-\alpha a} = Fe^{ika} \\ \alpha\left(Ce^{\alpha a} - De^{-\alpha a}\right) = ikFe^{ika}. \end{cases} \end{aligned} \tag{2.98}$$

Solving for $\frac{F}{A}$, we obtain

$$\frac{F}{A} = \frac{4i\alpha k}{e^{ika}\left[4i\alpha k \cosh \alpha a - (\alpha^2 - k^2)2 \sinh \alpha a\right]}$$
$$\left(\cosh \alpha a = \frac{e^{\alpha a} + e^{-\alpha a}}{2}, \quad \sinh \alpha a = \frac{e^{\alpha a} - e^{-\alpha a}}{2} \right). \tag{2.99}$$

The number of particles transmitted through the barrier is proportional to $\left|\frac{F}{A}\right|^2$, i.e.

$$\left|\frac{F}{A}\right|^2 = \frac{4}{\left[4\cosh^2 \alpha a + \left(\frac{\alpha}{k} - \frac{k}{\alpha}\right)^2 \sinh^2 \alpha a\right]}. \tag{2.100}$$

If αa (α and a represent height and width of the potential barrier respectively) is large, then both $\cosh \alpha a$ and $\sinh \alpha a$ behave as $e^{\alpha a}/2$ and

$$\left|\frac{F}{A}\right|^2 = \frac{16\alpha^2 k^2}{(\alpha^2 + k^2)^2}e^{-2\alpha a}. \tag{2.101}$$

The number of transmitted particles falls off exponentially as

$$\exp\left\{ -2\left[\frac{2m}{\hbar^2}(V_0 - E)\right]^{\frac{1}{2}} a \right\}. \tag{2.102}$$

If $h \to 0$, the transmission number goes to 0 as in CM.

Tunneling of particles occurs in radioactive decay. For example in the case of $^{238}_{92}$U, the height of the potential barrier is about 30 MeV whereas the kinetic energy of the emitted α-particles is only 4.2 MeV, so classically the decay cannot take place, QM is required to explained the phenomenon.

A summary of the steps for solving one-dimensional quantum mechanical problems:

(1) Write down the Hamiltonian H of the system.
(2) Convert the Hamiltonian H to the operator form $\hat{H}$ $(p \to -i\hbar\frac{\partial}{\partial x})$.
(3) Write down the Schrödinger equation $\hat{H}\Psi = i\hbar\frac{\partial\Psi}{\partial t}$. If $\hat{H}$ is time-independent, separate the variables:

$$\hat{H}\psi = E\psi \quad \text{(time-independent Schrödinger equation)} \qquad (2.103)$$

$$\Psi = \psi e^{-\frac{iEt}{\hbar}}. \qquad (2.104)$$

(4) Solve the Schrödinger equation by using the standard conditions of wavefunction: single-valued, square-integrable and continuous.

For more cases of one-dimensional QM problems, see for example Refs. 1 and 2.

2.2.5 *Linear harmonic oscillator*

For any system near its equilibrium position x_0, the potential energy can be expanded in a Taylor series

$$V(x) = V(x_0) + V'(x_0)(x - x_0) + \frac{1}{2!}V''(x_0)(x - x_0)^2 + \ldots \qquad (2.105)$$

Since at the equilibrium position $V(x_0)$ is an extremum, $V'(x_0) = 0$. The term $V(x_0)$ can be chosen as zero and the equilibrium position x_0 can be taken as $x_0 = 0$. Furthermore, if the second-order term is not zero and we drop the higher-order terms, we can write the potential as

$$V(x) = \frac{1}{2}kx^2, \quad k = \left.\frac{d^2V}{dx^2}\right|_{x=0} > 0. \qquad (2.106)$$

Classically, the dynamic equation of harmonic oscillator is

$$m\ddot{x} + kx = 0, \qquad (2.107)$$

and it can be rewritten as

$$\ddot{x} + \omega^2 x = 0, \quad \omega = \sqrt{\frac{k}{m}}, \qquad (2.108)$$

while the solution is

$$x = A\sin(\omega t + \phi). \tag{2.109}$$

In QM, the problem of linear harmonic oscillator is solving the Schrödinger equation for $V(x) = \frac{1}{2}m\omega^2 x^2$.

2.2.5.1   *Analytic solution*

The time-independent Schrödinger equation is

$$\frac{\mathrm{d}^2\psi}{\mathrm{d}x^2} + \frac{2m}{\hbar^2}\left(E - \frac{1}{2}m\omega^2 x^2\right)\psi = 0. \tag{2.110}$$

Let

$$y = \sqrt{\frac{m\omega}{\hbar}}\,x = \beta x, \quad \lambda = \frac{2E}{\hbar\omega}, \tag{2.111}$$

where

$$\beta = \sqrt{\frac{m\omega}{\hbar}}, \tag{2.112}$$

then we have

$$\frac{\mathrm{d}^2\psi}{\mathrm{d}y^2} + (\lambda - y^2)\psi = 0. \tag{2.113}$$

When y^2 is very large,

$$\frac{\mathrm{d}^2\psi}{\mathrm{d}y^2} - y^2\psi = 0, \tag{2.114}$$

the approximate solution is

$$\psi \sim \mathrm{e}^{\pm y^2/2}. \tag{2.115}$$

Since the boundary condition requires $\psi \to 0$ when $y \to \infty$, the solution $\mathrm{e}^{+y^2/2}$ is not permissible, therefore we obtain

$$\psi \sim \mathrm{e}^{-y^2/2}. \tag{2.116}$$

Let

$$\psi = H(y)\mathrm{e}^{-y^2/2}. \tag{2.117}$$

Then Eq. (2.113) becomes

$$\frac{\mathrm{d}^2 H}{\mathrm{d}y^2} - 2y\frac{\mathrm{d}H}{\mathrm{d}y} + (\lambda - 1)H = 0, \tag{2.118}$$

which is actually the Hermite equation.

Expand $H(y)$ in a series,

$$H(y) = \sum_r a_r y^r, \tag{2.119}$$

and substitute this series (2.119) back into the Hermite equation (2.118), then we get a recurrence relation

$$a_{r+2} = \frac{2r + 1 - \lambda}{(r+1)(r+2)} a_r. \tag{2.120}$$

This means that

$$a_2, a_4, \ldots \text{ can be expressed by } a_0,$$

$$a_3, a_5, \ldots \text{ can be expressed by } a_1,$$

so r even and r odd yield two independent solutions.

$$\lim_{r \to \infty} \frac{a_{r+2}}{a_r} = \lim_{r \to \infty} \frac{2r + 1 - \lambda}{(r+1)(r+2)} = \mathcal{O}\left(\frac{2}{r}\right). \tag{2.121}$$

Consider the series expansion of $\exp(y^2)$,

$$e^{y^2} = 1 + y^2 + \frac{y^4}{2!} + \cdots + \frac{(y^2)^s}{s!} + \frac{(y^2)^{s+1}}{(s+1)!} + \cdots$$

$$= 1 + y^2 + \frac{y^4}{2!} + \cdots + \frac{y^r}{\left(\frac{r}{2}\right)!} + \frac{y^{r+2}}{\left(\frac{r}{2}+1\right)!} + \cdots \qquad (r = 2s). \tag{2.122}$$

Since

$$\lim_{r \to \infty} \frac{\left(\frac{r}{2}\right)!}{\left(\frac{r}{2}+1\right)!} = \lim_{r \to \infty} \frac{1}{\frac{r}{2}+1} = \mathcal{O}\left(\frac{2}{r}\right), \tag{2.123}$$

this shows that $\exp(y^2)$ has the same behaviour for large y as $H(y) = \sum_r a_r y^r$, i.e.

$$y \to \infty, \qquad H(y) \to \exp(y^2), \tag{2.124}$$

$$\psi(y) \to e^{-y^2/2} e^{y^2} = e^{y^2/2}. \tag{2.125}$$

It is divergent, which is not allowed. This means that the series must terminate at certain term, i.e. $H(y)$ must be a polynomial of the variable y rather than an infinite series. To terminate the series of $H(y)$ (2.119), we should have

$$\lambda = 2r + 1. \tag{2.126}$$

The corresponding polynomial is called Hermite polynomial of degree r.

Substituting (2.126) into (2.118), we get

$$H_r'' - 2yH_r' + 2rH_r = 0, \tag{2.127}$$

where $H_r' = \frac{\mathrm{d}H_r}{\mathrm{d}y}, H_r'' = \frac{\mathrm{d}^2 H_r}{\mathrm{d}y^2}$. A general expression of Hermite polynomial is

$$H_r(y) = (-1)^r e^{y^2} \frac{\mathrm{d}^r}{\mathrm{d}y^r}(e^{-y^2}). \tag{2.128}$$

Proof. If (2.128) satisfies (2.127), we should have

$$\left[\frac{\mathrm{d}^{r+2}}{\mathrm{d}y^{r+2}} + 2y \frac{\mathrm{d}^{r+1}}{\mathrm{d}y^{r+1}} + (2+2r) \frac{\mathrm{d}^r}{\mathrm{d}y^r} \right] e^{-y^2} = 0. \tag{2.129}$$

We use mathematical induction.

Verify for $r = 0$: from Eq. (2.129) we obtain

$$\left[\frac{\mathrm{d}^2}{\mathrm{d}y^2} + 2y \frac{\mathrm{d}}{\mathrm{d}y} + 2 \right] e^{-y^2} = 0, \tag{2.130}$$

which is true.

If $r = n - 1$ is true, i.e.

$$\left[\frac{\mathrm{d}^{n+1}}{\mathrm{d}y^{n+1}} + 2y \frac{\mathrm{d}^n}{\mathrm{d}y^n} + 2n \frac{\mathrm{d}^{n-1}}{\mathrm{d}y^{n-1}} \right] e^{-y^2} = 0. \tag{2.131}$$

Differentiate both sides of Eq. (2.131) with respect to y, we get

$$\left[\frac{\mathrm{d}^{n+2}}{\mathrm{d}y^{n+2}} + 2y \frac{\mathrm{d}^{n+1}}{\mathrm{d}y^{n+1}} + (2n+2) \frac{\mathrm{d}^n}{\mathrm{d}y^n} \right] e^{-y^2} = 0, \tag{2.132}$$

thus $r = n$ is also true. $\qquad\square$

The first few Hermite polynomials $H_r(y)$ are as follows (the coefficient of the highest power of y is chosen to be 2^r):

$$H_0(y) = 1$$
$$H_1(y) = 2y$$
$$H_2(y) = 4y^2 - 2$$
$$H_3(y) = 8y^3 - 12y$$
$$H_4(y) = 16y^4 - 48y^2 + 12$$
$$H_5(y) = 32y^5 - 160y^3 + 120y.$$

Here we note some useful properties.

- Derivative

$$H'_r = 2rH_{r-1}. \tag{2.133}$$

- Recurrence relation

$$H_{r+1}(y) - 2yH_r(y) + 2rH_{r-1}(y) = 0. \tag{2.134}$$

- Orthogonal relation

$$\int_{-\infty}^{+\infty} H_m(y)H_n(y)\mathrm{e}^{-y^2}\,\mathrm{d}y = \sqrt{\pi}2^n n!\delta_{mn}. \tag{2.135}$$

For $m = n$ in Eq. (2.135), we get the normalization property

$$\int_{-\infty}^{+\infty} H_n^2(y)\mathrm{e}^{-y^2}\,\mathrm{d}y = \sqrt{\pi}2^n n!. \tag{2.136}$$

Proof. We use the method of mathematical induction.

For $n = 0$,

$$\int_{-\infty}^{+\infty} \mathrm{e}^{-y^2}\,\mathrm{d}y = \sqrt{\pi}, \tag{2.137}$$

which is true.

If for $n = r - 1$, Eq. (2.136) is true, i.e.

$$\int_{-\infty}^{+\infty} H_{r-1}^2(y)\mathrm{e}^{-y^2}\,\mathrm{d}y = \sqrt{\pi}2^{r-1}(r-1)! \tag{2.138}$$

then

$$\int_{-\infty}^{+\infty} H_r^2 \mathrm{e}^{-y^2}\,\mathrm{d}y$$

$$= (-1)^r \int_{-\infty}^{+\infty} \left(\frac{\mathrm{d}^r}{\mathrm{d}y^r}\mathrm{e}^{-y^2}\right) H_r\,\mathrm{d}y$$

$$= (-1)^{r-1}(-1) \int_{-\infty}^{+\infty} \left[\frac{\mathrm{d}}{\mathrm{d}y}\left(\frac{\mathrm{d}^{r-1}}{\mathrm{d}y^{r-1}}\mathrm{e}^{-y^2}\right)\right] H_r\,\mathrm{d}y$$

$$= (-1)^{r-1}\left\{(-1)H_r \frac{\mathrm{d}^{r-1}}{\mathrm{d}y^{r-1}}\mathrm{e}^{-y^2}\Big|_{-\infty}^{+\infty} + \int_{-\infty}^{+\infty}\left(\frac{\mathrm{d}^{r-1}}{\mathrm{d}y^{r-1}}\mathrm{e}^{-y^2}\right)\frac{\mathrm{d}H_r}{\mathrm{d}y}\right\}\,\mathrm{d}y$$

$$= (-1)^{r-1} \int_{-\infty}^{+\infty} \left(\frac{\mathrm{d}^{r-1}}{\mathrm{d}y^{r-1}} \mathrm{e}^{-y^2} \right) H'_r \, \mathrm{d}y$$

$$= (-1)^{r-1} \int_{-\infty}^{+\infty} \left(\frac{\mathrm{d}^{r-1}}{\mathrm{d}y^{r-1}} \mathrm{e}^{-y^2} \right) 2r H_{r-1} \, \mathrm{d}y$$

$$= 2r \int_{-\infty}^{+\infty} H_{r-1}^2 \mathrm{e}^{-y^2} \, \mathrm{d}y$$

$$= 2r[\sqrt{\pi} 2^{r-1}(r-1)!]$$

$$= \sqrt{\pi} 2^r r!.$$

Thus when $n = r$, Eq. (2.136) is also true. In the process of calculation we have used the method of integration by parts, the fact that $\mathrm{e}^{-y^2} = 0$ at the point $y = \pm\infty$, and also Eq. (2.133). $\qquad\square$

Normalize the wavefunction $\psi_n = A_n \mathrm{e}^{-y^2/2} H_n(y)$,

$$\int \psi_n^* \psi_n \, \mathrm{d}x = 1 \implies \int |A_n|^2 H_n^2(y) \mathrm{e}^{-y^2} \, \mathrm{d}x = 1$$

$$\implies \frac{|A_n|^2}{\beta} \int H_n^2(y) \mathrm{e}^{-y^2} \, \mathrm{d}y = \sqrt{\pi} 2^n n! \frac{|A_n|^2}{\beta} = 1$$

$$\implies A_n = \beta^{\frac{1}{2}} \sqrt{\frac{1}{\sqrt{\pi} 2^n n!}} = \left(\frac{m\omega}{\pi\hbar} \right)^{\frac{1}{4}} \frac{1}{\sqrt{2^n n!}}.$$

Now we have the stationary-state wavefunctions

$$\psi_n = \left(\frac{m\omega}{\pi\hbar} \right)^{\frac{1}{4}} \frac{1}{\sqrt{2^n n!}} \mathrm{e}^{-y^2/2} H_n(y), \quad n = 0, 1, 2, \tag{2.139}$$

while Eqs. (2.111) and (2.126) give the eigenvalues

$$E_n = \hbar\omega \left(n + \frac{1}{2} \right). \tag{2.140}$$

This shows that a linear harmonic oscillator has a zero-point energy of $\frac{1}{2}\hbar\omega$.

The zero-point energy can be observed in the scattering of light by crystal because the scattering is caused by oscillation of atoms. When $T \to 0$, the definite value of the intensity of scattering light is not zero, which means that there is a zero-point oscillation. The existence of this zero-point energy is directly related to the uncertainty principle.

2.2.5.2 *Algebraic solution*

We can also find the solution to the linear harmonic oscillator through algebraic calculations. To facilitate the calculations, we define the following "position" and "momentum" operators:

$$\hat{X} = \left(\frac{m\omega}{2\hbar}\right)^{\frac{1}{2}}\hat{x}, \quad \hat{P} = \frac{1}{(2m\hbar\omega)^{\frac{1}{2}}}\hat{p}. \tag{2.141}$$

$$[\hat{X}, \hat{P}] = \frac{1}{2\hbar}[\hat{x}, \hat{p}] = \frac{1}{2\hbar}i\hbar = \frac{i}{2}. \tag{2.142}$$

The Hamiltonian can then be written as

$$\begin{aligned}
\hat{H} &= \frac{1}{2m}\hat{p}^2 + \frac{1}{2}m\omega^2\hat{x}^2 \\
&= \hbar\omega(\hat{P}^2 + \hat{X}^2) \\
&= \hbar\omega\left(\hat{a}^\dagger\hat{a} + \frac{1}{2}\right),
\end{aligned} \tag{2.143}$$

with $\hat{a} = \hat{X} + i\hat{P}, \quad \hat{a}^\dagger = \hat{X} - i\hat{P}.$

$$[\hat{a}, \hat{a}^\dagger] = \hat{a}\hat{a}^\dagger - \hat{a}^\dagger\hat{a} = \left(\hat{X}^2 + \hat{P}^2 + \frac{1}{2}\right) - \left(\hat{X}^2 + \hat{P}^2 - \frac{1}{2}\right) = 1. \tag{2.144}$$

Use Dirac notation $|u_n\rangle$ to denote the eigenstates of the Hamiltonian,

$$\hat{H}|u_n\rangle = E_n|u_n\rangle. \tag{2.145}$$

Let $\epsilon_n = \frac{E_n}{\hbar\omega}$,

$$\left(\hat{a}^\dagger\hat{a} + \frac{1}{2}\right)|u_n\rangle = \frac{E_n}{\hbar\omega}|u_n\rangle = \epsilon_n|u_n\rangle \tag{2.146}$$

$$\Longrightarrow \hat{a}\left(\hat{a}^\dagger\hat{a} + \frac{1}{2}\right)|u_n\rangle = \hat{a}(\epsilon_n|u_n\rangle) \tag{2.147}$$

$$\Longrightarrow \left(\hat{a}^\dagger\hat{a} + \frac{3}{2}\right)\hat{a}|u_n\rangle = \epsilon_n\hat{a}|u_n\rangle \tag{2.148}$$

$$\Longrightarrow \left(\hat{a}^\dagger\hat{a} + \frac{1}{2}\right)\hat{a}|u_n\rangle = (\epsilon_n - 1)\hat{a}|u_n\rangle \tag{2.149}$$

Thus $\hat{a}|u_n\rangle$ is also an eigenstate of the Hamiltonian but with energy eigenvalue

$$E = \hbar\omega(\epsilon_n - 1) = E_n - \hbar\omega. \tag{2.150}$$

Similarly, we have

$$\left(\hat{a}^{\dagger}\hat{a} + \frac{1}{2}\right)\hat{a}^{\dagger}|u_n\rangle = (\epsilon_n + 1)\hat{a}^{\dagger}|u_n\rangle. \tag{2.151}$$

So $\hat{a}$ is called the lowering operator and $\hat{a}^{\dagger}$ is called the raising operator.

Since $V(x) \geq 0$, the eigenvalues of the Hamiltonian should be positive semidefinite, i.e. $E_n \geq 0$. There should be a lowest eigenstate $|u_0\rangle$ to prevent the eigenvalue from being negative,

$$\hat{a}|u_0\rangle = 0, \tag{2.152}$$

$$\hat{H}|u_0\rangle = E_0|u_0\rangle. \tag{2.153}$$

Then

$$E_0|u_0\rangle = \hbar\omega\left(\hat{a}^{\dagger}\hat{a} + \frac{1}{2}\right)|u_0\rangle = \frac{\hbar\omega}{2}|u_0\rangle$$

$$\implies E_0 = \frac{\hbar\omega}{2}.$$

Higher eigenstates can be generated by using raising operator $\hat{a}^{\dagger}$,

$$|u_1\rangle = A_1\hat{a}^{\dagger}|u_0\rangle, \quad |u_n\rangle = A_n(\hat{a}^{\dagger})^n|u_0\rangle, \tag{2.154}$$

where A_n is the normalization factor. The corresponding eigenvalues of energy are

$$E_n = \left(n + \frac{1}{2}\right)\hbar\omega, \quad n = 0, 1, 2, \ldots \tag{2.155}$$

Compare with Eq. (2.146), we have

$$\hat{a}^{\dagger}\hat{a}|u_n\rangle = n|u_n\rangle, \tag{2.156}$$

$$\hat{a}\hat{a}^{\dagger}|u_n\rangle = (1 + \hat{a}^{\dagger}\hat{a})|u_n\rangle = (n+1)|u_n\rangle. \tag{2.157}$$

To determine A_n, we note that $|u_{n+1}\rangle \propto \hat{a}^{\dagger}|u_n\rangle$ and that

$$\langle \hat{a}^{\dagger}u_n|\hat{a}^{\dagger}u_n\rangle = \langle u_n|\hat{a}\hat{a}^{\dagger}u_n\rangle$$

$$= (n+1)\langle u_n|u_n\rangle$$

$$= n+1,$$

since $|u_n\rangle$ is normalized. So the normalized eigenfunction should be

$$|u_{n+1}\rangle = \frac{1}{\sqrt{n+1}}\hat{a}^\dagger|u_n\rangle$$

$$= \frac{1}{\sqrt{n+1}\sqrt{n}}(\hat{a}^\dagger)^2|u_{n-1}\rangle$$

$$= \cdots = \frac{1}{\sqrt{(n+1)!}}(\hat{a}^\dagger)^{n+1}|u_0\rangle.$$

Hence $A_n = \frac{1}{\sqrt{n!}}$ and we have

$$|u_n\rangle = \frac{1}{\sqrt{n!}}(\hat{a}^\dagger)^n|u_0\rangle. \tag{2.158}$$

We can still get the wavefunctions $\psi_n(x)$ of eigenstates by writing $\hat{a}$ and $\hat{a}^\dagger$ explicitly in x representation:

$$\hat{a} = \frac{1}{\sqrt{2\hbar m\omega}}\left(m\omega x + \hbar\frac{\mathrm{d}}{\mathrm{d}x}\right), \tag{2.159}$$

$$\hat{a}^\dagger = \frac{1}{\sqrt{2\hbar m\omega}}\left(m\omega x - \hbar\frac{\mathrm{d}}{\mathrm{d}x}\right). \tag{2.160}$$

We first find $\psi_0(x)$ from $|u_0\rangle$.

$$\hat{a}|u_0\rangle = 0 \implies \sqrt{\frac{\hbar}{2m\omega}}\left(\frac{\mathrm{d}}{\mathrm{d}x} + \frac{m\omega}{\hbar}x\right)\psi_0(x) = 0 \tag{2.161}$$

$$\implies \psi_0(x) = \left(\frac{m\omega}{\pi\hbar}\right)^{\frac{1}{4}} e^{-\frac{1}{2}\frac{m\omega}{\hbar}x^2}. \tag{2.162}$$

Applying the raising operator,

$$\psi_n(x) = \frac{1}{\sqrt{n!}}(\hat{a}^\dagger)^n\psi_0(x) = \left(\frac{m\omega}{\pi\hbar}\right)^{\frac{1}{4}}\frac{1}{\sqrt{2^n n!}}e^{-y^2/2}H_n(y). \tag{2.163}$$

Chapter 3

Matrix Mechanics

3.1 Introduction

Matrix mechanics was developed by Werner Heisenberg, Max Born, and Pascual Jordan. It is an equivalent, alternative formulation of quantum mechanics.

For the Schrödinger formulation, the mathematical tool is differential operators, the fundamental law is expressed by differential equations; whereas for the Heisenberg formulation, the mathematical tool is matrices, the fundamental law is expressed by matrix equations. This is summarized in Table 3.1.

Consider a differential equation

$$\hat{A}|\psi\rangle = a|\psi\rangle. \tag{3.1}$$

In some α representation with orthonormal basis vectors $|\psi_i\rangle$ satisfying

$$\hat{\alpha}|\psi_i\rangle = \alpha_i|\psi_i\rangle, \tag{3.2}$$

the wave function $|\psi\rangle$ can be expanded as

$$|\psi\rangle = \sum_i c_i|\psi_i\rangle. \tag{3.3}$$

In the matrix language, $|\psi\rangle$ is represented by a column vector:

$$\begin{pmatrix} c_1 \\ c_2 \\ \vdots \end{pmatrix}. \tag{3.4}$$

Table 3.1   Different formulation of QM.

Formulation	Mathematical tool	Law
Schrödinger formulation	Differential operator	Differential equation
Heisenberg formulation	Matrix	Matrix equation

Substituting Eq. (3.3) into Eq. (3.1), and using the closure relation $\sum_j |\psi_j\rangle\langle\psi_j| = 1$, we obtain

$$\sum_i c_i \sum_j |\psi_j\rangle\langle\psi_j|\hat{A}|\psi_i\rangle = a \sum_i c_i|\psi_i\rangle. \tag{3.5}$$

Take inner product with $\langle\psi_l|$,

$$\langle\psi_l|\psi_j\rangle = \delta_{lj} \implies \sum_{i,j}\langle\psi_j|\hat{A}|\psi_i\rangle c_i \delta_{lj} = a\sum_i c_i \delta_{il} \tag{3.6}$$

$$\implies \sum_i c_i\langle\psi_l|\hat{A}|\psi_i\rangle = ac_l. \tag{3.7}$$

Let $a_{li} = \langle\psi_l|\hat{A}|\psi_i\rangle$, then Eq. (3.7) becomes

$$\sum_i a_{li}c_i = ac_l, \tag{3.8}$$

or

$$\underline{\underline{A}}\,\underline{c} = a\underline{c}, \tag{3.9}$$

where $\underline{\underline{A}}$ is the matrix (a_{ij}), $\underline{c}$ is a column vector.

So, in the α representation,

$$\hat{A} \longleftarrow \underline{\underline{A}} \qquad (\longleftarrow\text{: is represented by})$$
$$|\psi\rangle \longleftarrow \underline{c}$$
$$\hat{A}|\psi\rangle = a|\psi\rangle \longleftarrow \underline{\underline{A}}\,\underline{c} = a\underline{c}.$$

The basis $|\psi_1\rangle,\ldots,|\psi_i\rangle,\ldots$ are eigenfunctions of operator $\hat{\alpha}$. $\underline{\underline{A}}$ and $\underline{c}$ depend on the basis $|\psi_i\rangle$, i.e. different set of $|\psi_i\rangle$ gives different representative, just as a vector $\vec{A}$'s components (a_x, a_y, a_z) depend on the coordinate system chosen.

3.2 Matrix algebra

Before we go deep into matrix representation of quantum mechanics, let's review some basics of matrix algebra.

Definition 3.1. A matrix with m rows and n columns is called an $m \times n$ or "m by n" matrix. Given matrices $\underline{\underline{A}} = (a_{ij})$, $\underline{\underline{B}} = (b_{ij})$ and c a constant, we have:

(1) *Addition, subtraction and scalar multiplication:*

$$\underline{\underline{A}} \pm \underline{\underline{B}} = (a_{ij} \pm b_{ij}), \tag{3.10}$$

$$c\underline{\underline{A}} = (ca_{ij}). \tag{3.11}$$

(2) *Matrix multiplication:* Multiplication between two matrices is only possible when the number of columns of the first matrix equals the number of rows of the second matrix. In this case,

$$\underline{\underline{A}}\,\underline{\underline{B}} = (a_{ik}b_{kj}), \tag{3.12}$$

where we have adopted the Einstein's summation convention: repeated indices are to be summed. The repeated index k is called a dummy index, here we must sum for all possible k, i.e. the number of columns of the matrix $\underline{\underline{A}}$.

(3) *Transpose:*

$$\widetilde{\underline{\underline{A}}} = (a_{ji}), \tag{3.13}$$

$$\widetilde{(\underline{\underline{A}}\,\underline{\underline{B}})} = \widetilde{\underline{\underline{B}}}\,\widetilde{\underline{\underline{A}}}. \tag{3.14}$$

Definition 3.2. A square matrix is a matrix with the same number of rows and columns, i.e. an $n \times n$ matrix. Given square matrices $\underline{\underline{A}} = (a_{ij})$, $\underline{\underline{B}} = (b_{ij})$, we have:

(1) *Trace:*

$$\text{Trace } \underline{\underline{A}} = \sum_i a_{ii}, \tag{3.15}$$

$$\text{Trace } (\underline{\underline{A}}\,\underline{\underline{B}}) = \text{Trace } (\underline{\underline{B}}\,\underline{\underline{A}}). \tag{3.16}$$

(2) *Inverse:* When $\underline{\underline{A}}$ is nonsingular (i.e. $|\underline{\underline{A}}| \neq 0$), $\underline{\underline{A}}^{-1}$ is obtained from $\underline{\underline{A}}$ by replacing each element by its cofactor and then transposing the resulting matrix and finally dividing it by $|\underline{\underline{A}}|$.

$$(\underline{\underline{A}}\,\underline{\underline{B}})^{-1} = \underline{\underline{B}}^{-1}\underline{\underline{A}}^{-1}. \tag{3.17}$$

Example 3.1. Examples of square matrices.

(1) Symmetric matrix: $\widetilde{\underline{\underline{A}}} = \underline{\underline{A}}$ $(a_{ij} = a_{ji})$.
(2) Diagonal matrix: $\underline{\underline{A}} = (a_i \delta_{ij})$.
(3) Unit matrix: $\underline{\underline{I}} = (\delta_{ij})$.

Definition 3.3. A matrix $\underline{\underline{A}} = (a_{ij})$ is a Hermitian matrix if it is equal to its adjoint, i.e.

$$\underline{\underline{A}} = \underline{\underline{A}}^{\dagger}, \quad a_{ij} = (\underline{\underline{A}}^{\dagger})_{ij} \equiv a_{ji}^{*}. \tag{3.18}$$

Definition 3.4. A matrix $\underline{\underline{A}} = (a_{ij})$ is a unitary matrix if its adjoint is its inverse, i.e.

$$\underline{\underline{A}} = (\underline{\underline{A}}^{\dagger})^{-1}, \quad (\underline{\underline{A}}^{\dagger}\underline{\underline{A}})_{ij} = \delta_{ij}. \tag{3.19}$$

To find the eigenvalues of a matrix $\underline{\underline{A}} = (a_{ij})$,

$$\underline{\underline{A}}\,\underline{c} = a\underline{c}$$
$$\Longrightarrow |\underline{\underline{A}} - a\underline{\underline{I}}| = 0. \tag{3.20}$$

This is called the characteristic equation, whose n roots are the eigenvalues.

Theorem 3.1. *The eigenvalues of a Hermitian matrix are real.*

Theorem 3.2. *The eigenvectors of a Hermitian matrix corresponding to different eigenvalues are orthogonal.*

Proof. See Sec. 2.1.3.   □

3.3   Matrix mechanics

Now we write the Schrödinger equation in matrix form.
    With

$$|\psi\rangle = \sum_{j} c_j |\psi_j\rangle, \tag{3.21}$$

Schrödinger equation $\hat{H}|\psi\rangle = i\hbar\frac{\partial}{\partial t}|\psi\rangle$ becomes

$$\sum_j c_j \hat{H}|\psi_j\rangle = \sum_j i\hbar \dot{c}_j |\psi_j\rangle \tag{3.22}$$

$$\implies \langle\psi_i|\left(\sum_j c_j \hat{H}|\psi_j\rangle\right) = \langle\psi_i|\left(\sum_j i\hbar\dot{c}_j|\psi_j\rangle\right) \tag{3.23}$$

$$\implies \sum_j H_{ij}c_j = i\hbar\dot{c}_i \tag{3.24}$$

$$\implies \underline{\underline{H}}\,\underline{c} = i\hbar\underline{\dot{c}} \tag{3.25}$$

The time dependence of the wavefunction is reflected in the time dependence of the column vector.

To calculate the expectation value of an operator $\hat{A}$,

$$\langle\psi|\hat{A}|\psi\rangle = \sum_{i,j}\langle\psi|\psi_i\rangle\langle\psi_i|\hat{A}|\psi_j\rangle\langle\psi_j|\psi\rangle = \sum_{i,j} c_i^* a_{ij} c_j = \underline{c}^\dagger \underline{\underline{A}}\,\underline{c}. \tag{3.26}$$

To find the eigenvalues of an operator, we use the characteristic equation:

$$\hat{A}|\psi\rangle = a|\psi\rangle \longrightarrow \underline{\underline{A}}\,\underline{c} = a\underline{c} \longrightarrow |a_{ij} - a\delta_{ij}| = 0. \tag{3.27}$$

Example 3.2. For linear harmonic oscillator in the energy representation with basis $\{|u_n\rangle\}$, we can calculate the matrix of $\hat{x}$, $\hat{p}$, and $\hat{H}$.

Recall the discussion in Sec. 2.2.5.2, we have

$$\hat{a} = \frac{1}{\sqrt{2\hbar m\omega}}(m\omega\hat{x} + i\hat{p}), \tag{3.28}$$

$$\hat{a}^\dagger = \frac{1}{\sqrt{2\hbar m\omega}}(m\omega\hat{x} - i\hat{p}), \tag{3.29}$$

$$\hat{a}^\dagger|u_n\rangle = \sqrt{n+1}|u_{n+1}\rangle, \tag{3.30}$$

$$\hat{a}|u_n\rangle = \sqrt{n}|u_{n-1}\rangle. \tag{3.31}$$

From Eq. (3.30),

$$\langle u_i|\hat{a}^\dagger|u_j\rangle = \sqrt{j+1}\langle u_i|u_{j+1}\rangle = \sqrt{j+1}\,\delta_{i,j+1}. \tag{3.32}$$

Similarly from Eq. (3.31),

$$\langle u_i|\hat{a}|u_j\rangle = \sqrt{j}\langle u_i|u_{j-1}\rangle = \sqrt{j}\,\delta_{i,j-1}. \tag{3.33}$$

We can write $\hat{x}$ and $\hat{p}$ in terms of $\hat{a}$ and $\hat{a}^\dagger$,

$$\hat{x} = \sqrt{\frac{\hbar}{2m\omega}}(\hat{a}^\dagger + \hat{a}), \tag{3.34}$$

$$\hat{p} = i\sqrt{\frac{\hbar m\omega}{2}}(\hat{a}^\dagger - \hat{a}). \tag{3.35}$$

Set $\beta = \sqrt{\frac{m\omega}{\hbar}}$, then

$$x_{ij} = \langle u_i|\hat{x}|u_j\rangle = \frac{1}{\beta}\left[\sqrt{\frac{j+1}{2}}\delta_{i,j+1} + \sqrt{\frac{j}{2}}\delta_{i,j-1}\right]. \tag{3.36}$$

In matrix form,

$$(x_{ij}) = \frac{1}{\beta}\begin{pmatrix} 0 & \frac{1}{\sqrt{2}} & 0 & 0 & \dots \\ \frac{1}{\sqrt{2}} & 0 & \sqrt{\frac{2}{2}} & 0 & \dots \\ 0 & \sqrt{\frac{2}{2}} & 0 & \sqrt{\frac{3}{2}} & \dots \\ 0 & 0 & \sqrt{\frac{3}{2}} & 0 & \dots \\ \vdots & \vdots & \vdots & \vdots & \ddots \end{pmatrix}. \tag{3.37}$$

Similarly,

$$p_{ij} = \langle u_i|\hat{p}|u_j\rangle = i\hbar\beta\left[\sqrt{\frac{j+1}{2}}\delta_{i,j+1} - \sqrt{\frac{j}{2}}\delta_{i,j-1}\right], \tag{3.38}$$

$$(p_{ij}) = i\hbar\beta\begin{pmatrix} 0 & -\frac{1}{\sqrt{2}} & 0 & 0 & \dots \\ \frac{1}{\sqrt{2}} & 0 & -\sqrt{\frac{2}{2}} & 0 & \dots \\ 0 & \sqrt{\frac{2}{2}} & 0 & -\sqrt{\frac{3}{2}} & \dots \\ 0 & 0 & \sqrt{\frac{3}{2}} & 0 & \dots \\ \vdots & \vdots & \vdots & \vdots & \ddots \end{pmatrix}. \tag{3.39}$$

On the other hand,

$$H_{ij} = \langle u_i|\hat{H}|u_j\rangle = E_j\langle u_i|u_j\rangle = E_j\delta_{ij} = \left(j+\frac{1}{2}\right)\hbar\omega\delta_{ij}. \tag{3.40}$$

In matrix form,

$$(H_{ij}) = \hbar\omega \begin{pmatrix} \frac{1}{2} & 0 & 0 & \\ 0 & \frac{2}{3} & 0 & 0 \\ 0 & 0 & \frac{2}{5} & \\ & 0 & & \ddots \end{pmatrix}. \tag{3.41}$$

Remark 3.1. The matrix elements of a dynamical variable in its own representation is a diagonal matrix, and the diagonal elements are its eigenvalues.

Example 3.3. To find the matrix elements of the operator $\hat{x}^2$, we can do a matrix multiplication.

$$\begin{aligned} (x^2)_{ij} &= x_{ik}x_{kj} \\ &= \frac{1}{2\beta^2}\left[\sqrt{j(j-1)}\delta_{i,j-2} + (2j+1)\delta_{ij} + \sqrt{(j+1)(j+2)}\delta_{i,j+2}\right]. \end{aligned} \tag{3.42}$$

Here we summarize the general steps of solving QM problems in matrix representation.

(1) Choose a basis. This is the same as CM, first thing to do is to choose a basis $\{i, j, k\}$. Different basis will give different description. For some $\hat{\alpha}$ representation, the basis is the set of eigenvectors $\{|\psi_i\rangle\}$:

$$\hat{\alpha}|\psi_i\rangle = \alpha_i|\psi_i\rangle. \tag{3.43}$$

Some examples:

$$\hat{H}|\psi_i\rangle = E_n|\psi_i\rangle \qquad \text{energy representation} \tag{3.44}$$

$$\hat{p}|\psi_i\rangle = p'|\psi_i\rangle \qquad \text{momentum representation} \tag{3.45}$$

$$\hat{x}|\psi_i\rangle = x'|\psi_i\rangle \qquad \text{coordinate representation} \tag{3.46}$$

(2) Find the representation of $|\psi\rangle$, which is similar to finding components of a vector.

$$\begin{aligned} c_i &= \langle\psi_i|\psi\rangle & \boldsymbol{e}_i \cdot \boldsymbol{V} &= v_i \\ |\psi\rangle &= \sum_i c_i|\psi_i\rangle & \boldsymbol{V} &= \sum_i V_i\boldsymbol{e}_i \\ |\psi\rangle &\longleftarrow \underline{c} & \boldsymbol{V} &\longleftarrow (V_1, V_2, V_3) \end{aligned}$$

(3) Find the representation of the operators.

$$\hat{A} = (a_{ij}), \qquad a_{ij} = \langle\psi_i|\hat{A}|\psi_j\rangle. \tag{3.47}$$

(4) Write down the fundamental laws in matrix form. In $\hat{\alpha}$ representation,

   (a) $\hat{A}|\psi\rangle = a|\psi\rangle \longleftarrow \underline{\underline{A}}\,\underline{c} = a\underline{c}$.

   (b) $\hat{H}|\psi\rangle = i\hbar\frac{\partial}{\partial t}|\psi\rangle \longleftarrow \underline{\underline{H}}\,\underline{c} = i\hbar\underline{\dot{c}}$.

   (c) $\langle\psi|\hat{A}|\psi\rangle \longleftarrow \underline{c}^{\dagger}\underline{\underline{A}}\,\underline{c}$.

3.4   Transformation of representations

As mentioned, different representation will give different description. However, since they describe the same phenomena, there should be some connections between the different descriptions. Here we want to find the relationships between two different representations. With this relationship, we can transform the matrix form for one representation into the other.

3.4.1   *Transformation of basis*

Consider two representations with different bases: for $\hat{\alpha}$ representation the basis is $\{|\psi_i\rangle\}$, while for $\hat{\beta}$ representation the basis is $\{|\phi_j\rangle\}$. Using the closure relation $\sum_i |\psi_i\rangle\langle\psi_i| = 1$ we obtain

$$|\phi_j\rangle = \sum_i |\psi_i\rangle\langle\psi_i|\phi_j\rangle = \sum_i |\psi_i\rangle u_{ij}, \quad u_{ij} = \langle\psi_i|\phi_j\rangle. \tag{3.48}$$

Since

$$u_{jk}^{\dagger} = u_{kj}^{*} = \langle\psi_k|\phi_j\rangle^{*} = \langle\phi_j|\psi_k\rangle, \tag{3.49}$$

we have

$$\sum_j u_{ij} u_{jk}^{\dagger} = \sum_j \langle\psi_i|\phi_j\rangle\langle\phi_j|\psi_k\rangle = \langle\psi_i|\psi_k\rangle = \delta_{ik}, \tag{3.50}$$

i.e.

$$\underline{\underline{U}}\,\underline{\underline{U}}^{\dagger} = \underline{\underline{I}}, \tag{3.51}$$

where $\underline{\underline{U}} = (u_{ij})$. Therefore, $\underline{\underline{U}}$ is a unitary matrix.

3.4.2 *Transformation of a state vector*

Continuing with the same notation, consider a state vector in the two representations.

$$\text{For } \hat{\alpha} \text{ representation,} \quad |\psi\rangle = \sum_i c_i|\psi_i\rangle, \quad |\psi\rangle \longleftarrow \underline{c},$$

$$\text{for } \hat{\beta} \text{ representation,} \quad |\psi\rangle = \sum_i f_i|\phi_i\rangle, \quad |\psi\rangle \longleftarrow \underline{f}.$$

Then we have

$$\langle\psi_l|\sum_i c_i|\psi_i\rangle = \langle\psi_l|\sum_i f_i|\phi_i\rangle, \tag{3.52}$$

$$c_l = \sum_i \langle\psi_l|\phi_i\rangle f_i = \sum_i u_{li}f_i, \tag{3.53}$$

i.e. $\underline{c} = \underline{\underline{U}}\,\underline{f}$.

3.4.3 *Transformation of an operator*

Again with the same notation, consider an operator $\hat{A}$.

$$\text{For } \hat{\alpha} \text{ representation} \quad \hat{A} \longleftarrow (a_{ij}^{(\alpha)}), \quad a_{ij}^{(\alpha)} = \langle\psi_i|\hat{A}|\psi_j\rangle,$$

$$\text{for } \hat{\beta} \text{ representation} \quad \hat{A} \longleftarrow (a_{ij}^{(\beta)}), \quad a_{ij}^{(\beta)} = \langle\phi_i|\hat{A}|\phi_j\rangle.$$

Using the closure relation of $|\phi_i\rangle$,

$$a_{ij}^{(\alpha)} = \langle\psi_i|\hat{A}|\psi_j\rangle = \sum_{l,m}\langle\psi_i|\phi_l\rangle\langle\phi_l|\hat{A}|\phi_m\rangle\langle\phi_m|\psi_j\rangle = u_{il}a_{lm}^{(\beta)}u_{mj}^{\dagger}. \tag{3.54}$$

In other words,

$$\underline{\underline{A}}^{(\alpha)} = \underline{\underline{U}}\,\underline{\underline{A}}^{(\beta)}\underline{\underline{U}}^{-1} \qquad (\underline{\underline{U}}^{-1} = \underline{\underline{U}}^{\dagger}). \tag{3.55}$$

3.5 Coordinate representation and momentum representation

At last, we focus on two representations that are important but not easy to understand: coordinate representation and momentum representation, as an application of the general theory of representation explained above.

Firstly, recall the eigenvalue equations

$$\hat{p}|p'\rangle = p'|p'\rangle, \tag{3.56}$$

$$\hat{x}|x'\rangle = x'|x'\rangle. \tag{3.57}$$

If we choose $|x\rangle$ as the basis (coordinate representation), the representative of $|\psi\rangle$ is

$$\langle x|\psi\rangle = \psi(x). \tag{3.58}$$

The difficulty of understanding these two representations lies in the fact that the eigenvalue of coordinate or momentum operator is continuous. The representative of state $|\psi\rangle$ is then a continuous function rather than a column vector.

For continuous functions, the sum in inner product is replaced by integration. In this case, the closure relation is given by

$$\int |x\rangle\langle x|\,\mathrm{d}x = 1. \tag{3.59}$$

The previously given interpretation of inner product in Dirac notation can be seen more clearly now,

$$\langle\psi_1|\psi_2\rangle = \int \langle\psi_1|x\rangle\langle x|\psi_2\rangle\,\mathrm{d}x = \int \psi_1^*(x)\psi_2(x)\,\mathrm{d}x. \tag{3.60}$$

Proposition 3.1. *If* $|x\rangle = \hat{A}|\psi\rangle$, *then* $\langle x| = \langle\psi|\hat{A}^\dagger$.

Proof. From $\langle\phi|x\rangle = \langle x|\phi\rangle^*$ we find the corresponding $\langle x|$.

$$\langle\phi|x\rangle = \langle\phi|\hat{A}|\psi\rangle = \langle\hat{A}^\dagger\phi|\psi\rangle = \langle\psi|\hat{A}^\dagger|\phi\rangle^* = \langle x|\phi\rangle^*. \tag{3.61}$$

$\square$

The coordinate wavefunction $\psi(x)$ or momentum wavefunction $\phi(p)$ is just the representative of the corresponding representation,

$$\langle x|\psi\rangle = \psi(x), \tag{3.62}$$

$$\langle p|\psi\rangle = \phi(p). \tag{3.63}$$

The momentum eigenfunction in coordinate representation is

$$\psi_p(x) = \langle x|p\rangle = \frac{1}{\sqrt{2\pi\hbar}}e^{\frac{i}{\hbar}px}, \tag{3.64}$$

and $\langle x|p\rangle$ is just element of the "unitary matrix" between coordinate representation and momentum representation which enables us to transform the representative from one representation to the other. Similarly,

$$\langle p|x\rangle = \langle x|p\rangle^* = \frac{1}{\sqrt{2\pi\hbar}} \mathrm{e}^{-\frac{i}{\hbar}px}. \tag{3.65}$$

Using the closure relation, we have

$$\langle x|\psi\rangle = \int \langle x|p\rangle\langle p|\psi\rangle \,\mathrm{d}p, \tag{3.66}$$

$$\implies \psi(x) = \frac{1}{\sqrt{2\pi\hbar}} \int \mathrm{e}^{\frac{i}{\hbar}px}\phi(p) \,\mathrm{d}p. \tag{3.67}$$

$$\langle p|\psi\rangle = \int \langle p|x\rangle\langle x|\psi\rangle \,\mathrm{d}x, \tag{3.68}$$

$$\implies \phi(p) = \frac{1}{\sqrt{2\pi\hbar}} \int \mathrm{e}^{-\frac{i}{\hbar}px}\psi(x) \,\mathrm{d}x, \tag{3.69}$$

where we have recovered the Fourier transform relation between x and p representations.

Chapter 4

Central Forces and Angular Momentum

4.1 Spherically symmetric potential

Central forces correspond to a potential whose value depends only on the distance between the origin and the point of interest; it is indifferent to the direction, i.e. $V = V(r)$ rather than $V = V(\boldsymbol{r})$. The potential is spherically symmetric and it produces no torque about the origin, which means the orbital angular momentum is conserved.

4.1.1 *Angular momentum*

The operator of angular momentum is given by

$$\hat{L} = \hat{r} \times \hat{p}. \tag{4.1}$$

If we break it down into components, we have

$$\hat{L}_x = \frac{\hbar}{i}\left(y\frac{\partial}{\partial z} - z\frac{\partial}{\partial y} \right), \tag{4.2}$$

$$\hat{L}_y = \frac{\hbar}{i}\left(z\frac{\partial}{\partial x} - x\frac{\partial}{\partial z} \right), \tag{4.3}$$

$$\hat{L}_z = \frac{\hbar}{i}\left(x\frac{\partial}{\partial y} - y\frac{\partial}{\partial x} \right). \tag{4.4}$$

Using the spherical coordinate (r, θ, ϕ) (see Fig. 4.1),

$$x = r\sin\theta\cos\phi, \tag{4.5}$$

$$y = r\sin\theta\sin\phi, \tag{4.6}$$

$$z = r\cos\theta, \tag{4.7}$$

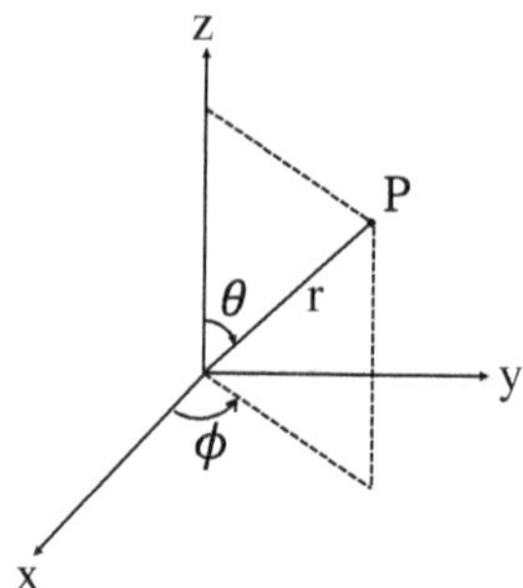

Fig. 4.1   Spherical coordinates (r, θ, ϕ) of an arbitrary point in space. The range of the coordinates are: $r \geq 0$, $0 \leq \theta \leq \pi$, $0 \leq \phi < 2\pi$.

we can obtain the angular momentum in z direction

$$\hat{L}_z = \frac{\hbar}{i} \frac{\partial}{\partial \phi}. \tag{4.8}$$

We can also calculate the square of angular momentum

$$\hat{L}^2 = (\hat{r} \times \hat{p}) \cdot (\hat{r} \times \hat{p}) = -\hbar^2 \left[\frac{1}{\sin \theta} \frac{\partial}{\partial \theta} \left(\sin \theta \frac{\partial}{\partial \theta} \right) + \frac{1}{\sin^2 \theta} \frac{\partial^2}{\partial \phi^2} \right]. \tag{4.9}$$

It can be easily proved that $\hat{L}^2$ and $\hat{L}_z$ commute, i.e. $[\hat{L}^2, \hat{L}_z] = 0$, hence they have common eigenfunctions.

4.1.2   *Hamiltonian*

The Hamiltonian is still the summation of kinetic energy and potential energy, but now it is convenient to use spherical coordinates. Then the angular part of the Laplace operator can be written in terms of the square of the angular momentum.

$$\hat{H} = \frac{\hat{p}^2}{2m} + V(r), \quad \hat{p}^2 = -\hbar^2 \nabla^2 = \hbar^2 \left[\frac{1}{r^2} \frac{\partial}{\partial r} \left(r^2 \frac{\partial}{\partial r} \right) - \frac{\hat{L}^2}{\hbar^2 r^2} \right]. \tag{4.10}$$

4.1.3   *Separation of variables*

To solve the time-independent Schrödinger equation,

$$\hat{H}\psi = E\psi, \tag{4.11}$$

we separate the variables

$$\psi = R(r)Y(\theta, \phi). \tag{4.12}$$

Then the differential equation is separated into a radial part and an angular part, corresponding to $R(r)$ and $Y(\theta, \phi)$ respectively:

$$\frac{1}{Y}\left[\frac{1}{\sin\theta}\frac{\partial}{\partial\theta}\left(\sin\theta\frac{\partial Y}{\partial\theta}\right) + \frac{1}{\sin^2\theta}\frac{\partial^2 Y}{\partial\phi^2}\right] = -\beta,\tag{4.13}$$

$$\frac{1}{R}\left[\frac{d}{dr}\left(r^2\frac{dR}{dr}\right) + r^2\frac{2m}{\hbar^2}[E - V(r)]\right] = \beta.\tag{4.14}$$

From Eq. (4.9), the angular equation (4.13) is equivalent to

$$\hat{L}^2 Y = \beta\hbar^2 Y.\tag{4.15}$$

Again we separate the variables of $Y(\theta, \phi)$,

$$Y = \Theta(\theta)\Phi(\phi).\tag{4.16}$$

Then the angular equation becomes

$$\frac{1}{\Phi}\frac{d^2\Phi}{d\phi^2} = -\alpha^2,\tag{4.17}$$

$$\frac{1}{\Theta}\left[\sin\theta\frac{d}{d\theta}\left(\sin\theta\frac{d\Theta}{d\theta}\right) + \beta\sin^2\theta\Theta\right] = \alpha^2.\tag{4.18}$$

From Eq. (4.8), Eq. (4.17) is equivalent to

$$L_z^2\Phi = \alpha^2\hbar^2\Phi.\tag{4.19}$$

4.1.4 *Solution of the angular equation*

For Eq. (4.17),

$$\frac{d^2\Phi}{d\phi^2} = -\alpha^2\Phi,\tag{4.20}$$

the solution is

$$\Phi = \frac{1}{\sqrt{2\pi}}e^{i\alpha\phi},\tag{4.21}$$

where $\frac{1}{\sqrt{2\pi}}$ is the normalization factor. As a wavefunction, Φ should be well defined, i.e.

$$\Phi(\phi) = \Phi(\phi + 2\pi)\tag{4.22}$$

$$\implies \frac{1}{\sqrt{2\pi}}e^{i\alpha\phi} = \frac{1}{\sqrt{2\pi}}e^{i\alpha(\phi+2\pi)}\tag{4.23}$$

$$\implies e^{i\alpha\cdot 2\pi} = 1\tag{4.24}$$

$$\implies \alpha = 0, \pm 1, \pm 2, \ldots\tag{4.25}$$

We can denote α as m. It can be easily verified that

$$\int_0^{2\pi} \Phi_m^* \Phi_m \, \mathrm{d}\phi = 1. \tag{4.26}$$

With $\alpha = m$, Eq. (4.18) becomes

$$\frac{1}{\sin\theta} \frac{\mathrm{d}}{\mathrm{d}\theta} \left(\sin\theta \frac{\mathrm{d}\Theta}{\mathrm{d}\theta} \right) - \frac{m^2}{\sin^2\theta} \Theta + \beta\Theta = 0. \tag{4.27}$$

Set $z = \cos\theta$, $\rho(z) = \Theta(\theta)$, then we have the associated Legendre equation:

$$\frac{\mathrm{d}}{\mathrm{d}z} \left[(1 - z^2) \frac{\mathrm{d}\rho(z)}{\mathrm{d}z} \right] + \left[\beta - \frac{m^2}{1 - z^2} \right] \rho(z) = 0. \tag{4.28}$$

Combining the two, the solutions to the angular equation (4.15) are the spherical harmonics $Y_l^m(\theta, \phi)$,

$$\beta = l(l + 1), \quad l = 0, 1, 2, \ldots \tag{4.29}$$

$$\alpha = m, \quad m = 0, \pm 1, \pm 2, \ldots, \pm l, \tag{4.30}$$

where l is the azimuthal quantum number, m is the magnetic quantum number.

A brief derivation of the solution is given over the next two subsections.

4.1.5 *Associated Legendre equation: Regular point and indicial equation*

Here we digress a little for a review of some mathematics.

Consider a second-order linear ordinary differential equation,

$$\frac{\mathrm{d}^2 F}{\mathrm{d}\xi^2} + p(\xi) \frac{\mathrm{d}F}{\mathrm{d}\xi} + q(\xi) F = 0. \tag{4.31}$$

The singular point, say $\xi = \xi_0$, is a regular point inasmuch as $(\xi - \xi_0)p(\xi)$ and $(\xi - \xi_0)^2 q(\xi)$ are analytic at the point ξ_0 and its neighborhood. Then the solution is of the form

$$F = (\xi - \xi_0)^s \sum_{\nu=0}^{\infty} a_\nu (\xi - \xi_0)^\nu. \tag{4.32}$$

Substitute this solution back into the differential equation will lead to an indicial equation by which the index s can be determined.

Now we go back to Eq. (4.28), for which $z = \pm 1$ are two regular points. Let $x = 1 - z$, $y = 1 + z$, then in both cases we find the same value of the index $s = |m|/2$. So we let

$$\rho(z) = x^{|m|/2} y^{|m|/2} G(z) = (1 - z^2)^{|m|/2} G(z). \tag{4.33}$$

Substitute this equation into Eq. (4.28), we get

$$(1 - z^2)G'' - 2(|m| + 1)zG' + [\beta - |m|(|m| + 1)]G = 0, \tag{4.34}$$

where $G' = \frac{\mathrm{d}G}{\mathrm{d}z}$, $G'' = \frac{\mathrm{d}^2 G}{\mathrm{d}z^2}$. Let

$$G = \sum_{\nu=0}^{\infty} a_\nu z^\nu, \tag{4.35}$$

then from Eq. (4.34) we obtain

$$a_{\nu+2} = \frac{(\nu + |m|)(\nu + |m| + 1) - \beta}{(\nu + 1)(\nu + 2)} a_\nu. \tag{4.36}$$

It is found that an infinite series with this relation diverges for $z = \pm 1$, which is not permissible. So G must contain only a finite number of terms, which request that

$$\beta = (\nu' + |m|)(\nu' + |m| + 1), \quad \nu' = 0, 1, 2, \ldots \tag{4.37}$$

Let

$$l = \nu' + |m|, \tag{4.38}$$

then

$$\beta = l(l + 1), \quad |m| \leq l. \tag{4.39}$$

4.1.6 *Spherical harmonics*

The explicit expression of the spherical harmonics $Y_l^m(\theta, \phi)$ is

$$Y_l^m(\theta, \phi) = (-1)^{\frac{m+|m|}{2}} \sqrt{\frac{2l + 1}{4\pi} \frac{(l - |m|)!}{(l + |m|)!}} P_l^m(\theta) e^{im\phi}, \tag{4.40}$$

where $P_l^m(\theta)$ are the associated Legendre polynomials

$$P_l^m(x) = (1 - x^2)^{\frac{|m|}{2}} \frac{d^{|m|}}{dx^{|m|}} P_l(x), \tag{4.41}$$

and $P_l(x)$ is the Legendre polynomials

$$P_l(x) = \frac{1}{2^l l!} \frac{d^l}{dx^l} (x^2 - 1)^l. \tag{4.42}$$

Since $\hat{L}^2$ and $\hat{L}_z$ commute, it is expected that they have the same eigenfunctions. $Y_l^m(\theta, \phi)$ is one such common eigenfunction,

$$\hat{L}^2 Y_l^m(\theta, \phi) = l(l + 1)\hbar Y_l^m(\theta, \phi), \tag{4.43}$$

$$\hat{L}_z Y_l^m(\theta, \phi) = m\hbar Y_l^m(\theta, \phi). \tag{4.44}$$

A graphical representation of L and L_z with $l = 2$ is given in Fig. 4.2.

The five arrows, with L_z varies from $-2\hbar$ to $2\hbar$, have the same length $\sqrt{6}\hbar$. This shows that l reflects the magnitude of the angular momentum and it is called the azimuthal quantum number. On the other hand m reflects the z component of the angular momentum, i.e. the direction of the angular momentum, and it is called the magnetic quantum number.

4.1.7 *Solution of the Schrödinger equation*

The solution of the Schrödinger equation is

$$\psi = R_{El}(r) Y_l^m(\theta, \phi) \tag{4.45}$$

The function $R_{El}(r)$ does not depend on m, so the states in a central potential are intrinsically $(2l+1)$-fold degenerate. This degeneration comes from

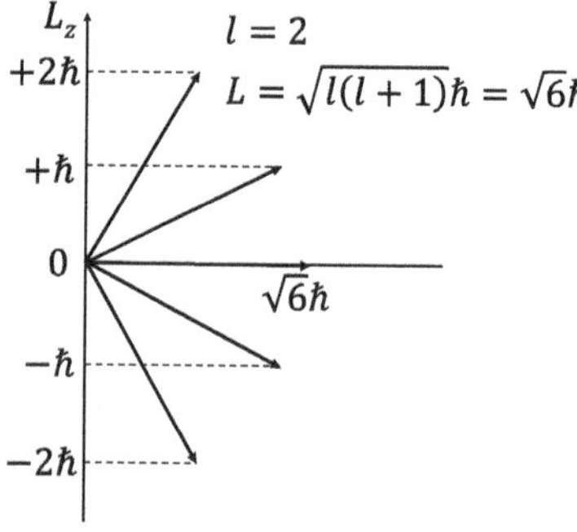

Fig. 4.2   A graphical representation of L and L_z with $l = 2$.

the fact that the choice of z direction is arbitrary, and having chosen a particular z direction, different directions of the angular momentum (different m) do not affect the energy.

The probability for finding the electron at (r, θ, ϕ) in the differential volume element $d^3r = r^2\, dr\, d\Omega$ is

$$|\psi_{Elm}|^2 r^2\, dr\, d\Omega, \tag{4.46}$$

where $d\Omega = \sin\theta\, d\theta\, d\phi$ is the solid angle, i.e. the element of area on a unit sphere. The normalization condition is:

$$\int |\psi_{Elm}|^2\, d^3r = 1, \tag{4.47}$$

$$\implies \int |R_{El}|^2 |Y_l^m|^2 r^2\, dr\, d\Omega = \int |R_{El}|^2 r^2\, dr \int |Y_l^m|^2\, d\Omega = 1, \tag{4.48}$$

where $|R_{El}|^2 r^2\, dr$ is the probability for finding the electron at a distance r in the interval dr, $|Y_l^m|^2\, d\Omega$ is the probability for finding the electron at the direction (θ, ϕ) in the element $d\Omega$.

Since the spherical harmonics Y_l^m are already normalized, the normalization condition is

$$\int_0^\infty |R_{El}|^2 r^2\, dr = 1. \tag{4.49}$$

4.2　Radial equation for the Coulomb potential

4.2.1　*Radial part of the wavefunction: Principle quantum number*

Spherical harmonics $Y_l^m(\theta, \phi)$ is the angular part of the wavefuntion. The radial part of the wavefunction, $R(r)$, is determined by Eq. (4.14). Substituting $\beta = l(l+1)$ into Eq. (4.14), we get

$$\frac{1}{R}\left[\frac{d}{dr}\left(r^2 \frac{dR}{dr}\right) + r^2 \frac{2m}{\hbar^2}[E - V(r)]\right] = l(l+1), \tag{4.50}$$

or equivalently

$$\frac{1}{r^2}\frac{d}{dr}\left(r^2 \frac{dR}{dr}\right) + \left[\frac{2\mu}{\hbar^2}(E - V) - \frac{l(l+1)}{r^2}\right] R = 0. \tag{4.51}$$

We have changed the notation of mass from m to μ to prevent confusion with the m in $Y_l^m(\theta, \phi)$.

From now on we solve this equation for the Coulomb potential

$$V = \frac{-e^2}{4\pi\epsilon_0 r}.$$ (4.52)

Let

$$\gamma = \sqrt{-\frac{2\mu E}{\hbar^2}},$$ (4.53)

where $E < 0$ corresponding to a total energy insufficient to ionize the atom. Let

$$\lambda = \frac{\mu e^2}{4\pi\epsilon_0 \hbar^2 \gamma}, \quad \rho = 2\gamma r, \quad S(\rho) = R(r),$$ (4.54)

then the equation becomes

$$\frac{1}{\rho^2}\frac{\mathrm{d}}{\mathrm{d}\rho}\left(\rho^2\frac{\mathrm{d}S}{\mathrm{d}\rho}\right) + \left[-\frac{1}{4} - \frac{l(l+1)}{\rho^2} + \frac{\lambda}{\rho}\right]S = 0.$$ (4.55)

We first discuss the asymptotic equation. For ρ large,

$$\frac{\mathrm{d}^2 S}{\mathrm{d}\rho^2} = \frac{1}{4}S.$$

The solution is

$$S = \mathrm{e}^{+\rho/2} \text{ and } S = \mathrm{e}^{-\rho/2},$$ (4.56)

while $S = \mathrm{e}^{+\rho/2}$ is not permissible.

So let $S(\rho) = \mathrm{e}^{-\rho/2}F(\rho)$ and substitute this into Eq. (4.55), then we get the equation of $F(\rho)$,

$$F'' + \left(\frac{2}{\rho} - 1\right)F' + \left[\frac{\lambda}{\rho} - \frac{l(l+1)}{\rho^2} - \frac{1}{\rho}\right]F = 0, \quad 0 \le \rho < \infty.$$ (4.57)

Since $\rho = 0$ is a regular point, let $F(\rho) = \rho^s L(\rho)$. The solutions of indicial equation $s(s+1) - l(l+1) = 0$ are $s = l, -(l+1)$, where $s = -(l+1)$ is not permissible. Therefore

$$F(\rho) = \rho^l L(\rho),$$ (4.58)

then we have

$$\rho L'' + [2(l+1) - \rho]L' + (\lambda - l - 1)L = 0.$$ (4.59)

Assume that $L(\rho)$ can be expressed by a power series in ρ,

$$L(\rho) = \sum_{\nu=0}^{\infty} a_\nu \rho^\nu, \tag{4.60}$$

then Eq. (4.59) leads to

$$(\lambda - l - 1 - \nu)a_\nu + [2(\nu + 1)(l + 1) + \nu(\nu + 1)]a_{\nu+1} = 0. \tag{4.61}$$

For an acceptable wavefunction, the series must break off after a finite number of terms. The condition that it breaks off after the term in $\rho^{n'}$ is

$$\lambda - l - 1 - n' = 0, \tag{4.62}$$

or

$$\lambda = n, \quad \text{where } n = n' + l + 1. \tag{4.63}$$

n' is called the radial quantum number and n is called the principal quantum number.

$$n' = 0, 1, 2, 3, \ldots$$

$$l = 0, 1, 2, \ldots, (n - 1).$$

Combine

$$\lambda = n, \quad \lambda = \frac{\mu e^2}{4\pi\epsilon_0 \hbar^2 \gamma}, \quad \gamma = \sqrt{-\frac{2\mu E}{\hbar^2}},$$

we get

$$E = -\frac{m}{2\hbar^2} \left(\frac{e^2}{4\pi\epsilon_0}\right)^2 \frac{1}{n^2}. \tag{4.64}$$

Introduce the dimensionless variable

$$\rho = r/a_0, \quad \epsilon = E/\mathrm{R_y}, \tag{4.65}$$

where

$$a_0 = 4\pi\epsilon_0 \frac{\hbar^2}{me^2} = 0.529 \times 10^{-8} \text{ cm}, \quad \mathrm{R_y} = \frac{me^4}{2\hbar^2} \frac{1}{(4\pi\epsilon_0)^2} = 13.6 \text{ eV}. \tag{4.66}$$

We have

$$E = -\frac{\mathrm{R_y}}{n^2}, \quad \epsilon = -\frac{1}{n^2}, \quad n = 1, 2, 3, \ldots \tag{4.67}$$

4.2.2 *Eigenstates of the hydrogen atom*

The wavefunction ψ_{Elm} can be denoted as ψ_{nlm}. Combine the angular part $Y_l^m(\theta,\phi)$ and the radial part $R(r)$, expressions of a few stationary wavefunctions are given below:

$$\psi_{100} = \frac{1}{(\pi a_0^3)^{3/2}} e^{-r/a_0},$$

$$\psi_{200} = \frac{1}{4(2\pi a_0^3)^{1/2}} \left(2 - \frac{r}{a_0}\right) e^{-r/2a_0},$$

$$\psi_{211} = \frac{1}{8(\pi a_0^3)^{1/2}} \frac{r}{a_0} e^{-r/2a_0} \cdot (-\sin\theta) \cdot e^{i\phi},$$

$$\psi_{210} = \frac{1}{8(\pi a_0^3)^{1/2}} \frac{r}{a_0} e^{-r/2a_0} \cdot \sqrt{2}\cos\theta,$$

$$\psi_{21-1} = \frac{1}{8(\pi a_0^3)^{1/2}} \frac{r}{a_0} e^{-r/2a_0} \cdot \sin\theta \cdot e^{-i\phi}.$$

These states are the eigenstates of hydrogen atom. Conventionally, the state with $l = 0, 1, 2, 3, \ldots$ is noted respectively as $s, p, d, f, \ldots$

Example 4.1. Consider the state $|\psi\rangle = \frac{1}{\sqrt{30}}(2|100\rangle - 3|211\rangle + 4|200\rangle - |320\rangle)$, where $|nlm\rangle$ are the eigenstates of hydrogen atom. If E_1 is the ground state energy, we can calculate the expectation value of the energy:

$$\langle H \rangle = \frac{1}{30}\left(2^2 + \frac{9}{2^2} + \frac{16}{2^2} + \frac{1}{3^2}\right)E_1 = \frac{373}{1080}E_1, \tag{4.68}$$

where the four terms correspond to the four eigenstates.

We can also calculate the expectation value of $\hat{L}^2$:

$$\langle L^2 \rangle = \frac{1}{30}(9 \times 2 + 6)\hbar^2 = \frac{4}{5}\hbar^2. \tag{4.69}$$

The expectation value is the average result of measurement. However, if we actually carry out the measurement, the result can only be one of several possible values with different probabilities. For example, consider the angular momentum in the z direction. The first, third and fourth terms correspond to

$$L_z = 0, \quad \text{probability} = \frac{1}{30}(4 + 16 + 1) = \frac{21}{30} = \frac{7}{10}.$$

For the second term,

$$L_z = \hbar, \quad \text{probability} = \frac{1}{30} \times 9 = \frac{3}{10}.$$

This means if we measure L_z, there is $7/10$ chance that it is 0 and $3/10$ chance that it is $\hbar$.

Example 4.2. For the electron of hydrogen, we know $L = \sqrt{6}\hbar$, $L_z = \pm\hbar$, so the probability per solid angle in the neighbourhood of (θ, ϕ) can be calculated.

$$L = \sqrt{6}\hbar \quad \rightarrow \quad l = 2,$$
$$L_z = \pm\hbar \quad \rightarrow \quad m = \pm 1,$$
$$Y_2^{\pm 1} = \mp \left(\frac{15}{8\pi}\right)^{1/2} \sin\theta \cos\theta\, e^{\pm i\phi},$$
$$|Y_2^{\pm 1}|^2 = \frac{15}{8\pi} \sin^2\theta \cos^2\theta.$$

To find the angles with the maximum probability, consider the equation

$$\frac{d}{d\theta}(|Y_2^{\pm 1}|^2) = 0,$$

which gives

$$\theta = (2n+1)\frac{\pi}{8}, \quad n = 0, \pm 1, \pm 2, \ldots$$

Chapter 5

Summary, Problems and Solutions

5.1 A brief summary of this course

To help you remember the important points of Part 1, we give a summary labeled from "one" to "nine".

One. One wave equation, namely the Schrödinger equation,

$$i\hbar\frac{\partial \Psi}{\partial t} = \hat{H}\Psi. \tag{5.1}$$

When $\hat{H}$ is independent of time, we can separate the variables and consider the time-independent Schrödinger equation,

$$\hat{H}\psi = E\psi, \tag{5.2}$$

where

$$\Psi = \psi \mathrm{e}^{-\frac{i}{\hbar}Et}. \tag{5.3}$$

Two. Two important types of operators to take note.

(1) Hermitian operator:

$$\hat{A} = \hat{A}^{\dagger}. \tag{5.4}$$

In general,

$$\langle \psi|\hat{A}|\phi\rangle = \langle \hat{A}^{\dagger}\psi|\phi\rangle, \tag{5.5}$$

then if $\hat{A}$ is a Hermitian operator,

$$\langle \psi|\hat{A}|\phi\rangle = \langle \hat{A}\psi|\phi\rangle. \tag{5.6}$$

87

(2) Unitary operator:

$$\hat{U}\hat{U}^\dagger = \underline{I}, \text{ or } \hat{U}^\dagger = \hat{U}^{-1}. \tag{5.7}$$

For the change of basis from $|\psi_i\rangle$ to $|\phi_j\rangle$, the unitary matrix connecting these two representations is

$$U_{ij} = \langle\psi_i|\phi_j\rangle. \tag{5.8}$$

Three. Three types of scalar value.

(1) Eigenvalue: $\hat{\alpha}|\psi_i\rangle = \alpha_i|\psi_i\rangle$.
(2) Expectation value: $\langle\psi|\hat{\alpha}|\psi\rangle$.
(3) Possible value: For a quantum state $|\psi\rangle = \sum_i c_i|\psi_i\rangle$, if the physical quantity corresponding to $\hat{\alpha}$ is measured, the probabilities of getting the results $\alpha_1, \alpha_2, \ldots$ are respectively $|c_1|^2, |c_2|^2, \ldots$

Four. Four steps to solve a quantum system.

(1) Write down the Hamiltonian H from CM.
(2) Transform it into the operator form $\hat{H}$.
(3) Write down the Schrödinger equation $\hat{H}\Psi = i\hbar\frac{\partial\Psi}{\partial t}$.
(4) Solve for Ψ by the boundary condition, initial condition and standard condition of wavefunction. For stationary states, $\Psi_n = \psi_n e^{-\frac{i}{\hbar}E_n t}$, $\hat{H}\psi_n = E_n\psi_n$, no initial condition is needed.

Five. Five postulates of QM.

(1) Quantum state: Any state of a system with n degrees of freedom is described as completely as possible by a wavefunction $\Psi(q_1, q_2, \ldots, q_n, t)$ which depends upon the coordinates q_i and the time t.
(2) Physical observable: With every physical observable there is associated a linear operator, $A \to \hat{A}$, and it is Hermitian, $\hat{A} = \hat{A}^\dagger$.
(3) Time evolution of quantum state: The state wavefunction Ψ satisfies Schrödinger's time-dependent equation, $\hat{H}\Psi = i\hbar\frac{\partial\Phi}{\partial t}$.
(4) Measurement: Suppose the state Ψ is an eigenstate of the operator $\hat{A}$, i.e. $\hat{A}\Psi = a\Psi$, then a measurement of the observable A will yield precisely the eigenvalue a.
(5) Expectation value: The expectation value of a physical observable A for a state $|\Psi\rangle$ is given by its corresponding operator $\hat{A}$, $\langle A\rangle = \int \Psi^*\hat{A}\Psi\,d\tau$.

Six. Six important cases.

(1) Free particle;
(2) Potential well;
(3) Potential barrier;
(4) Tunneling effect;
(5) Linear harmonic oscillator;
(6) Hydrogen atom.

Seven. Seven important relations.

(1) Commutator and uncertainty relation;
(2) Commutator and common eigenfunctions;
(3) Hermiticity and real eigenvalues;
(4) Hermiticity and orthogonality of eigenfunctions;
(5) Unitary transformation and expectation values;
(6) Unitary transformation and eigenvalues;
(7) Boundary conditions, standard conditions and quantum numbers.

Eight. Eight important items.

(1) Methods of theoretical research: Generalization, idealization, mathematical induction.
(2) Important physical quantities: Lagrangian, Hamiltonian, Poisson brackets.
(3) Separation of variables, ordinary point, regular point.
(4) Dirac delta function, Fourier transform.
(5) Phase velocity and group velocity.
(6) Probabilities and uncertainties.
(7) Hilbert space, representations, transformation of representations.
(8) Bra, ket, and closure relations.

Nine. Nine pioneers of QM.

(1) 1900 Planck;
(2) 1905 Einstein;
(3) 1913 Bohr;
(4) 1922 Compton;
(5) 1924 de Broglie;
(6) 1925 Schrödinger;
(7) 1926 Heisenberg;

(8) 1927 Davison and Germer;
(9) 1927 Born.

5.2   Problems

Problem 5.1. *A particle of mass m at a distance r from the origin, is moving in a plane under an attractive force $\mu m/r^2$ directed towards the origin. Write down the kinetic and potential energies and verify that Lagrange's equations may be written as*

$$m\ddot{r} - mr\dot{\theta}^2 + \frac{\mu m}{r^2} = 0,$$

$$m\frac{\mathrm{d}}{\mathrm{d}t}\left(r^2\dot{\theta}\right) = 0.$$

Problem 5.2. *A particle of mass m undergoes simple harmonic oscillations along the x-axis. The controlling force is directed towards the origin and has value $-kx$. Write down the kinetic and potential energies and hence the Lagrangian. Find the momentum conjugate to x and show that the Hamiltonian of the system is*

$$H = \frac{1}{2m}p^2 + \frac{1}{2}kx^2.$$

Problem 5.3. *Prove the following properties of Poisson brackets:*

(1) $\{E+F,G\} = \{E,G\} + \{F,G\}$,
(2) $\{F,\{G,K\}\} + \{G,\{K,F\}\} + \{K,\{F,G\}\} = 0$,

where E, F, G, K are all functions of q_i and p_i.

Problem 5.4. *A photon of energy $E = 100\ keV$ collides with a free electron at rest. It is scattered through $90°$. Find the energy of the scattered photon.*

Hint. For the electron mass use $mc^2 = 500$ keV and recall that $E = [(pc)^2 + (mc^2)^2]^{\frac{1}{2}}$.

Problem 5.5. *Use the Bohr theory of the hydrogen atom to show that the velocity v_0 of the electron in the first Bohr orbit is given by*

$$v_0 = \frac{e^2}{4\pi\epsilon_0\hbar} = \alpha c,$$

where

$$\alpha = \frac{e^2}{4\pi\epsilon_0 \hbar c} \approx \frac{1}{137}$$

is known as the fine-structure constant. The constants are $\epsilon_0 = 8.854 \times 10^{-12}$ C/V-m (permittivity of free space), $\hbar = \frac{h}{2\pi} = 6.58 \times 10^{-16}$ eV-s, and $\hbar c = 197.328$ MeV-F, where 1 F (Fermi) $= 10^{-15}$ m $= 10^{-13}$ cm.

Problem 5.6. *In a demonstration of the diffraction from two slits it was found that the spacing between interference minima was $\Delta x_I = 5.5$ mm and the first diffraction minimum was located at $\Delta x_D = 14.2$ mm from the beam center. The distance of the screen from the slits was $L = 1$ m and the wavelength (neon laser) was $\lambda = 6328$ Å.*

(1) *Find the width of the slits and their separation.*
(2) *Make a quantitative plot (on graph paper) of the observed pattern.*

Problem 5.7. *Low energy electrons $(E = 200$ eV) were reflected from the face of a nickel crystal (lattice constant $a = 3.52 \times 10^{-8}$ cm). Use the condition for constructive interference*

$$\Delta = 2a\cos\theta = n\lambda$$

to find the first three $(n = 1, 2, 3)$ interference maximum θ_1, θ_2, θ_3.

Problem 5.8. *Consider the potential*

$$V(x) = \begin{cases} \infty, & x \leq -L, \ x \geq L, \\ 0, & -L < x < L, \end{cases}$$

as depicted in Fig. 5.1. Find the wavefunction of stationary states and the corresponding energy eigenvalues.

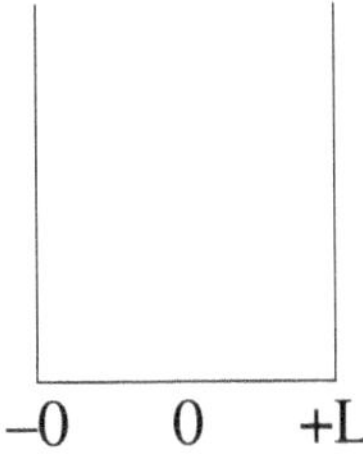

Fig. 5.1   Potential barrier.

Problem 5.9. *Consider the wavefunction*

$$\psi(x) = \begin{cases} 0, & x < -\dfrac{L}{2}, \ x > \dfrac{L}{2}, \\ Ae^{ikx}\cos\dfrac{3\pi x}{L}, & -\dfrac{L}{2} \le x \le \dfrac{L}{2}. \end{cases}$$

(1) *Find the value of the normalization constant A.*

(2) *What is the probability that the particle is found between $x = 0$ and $x = \frac{L}{4}$?*

Problem 5.10. *Show that the reflection coefficients for the two cases depicted in Fig. 5.2 are equal.*

Problem 5.11. *Calculate the reflection coefficient R and transmission coefficient T for the case as depicted in Fig. 5.3. Show that $T + R = 1$.*

Problem 5.12. *Show that the current density j may be written as*

$$j = \frac{1}{2m}[\psi^*\hat{p}\psi + (\psi^*\hat{p}\psi)^*],$$

where $\hat{p}$ is the momentum operator.

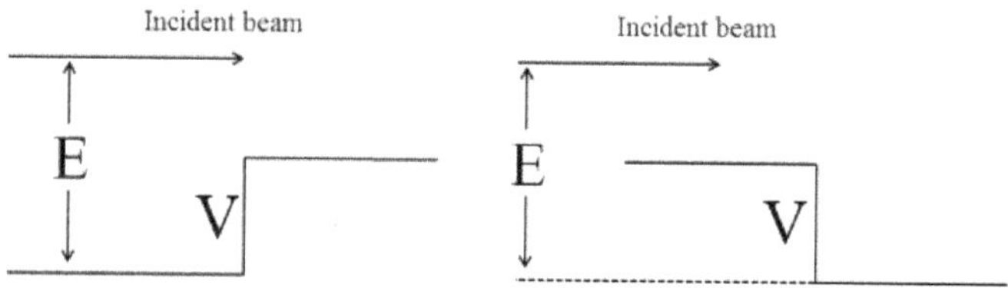

Fig. 5.2   Beam reflection in the face of two different potentials.

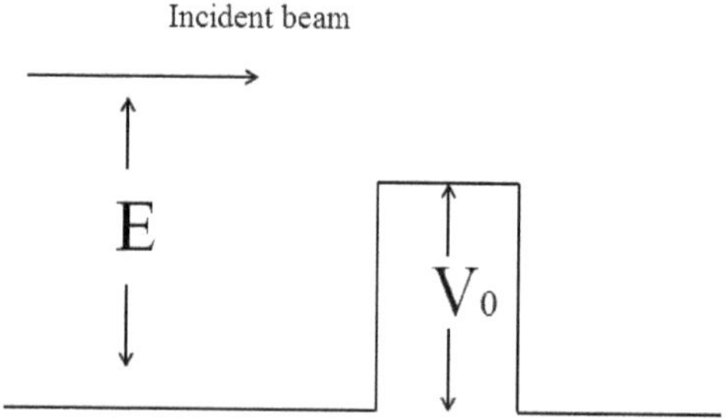

Fig. 5.3   Beam reflection and transmission.

Problem 5.13. *Show that for a one-dimensional wavefunction of the form*

$$\psi(x,t) = A\exp[i\phi(x,t)],$$

the current density is given by

$$j = \frac{\hbar}{m}|A|^2\frac{\partial\phi}{\partial x}.$$

Problem 5.14. *From*

$$\psi(x) = \frac{1}{\sqrt{2\pi\hbar}}\int_{-\infty}^{\infty}\phi(p)e^{\frac{i}{\hbar}px}\,\mathrm{d}p$$

prove

$$\phi(p) = \frac{1}{\sqrt{2\pi\hbar}}\int_{-\infty}^{\infty}\psi(x)e^{-\frac{i}{\hbar}px}\,\mathrm{d}x. \tag{5.9}$$

Then show that

$$\int_{-\infty}^{\infty}\psi^*(x)\psi(x)\,\mathrm{d}x = \int_{-\infty}^{\infty}\phi^*(p)\phi(p)\,\mathrm{d}p. \tag{5.10}$$

Problem 5.15. *If the particle is in the position eigenstate*

$$\psi_{x'}(x) = \delta(x - x'),$$

find its Fourier transform $\phi(p)$, and prove that

$$\phi^*(p)\phi(p) = const.$$

What is the physical meaning of this result?

Problem 5.16. *A single particle is confined to move in one dimension, whose wavefunction is*

$$\psi(x) = \begin{cases} 2\lambda^{\frac{3}{2}}xe^{-\lambda x}, & x \geq 0, \quad (\lambda > 0) \\ 0, & x < 0. \end{cases}$$

Find the wavefunction in momentum space $\phi(p)$. What is the expectation value of x and p?

Hint. $\int_0^{\infty} x^n e^{-ax}\,\mathrm{d}x = \frac{n!}{a^{n+1}}.$

Problem 5.17. *The recurrence relation of Hermite polynomials is*

$$H_{n+1}(y) - 2yH_n(y) + 2nH_{n-1}(y) = 0.$$

Use this relation to prove that

$$x\psi_n(x) = \frac{1}{\beta}\left[\sqrt{\frac{n}{2}}\psi_{n-1}(x) + \sqrt{\frac{n+1}{2}}\psi_{n+1}(x)\right],$$

$$x^2\psi_n(x) = \frac{1}{2\beta^2}[\sqrt{n(n-1)}\psi_{n-2}(x) + (2n+1)\psi_n(x)$$

$$+ \sqrt{(n+1)(n+2)}\psi_{n+2}(x)],$$

where $\psi_n(x)$ is the nth eigenfunction of harmonic oscillator and $\beta = \sqrt{\frac{m\omega}{\hbar}}$, thence prove that $\langle x \rangle = 0$, $\langle V \rangle = \frac{1}{2}E_n$.

Problem 5.18. *The differentiation formula of Hermite polynomial is*

$$H_n'(y) = 2nH_{n-1}(y).$$

Use this relation to prove that

$$\frac{\mathrm{d}}{\mathrm{d}x}\psi_n(x) = \beta\left[\sqrt{\frac{n}{2}}\psi_{n-1}(x) - \sqrt{\frac{n+1}{2}}\psi_{n+1}(x)\right],$$

$$\frac{\mathrm{d}^2}{\mathrm{d}x^2}\psi_n(x) = \frac{\beta^2}{2}[\sqrt{n(n-1)}\psi_{n-2}(x) - (2n+1)\psi_n(x)$$

$$+ \sqrt{(n+1)(n+2)}\psi_{n+2}(x)],$$

where $\psi_n(x)$ is the nth eigenfunction of harmonic oscillator and $\beta = \sqrt{\frac{m\omega}{\hbar}}$, thence prove that $\langle p \rangle = 0$, $\langle T \rangle = \langle \frac{p^2}{2m} \rangle = \frac{1}{2}E_n$.

Problem 5.19. *Prove that*

$$[q, f(q)\hat{p}^2] = 2i\hbar f(q)\hat{p},$$

$$[\hat{p}, f(q)\hat{p}^2] = \frac{\hbar}{i}\left(\frac{\mathrm{d}f}{\mathrm{d}q}\right)\hat{p}^2.$$

Problem 5.20. *Prove that $i(\hat{p}_x^2 x^2 - x^2\hat{p}_x^2)$ is a Hermitian operator.*

Problem 5.21. *Three operators $\hat{A}$, $\hat{B}$ and $\hat{C}$ have the following relations:*

$$\hat{C} = \hat{A}\hat{B}, \quad [\hat{A}, \hat{B}] = 1.$$

If $|\psi\rangle$ is an eigenvector of $\hat{C}$ with eigenvalue λ, prove that $\hat{A}|\psi\rangle$ and $\hat{B}|\psi\rangle$ are also the eigenvectors of $\hat{C}$, and find their corresponding eigenvalues.

Problem 5.22. *Show that if $\hat{A}|\psi_n\rangle = a_n|\psi_n\rangle$ and $\hat{B}|\psi_n\rangle = b_n|\psi_n\rangle$ for all eigenvalues $\{a_n\}$ and $\{b_n\}$ of $\hat{A}$ and $\hat{B}$, respectively (i.e. $\hat{A}$ and $\hat{B}$ have completely common eigenstates), then $[\hat{A}, \hat{B}] = 0$ on the space of function spanned by the basis $\{|\psi_n\rangle\}$.*

Hint. Any element of this space $|\psi\rangle$ may be written as $|\psi\rangle = \sum_i c_i|\psi_i\rangle$ and one need merely to show that $[\hat{A}, \hat{B}] \sum_i c_i|\psi_i\rangle = 0$.

Problem 5.23. *Show that if $\hat{U}$ is unitary, then the eigenvalues u_n of $\hat{U}$ are of unit magnitude.*

Problem 5.24. *Let $\hat{U}$ be a unitary operator.*

(1) Show that if $\langle\psi|\psi\rangle = 1$, then $\langle\hat{U}\psi|\hat{U}\psi\rangle = 1$.
(2) If $|u_i\rangle$ is a complete orthonormal set $\langle u_i|u_j\rangle = \delta_{ij}$, show that $|v_i\rangle = \hat{U}|u_i\rangle$ is also an orthonormal set.

Problem 5.25. *A quantum mechanical system is in the state $|\psi\rangle = \sum_i c_i|\psi_i\rangle$, where $\{|\psi_i\rangle\}$ are energy eigenstates. Find the expectation value of energy in terms of the energy of the stationary states.*

Problem 5.26. *In the $\hat{\alpha}$ representation the basis are $\{|\psi_i\rangle\}$; in the $\hat{\beta}$ representation the basis are $\{|\phi_j\rangle\}$. Show that if $\hat{A}$ is a Hermitian operator, then $D^2 = \sum_i \sum_j |\langle\phi_j|\hat{A}|\psi_i\rangle|^2$ is independent of the choice of basis and equals the sum of the squares of all the eigenvalues of $\hat{A}$.*

Problem 5.27.

(1) Plot to scale the probability of finding the electron at a distance r from the proton for the 1s, 2s and 2p states of the hydrogen atom.
(2) Plot the probability of finding the electron at an angle θ with respect to the quantization axis for the five 3d states of the hydrogen atom. Also indicate the most probable direction of the orbital angular momentum.

Problem 5.28. *The angular part of the wavefunction of a particle in a central force field is*

$$F(\theta, \phi) = \frac{1}{\sqrt{12}}[(i - 1)Y_1^1 + (i + 1)Y_1^{-1} + 2\sqrt{2}Y_1^0].$$

Find

(1) *the probability for $L_z = -\hbar$;*
(2) *the expectation value of L^2 and L_z;*
(3) *the probability per solid angle in the neighbourhood of (θ, ϕ).*

Hint.

$$Y_1^{\pm 1} = \mp\sqrt{\frac{3}{8\pi}} \sin\theta e^{\pm i\phi}, \qquad Y_1^0 = \sqrt{\frac{3}{4\pi}} \cos\theta.$$

5.3 Quiz 1

(1) Use de Broglie's hypothesis to explain Bohr's quantum conditions.
(2) Write out the Schrödinger equation for a simple harmonic oscillator, and state the procedure to find the energy eigenvalues and the wavefunctions in this case.
(3) Consider the potential

$$V(x) = \begin{cases} 0, & x < 0, \\ V_0, & 0 \le x \le a, \\ \infty, & x > a, \end{cases}$$

as depicted in Fig. 5.4. The incoming wave has energy $E = \frac{4}{3}V_0$.

(a) Find the wavefunction in this case.
(b) Show that the reflection coefficients in region 1 and region 2 equal 1.

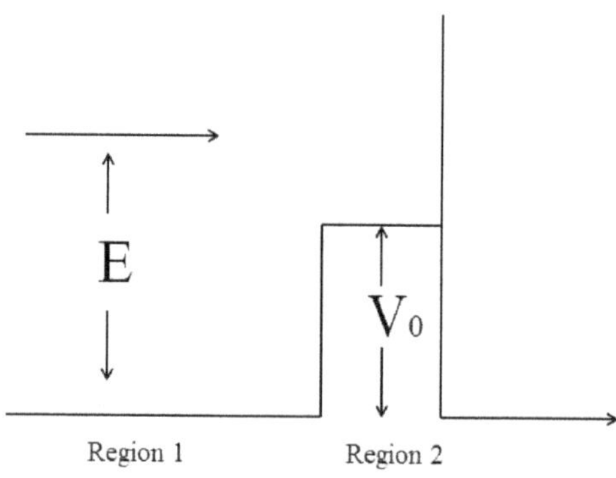

Fig. 5.4   Potential barrier.

5.4 Quiz 2

(1) Consider a particle with energy $E = p^2/2m$ moving in one dimension x. The uncertainty in its location is Δx. Show that if $\Delta x \Delta p > \hbar$, then $\Delta E \Delta t > \hbar$, where $\frac{p}{m}\Delta t = \Delta x$.

(2) Find the Hermitian adjoint of the following operators:

   (a) $\hat{A}\hat{B} - \hat{B}\hat{A}$;
   (b) $\hat{A}\hat{B} + \hat{B}\hat{A}$;
   (c) $i(\hat{A}\hat{B} - \hat{B}\hat{A})$;
   (d) $\hat{A}^\dagger$;
   (e) $\hat{A}^\dagger\hat{A}$.

   If $\hat{A}$ and $\hat{B}$ are both Hermitian, which of the above operators are Hermitian?

(3) Prove that for a linear harmonic oscillator the expectation value of the potential energy is equal to the expectation value of the kinetic energy. Since $(\Delta x)^2 = \langle x^2 \rangle - \langle x \rangle^2$, $(\Delta p)^2 = \langle p^2 \rangle - \langle p \rangle^2$, show that in the lowest energy state $\Delta x \Delta p = \frac{\hbar}{2}$.

5.5 Final examination

(1) Find the eigenfunction and eigenvalue of the momentum operator $\hat{p} = \frac{\hbar}{i}\frac{d}{dx}$. If ψ_{p_0} is the eigenstate of $\hat{p}$, prove that $e^{icx}\psi_{p_0}$ is also an eigenstate of $\hat{p}$ and find the corresponding eigenvalue.

(2) Show that unitary transformations preserve the Hermitian property of an operator and leave the eigenvalues invariant.

(3) Consider a hydrogen atom described by the following wavefunction

$$\psi(\boldsymbol{r}, t) = \sqrt{\frac{2}{3}}\psi_{100}\, e^{-iE_1 t/\hbar} + \sqrt{\frac{1}{3}}\psi_{211}\, e^{-iE_2 t/\hbar}.$$

   (a) What are the probabilities that measurement on this state yields the results $E = E_2$, $L^2 = 2\hbar^2$, $L_z = \hbar$?
   (b) What are the respective expectation value of E, L^2, and L_z?

(4) Let $|\psi_1\rangle$ and $|\psi_2\rangle$ be the first and second excited state of linear harmonic oscillator. Calculate the matrix elements $\langle \psi_2|\hat{x}|\psi_1 \rangle$, $\langle \psi_2|\hat{p}^2|\psi_2 \rangle$ and find the minimum uncertainty product $\Delta x \Delta p$ for the second excited state.

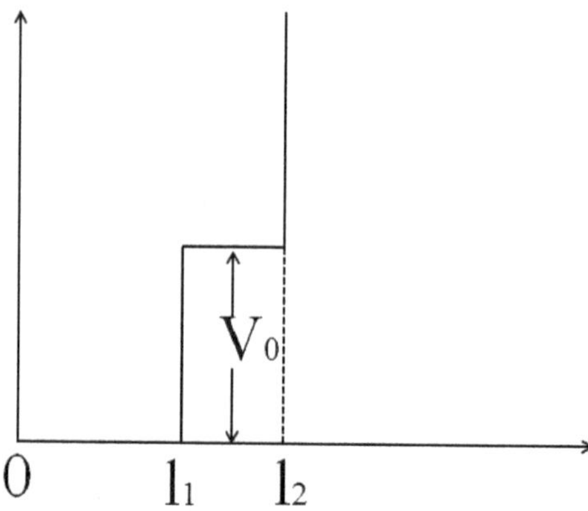

Fig. 5.5   Potential well.

(5) Consider the potential

$$V(x) = \begin{cases} 0, & 0 \le x < l_1, \\ \infty, & x < 0,\ x > l_2, \\ V_0, & l_1 \le x \le l_2, \end{cases}$$

as depicted in Fig. 5.5. Let the energy of the state be $E < V_0$.

(a) Prove that

$$k_2 \tan(k_1 l_1) = -k_1 \tanh[k_2(l_2 - l_1)],$$

where $k_1 \equiv \sqrt{2mE}/\hbar$, $k_2 \equiv \sqrt{2m(V_0 - E)}/\hbar$, $\tanh y \equiv (e^y - e^{-y})/(e^y + e^{-y})$.

(b) When $V_0 \gg E$, find the approximate values of E.

5.6   Solutions of problems

Solution 5.1. Kinetic energy:

$$E_k = \frac{1}{2}m(\dot{r}^2 + r^2\dot{\theta}^2).$$

Potential energy is the work done by the conservative force from the given position r to the position of zero potential energy we have assigned, in this case we choose $r = \infty$. Hence

$$E_p = \int_r^\infty -\frac{\mu m}{r^2}\, \mathrm{d}r = -\frac{\mu m}{r}.$$

Lagrangian:

$$L = T - V = \frac{1}{2}m(\dot{r}^2 + r^2\dot{\theta}^2) + \frac{\mu m}{r}.$$

Lagrange's equation:

$$\frac{\mathrm{d}}{\mathrm{dt}}\left(\frac{\partial L}{\partial \dot{q}_i}\right) - \frac{\partial L}{\partial q_i} = 0.$$

For $q_i = r$, we obtain

$$m\ddot{r} - mr\dot{\theta}^2 + \frac{\mu m}{r^2} = 0.$$

For $q_i = \theta$, we obtain

$$m\frac{\mathrm{d}}{\mathrm{dt}}(r^2\dot{\theta}) = 0.$$

Solution 5.2. Kinetic energy:

$$E_k = \frac{1}{2}m\dot{x}^2.$$

Potential energy:

$$E_p = \int_x^0 -kx\,\mathrm{d}x = \frac{1}{2}kx^2.$$

Lagrangian:

$$L = \frac{1}{2}m\dot{x}^2 - \frac{1}{2}kx^2.$$

Conjugate momentum:

$$p = \frac{\partial L}{\partial \dot{x}} = m\dot{x}.$$

Hamiltonian:

$$H = p \cdot \dot{x} - L = p \cdot \frac{p}{m} - \frac{p^2}{2m} + \frac{1}{2}kx^2 = \frac{p^2}{2m} + \frac{1}{2}kx^2.$$

Solution 5.3. Recall the definition of Poisson brackets:

$$\{E, F\} = \sum_k \left(\frac{\partial E}{\partial q_k}\frac{\partial F}{\partial p_k} - \frac{\partial E}{\partial p_k}\frac{\partial F}{\partial q_k}\right).$$

(1) By definition,

$$\{E+F,G\} = \sum_k \left[\frac{\partial(E+F)}{\partial q_k} \frac{\partial G}{\partial p_k} - \frac{\partial(E+F)}{\partial p_k} \frac{\partial G}{\partial q_k} \right]$$

$$= \sum_k \left[\left(\frac{\partial E}{\partial q_k} + \frac{\partial F}{\partial q_k} \right) \frac{\partial G}{\partial p_k} - \left(\frac{\partial E}{\partial p_k} + \frac{\partial F}{\partial p_k} \right) \frac{\partial G}{\partial q_k} \right]$$

$$= \sum_k \left[\frac{\partial E}{\partial q_k} \frac{\partial G}{\partial p_k} - \frac{\partial E}{\partial p_k} \frac{\partial G}{\partial q_k} \right] + \sum_k \left[\frac{\partial F}{\partial q_k} \frac{\partial G}{\partial p_k} - \frac{\partial F}{\partial p_k} \frac{\partial G}{\partial q_k} \right]$$

$$= \{E,G\} + \{F,G\}.$$

(2) Look at the equation to prove:

$$\{F,\{G,K\}\} + \{G,\{K,F\}\} + \{K,\{F,G\}\} = 0.$$

The left-hand side of the equation is a linear homogeneous function of the second derivatives of F, G, K. First let us collect the terms involving the second derivatives of F, which only come from the second and third brackets. Define linear differential operators D_1 and D_2 as

$$D_1 = \sum_k \left(\frac{\partial K}{\partial q_k} \frac{\partial}{\partial p_k} - \frac{\partial K}{\partial p_k} \frac{\partial}{\partial q_k} \right),$$

$$D_2 = \sum_k \left(\frac{\partial G}{\partial q_k} \frac{\partial}{\partial p_k} - \frac{\partial G}{\partial p_k} \frac{\partial}{\partial q_k} \right).$$

such that

$$\{K,F\} = D_1 F,$$
$$\{G,F\} = D_2 F.$$

Then

$$\{G,\{K,F\}\} + \{K,\{F,G\}\} = \{G,\{K,F\}\} - \{K,\{G,F\}\}$$

$$= D_2 D_1 F - D_1 D_2 F$$

$$= (D_2 D_1 - D_1 D_2)F.$$

The second derivatives in $D_2 D_1 - D_1 D_2$ cancel each other, thus there is no second derivatives of F in $\{F,\{G,K\}\} + \{G,\{K,F\}\} + \{K,\{F,G\}\}$. Similarly, there is no second derivatives of K or G. Since $\{F,\{G,K\}\} + \{G,\{K,F\}\} + \{K,\{F,G\}\}$ is a linear homogeneous function of the second derivatives of F, G, K, it should be identically zero.

Solution 5.4. For Compton scattering,

$$\lambda' - \lambda = \frac{h}{mc}(1 - \cos\theta).$$

Since $\lambda = \frac{hc}{E}$, $\theta = 90°$, we obtain

$$\frac{hc}{E'} - \frac{hc}{E} = \frac{h}{mc},$$

$$E' = 83 \text{ keV}.$$

Solution 5.5. From the centrifugal force we can find the radius in terms of v_0:

$$\frac{e^2}{4\pi\epsilon_0 r^2} = \frac{mv_0^2}{r} \implies r = \frac{e^2}{4\pi\epsilon_0 mv_0^2}.$$

Given the angular momentum,

$$mv_0 r = \hbar$$

$$\implies \frac{e^2}{4\pi\epsilon_0 v_0} = \hbar$$

$$\implies v_0 = \frac{e^2}{4\pi\epsilon_0 \hbar}.$$

Solution 5.6. Recall the formula from Sec. 1.3.4.2.

(1) Width:

$$a = \frac{\lambda}{\sin\theta} = \lambda / \frac{\Delta x_D}{\sqrt{\Delta x_D^2 + L^2}} = 4.46 \times 10^{-6} \text{ m}.$$

Separation:

$$d = \frac{\lambda L}{\Delta X_I} = 1.15 \times 10^{-4} \text{ m}.$$

(2) See Fig. 5.6.

Solution 5.7. First calculate the wavelength from the momentum.

$$p = \sqrt{2mE} \implies \lambda = \frac{h}{p} = 8.67 \times 10^{-11} \text{ m}.$$

Then from the formula, we can calculate θ.

$$n = 1, \quad \theta = 82.9°.$$

$$n = 2, \quad \theta = 75.7°.$$

$$n = 3, \quad \theta = 68.3°.$$

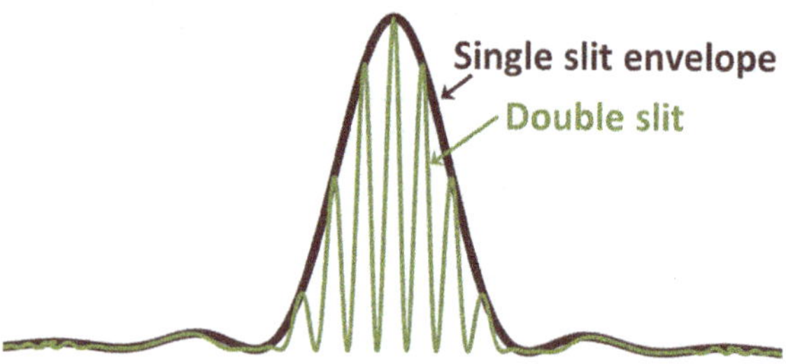

Fig. 5.6   Double-slit diffraction pattern. Credit: Brews ohare/Wikimedia Commons.

Solution 5.8. Write down the Schrödinger equation,

$$\hat{H}\psi = E\psi,$$

$$-\frac{\hbar^2}{2m}\frac{\mathrm{d}^2}{\mathrm{d}x^2}\psi = E\psi.$$

Set $k = \frac{\sqrt{2mE}}{\hbar}$, then

$$\left(\frac{\mathrm{d}^2}{\mathrm{d}x^2} + k^2\right)\psi = 0$$

$$\implies \psi = A\sin kx + B\cos kx.$$

From the standard conditions,

$$\psi(-L) = -A\sin kL + B\cos kL = 0,$$
$$\psi(L) = A\sin kL + B\cos KL = 0.$$

Thus we have either

$$\sin kL = 0, \quad k = \frac{n\pi}{2L}, \quad n = 2, 4, 6, \ldots$$

$$\implies \psi = A\sin kx$$

or

$$\cos kL = 0, \quad k = \frac{n\pi}{2L}, \quad n = 1, 3, 5, \ldots$$

$$\implies \psi = B\cos kx.$$

The energy eigenvalues are

$$E = \frac{\hbar^2 k^2}{2m} = \frac{\hbar^2 n^2 \pi^2}{8mL^2}.$$

Solution 5.9. Recall that the probability density is given by $|\psi(x)|^2$.

(1) Using the normalization condition,

$$1 = \int_{-\frac{L}{2}}^{\frac{L}{2}} |\psi(x)|^2 \, \mathrm{d}x = \int_{-\frac{L}{2}}^{\frac{L}{2}} A^2 \cos^2 \frac{3\pi x}{L} \, \mathrm{d}x = A^2 \frac{L}{2}$$

$$\implies A = \sqrt{\frac{2}{L}}.$$

(2) The probability can be calculated by integrating over the region:

$$P = \int_0^{\frac{L}{4}} |\psi(x)|^2 \, \mathrm{d}x$$

$$= \int_0^{\frac{L}{4}} \frac{2}{L} \cos^2 \frac{3\pi x}{L} \, \mathrm{d}x$$

$$= \frac{1}{4}\left(1 - \frac{2}{3\pi}\right)$$

$$= \frac{1}{4} - \frac{1}{6\pi}.$$

Solution 5.10. For the left case,

$$x < 0, \quad \psi = e^{ikx} + A e^{-ikx}, \quad k = \frac{\sqrt{2mE}}{\hbar};$$

$$x > 0, \quad \psi = B e^{ik'x}, \quad k' = \frac{\sqrt{2m(E-V)}}{\hbar}.$$

By the continuity of $\psi(x)$ and $\psi'(x)$ at 0, we obtain

$$1 + A = B,$$

$$ik(1 - A) = ik'B$$

$$\implies A = \frac{k - k'}{k + k'}.$$

Thus the reflection coefficient is given by

$$R = |A|^2 = \frac{(k - k')^2}{(k + k')^2} = \frac{(\sqrt{E} - \sqrt{E-V})^2}{(\sqrt{E} + \sqrt{E-V})^2}. \tag{5.11}$$

For the right case, it can be seen as the left case with $k = \frac{\sqrt{2m(E-V)}}{\hbar}$ and $k' = \frac{\sqrt{2mE}}{\hbar}$. Then we can still use Eq. (5.11) and obtain the same value of the reflection coefficient:

$$R' = \frac{(\sqrt{E-V} - \sqrt{E})^2}{(\sqrt{E-V} + \sqrt{E})^2} = \frac{(\sqrt{E} - \sqrt{E-V})^2}{(\sqrt{E} + \sqrt{E-V})^2} = R$$

Solution 5.11. The wavefunction is

$$e^{ikx} + Pe^{-ikx}, \quad x < 0,$$
$$Se^{ikx}, \quad x > a,$$
$$Ae^{ik'x} + Be^{-ik'x}, \quad 0 < x < a,$$

where

$$k = \frac{\sqrt{2mE}}{\hbar}, \quad k' = \frac{\sqrt{2m(E-V_0)}}{\hbar}.$$

Using the continuity of wavefunction and its derivative at $x = 0$ and $x = a$, we obtain

$$1 + P = A + B,$$
$$k(1 - P) = k'(A - B),$$
$$Se^{ika} = Ae^{ik'a} + Be^{-ik'a},$$
$$kSe^{ika} = k'Ae^{ik'a} - k'Be^{-ik'a}.$$

By these equations we can solve for P and S. Set $\beta = \frac{k}{k'}$, then

$$P = \frac{e^{2ik'a}(1 - \beta^2) - (1 - \beta^2)}{(1 + \beta)^2 - e^{2ik'a}(1 - \beta)^2},$$

$$S = e^{i(k'-k)a} \frac{(1 + \beta)^2 - (1 - \beta)^2}{(1 + \beta)^2 - e^{2ik'a}(1 - \beta)^2}.$$

So the reflection coefficient R and transmission coefficient T are

$$R = |S|^2 = \frac{(4\beta)^2}{(1 + \beta)^4 + (1 - \beta)^4 - 2(1 + \beta)^2(1 - \beta)^2 \cos 2k'a},$$

$$T = |P|^2 = \frac{(1 - \beta^2)^2(2 - 2\cos 2k'a)}{(1 + \beta)^4 + (1 - \beta)^4 - 2(1 + \beta)^2(1 - \beta)^2 \cos 2k'a},$$

and $R + T = 1$.

Solution 5.12. From $\boldsymbol{j} = -\frac{i\hbar}{2m}[\psi^*\nabla\psi - (\nabla\psi^*)\psi]$ and $\hat{p} = -i\hbar\nabla$, we obtain

$$\boldsymbol{j} = \frac{1}{2m}[\psi^*(-i\hbar\nabla)\psi - (-i\hbar\nabla\psi^*)\psi]$$

$$= \frac{1}{2m}[\psi^*(-i\hbar\nabla)\psi + \psi(i\hbar\nabla\psi^*)]$$

$$= \frac{1}{2m}[\psi^*(-i\hbar\nabla)\psi + \psi(-i\hbar\nabla\psi)^*]$$

$$= \frac{1}{2m}[\psi^*\hat{p}\psi + (\psi^*\hat{p}\psi)^*].$$

Solution 5.13. From definition,

$$j = \frac{1}{2m}\left[Ae^{-i\phi}(-ih)\left(i\frac{\partial\phi}{\partial x}\right)A^*e^{i\phi} + \text{c.c.}\right]$$

(c.c. means complex conjugate)

$$= \frac{1}{2m}\left[2|A|^2\hbar\frac{\partial\phi}{\partial x}\right]$$

$$= \frac{\hbar}{m}|A|^2\frac{\partial\phi}{\partial x}.$$

Solution 5.14. To show Eq. (5.9),

$$\frac{1}{\sqrt{2\pi\hbar}}\int_{-\infty}^{\infty}\psi(x)e^{-\frac{i}{\hbar}px}\,dx = \frac{1}{2\pi\hbar}\int_{-\infty}^{\infty}dx\,e^{-\frac{i}{\hbar}px}\int_{-\infty}^{\infty}dp'\,e^{\frac{i}{\hbar}p'x}\phi(p')$$

$$= \frac{1}{2\pi\hbar}\int_{-\infty}^{\infty}dp'\,\phi(p')\int_{-\infty}^{\infty}dx\,e^{\frac{i}{\hbar}(p'-p)x}$$

$$= \int_{-\infty}^{\infty}dp'\,\phi(p')\delta(p'-p)$$

$$= \phi(p).$$

Proof of Eq. (5.10) is equivalent to the proof of $\int_{-\infty}^{+\infty}dk\,g^*(k)g(k) = \int_{-\infty}^{+\infty}dx\,f^*(x)f(x)$ given in Sec. 1.3.7.

Solution 5.15. The Fourier transform $\phi(p)$ is given by

$$\phi(p) = \frac{1}{\sqrt{2\pi\hbar}}e^{-\frac{i}{\hbar}px'}.$$

It is clear that

$$\phi^*(p)\phi(p) = \frac{1}{2\pi\hbar}.$$

The physical meaning is that when a particle is in a position eigenstate its momentum is completely uncertain.

Solution 5.16. $\phi(p)$ is given by the Fourier transform of $\psi(x)$.

$$\phi(p) = \frac{1}{\sqrt{2\pi\hbar}} \int_0^{+\infty} 2\lambda^{\frac{3}{2}} x e^{-\lambda x} e^{-\frac{i}{\hbar}px}\, dx$$

$$= \frac{1}{\sqrt{2\pi\hbar}} (2\lambda^{\frac{3}{2}}) \frac{1}{(\lambda + \frac{ip}{\hbar})^2}.$$

The expectation values are

$$\langle x \rangle = \int_0^{+\infty} dx\, x \cdot 4\lambda^3 \cdot x^2 e^{-2\lambda x}$$

$$= \frac{3}{2\lambda},$$

$$\langle p \rangle = \int_{-\infty}^{+\infty} dp\, \phi(p)\phi^*(p)$$

$$= \int_{-\infty}^{+\infty} dp\, p \cdot \frac{1}{2\pi\hbar} \cdot 4\lambda^3 \cdot \frac{1}{(\lambda^2 + \frac{p^2}{\hbar^2})^2}$$

$$= 0.$$

Solution 5.17. Recall that

$$\psi_n(x) = \sqrt{\beta} \cdot \frac{1}{\sqrt{\sqrt{\pi}2^n n!}} e^{-\frac{\beta^2 x^2}{2}} H_n(\beta x).$$

To prove the equation given in the problem, we start from the right-hand side,

$$\sqrt{\frac{n}{2}}\psi_{n-1}(x) + \sqrt{\frac{n+1}{2}}\psi_{n+1}(x)$$

$$= \sqrt{\frac{n}{2}}\sqrt{\beta} \cdot \frac{1}{\sqrt{\sqrt{\pi}2^{n-1}(n-1)!}} e^{-\frac{\beta^2 x^2}{2}} H_{n-1}(\beta x)$$

$$+ \sqrt{\frac{n+1}{2}}\sqrt{\beta} \cdot \frac{1}{\sqrt{\sqrt{\pi}2^{n+1}(n+1)!}} e^{-\frac{\beta^2 x^2}{2}} H_{n+1}(\beta x)$$

$$= \sqrt{\beta} \cdot \frac{1}{\sqrt{\sqrt{\pi}2^n n!}} e^{-\frac{\beta^2 x^2}{2}} \left[n H_{n-1}(\beta x) + \frac{1}{2} H_{n+1}(\beta x) \right]$$

$$= \sqrt{\beta} \cdot \frac{1}{\sqrt{\sqrt{\pi}2^n n!}} e^{-\frac{\beta^2 x^2}{2}} \beta x H_n(\beta x)$$

$$= \beta x \psi_n(x).$$

Thus

$$x\psi_n(x) = \frac{1}{\beta} \left[\sqrt{\frac{n}{2}} \psi_{n-1}(x) + \sqrt{\frac{n+1}{2}} \psi_{n+1}(x) \right].$$

Also,

$$x^2 \psi_n(x) = \frac{1}{\beta} \left[\sqrt{\frac{n}{2}} x\psi_{n-1}(x) + \sqrt{\frac{n+1}{2}} x\psi_{n+1}(x) \right]$$

$$= \frac{1}{\beta} \left\{ \sqrt{\frac{n}{2}} \left[\frac{1}{\beta} \left(\sqrt{\frac{n-1}{2}} \psi_{n-2}(x) + \sqrt{\frac{n}{2}} \psi_n(x) \right) \right] \right.$$

$$\left. + \sqrt{\frac{n+1}{2}} \left[\frac{1}{\beta} \left(\sqrt{\frac{n+1}{2}} \psi_n(x) + \sqrt{\frac{n+2}{2}} \psi_{n+2}(x) \right) \right] \right\}$$

$$= \frac{1}{2\beta^2} [\sqrt{n(n-1)}\psi_{n-2}(x) + (2n+1)\psi_n(x)$$

$$+ \sqrt{(n+1)(n+2)}\psi_{n+2}(x)].$$

The expectation values are

$$\langle x \rangle = \langle \psi_n | x | \psi_n \rangle = \frac{1}{\beta} \left[\sqrt{\frac{n}{2}} \langle \psi_n | \psi_{n-1} \rangle + \sqrt{\frac{n+1}{2}} \langle \psi_n | \psi_{n+1} \rangle \right] = 0,$$

$$\langle V \rangle = \left\langle \psi_n \left| \frac{1}{2} m\omega^2 x^2 \right| \psi_n \right\rangle = \frac{1}{2} m\omega^2 \cdot \frac{2n+1}{2\beta^2} = \frac{1}{2} \left(n + \frac{1}{2} \right) \hbar\omega = \frac{1}{2} E_n.$$

Solution 5.18. Recall that

$$\psi_n(x) = \sqrt{\beta} \cdot \frac{1}{\sqrt{\sqrt{\pi}2^n n!}} e^{-\frac{\beta^2 x^2}{2}} H_n(\beta x) \tag{5.12}$$

$$\implies \frac{\mathrm{d}\psi}{\mathrm{d}x} = \beta \cdot \sqrt{\beta} \cdot \frac{1}{\sqrt{\sqrt{\pi}2^n n!}} e^{-\frac{\beta^2 x^2}{2}} (-\beta x H_n + H_n'). \tag{5.13}$$

To prove the equation given in the problem, we start from the right-hand side,

$$\sqrt{\frac{n}{2}}\psi_{n-1} - \sqrt{\frac{n+1}{2}}\psi_{n+1} = \sqrt{\beta}\cdot\frac{1}{\sqrt{\sqrt{\pi}2^n n!}}e^{-\frac{\beta^2 x^2}{2}}\left(nH_{n-1} - \frac{1}{2}H_{n+1}\right)$$

$$= \sqrt{\beta}\cdot\frac{e^{-\frac{\beta^2 x^2}{2}}}{\sqrt{\sqrt{\pi}2^n n!}}(nH_{n-1} - \beta x H_n + nH_{n-1})$$

$$= \sqrt{\beta}\cdot\frac{e^{-\frac{\beta^2 x^2}{2}}}{\sqrt{\sqrt{\pi}2^n n!}}(H_n' - \beta x H_n). \tag{5.14}$$

Combining Eqs. (5.13) and (5.14), we obtain

$$\frac{\mathrm{d}}{\mathrm{d}x}\psi_n(x) = \beta\left[\sqrt{\frac{n}{2}}\psi_{n-1}(x) - \sqrt{\frac{n+1}{2}}\psi_{n+1}(x)\right].$$

Differentiate both sides with respect to x,

$$\frac{\mathrm{d}^2}{\mathrm{d}x^2}\psi_n(x) = \beta\left[\sqrt{\frac{n}{2}}\frac{\mathrm{d}}{\mathrm{d}x}\psi_{n-1}(x) - \sqrt{\frac{n+1}{2}}\frac{\mathrm{d}}{\mathrm{d}x}\psi_{n+1}(x)\right]$$

$$= \frac{\beta^2}{2}[\sqrt{n(n-1)}\psi_{n-2}(x) - (2n+1)\psi_n(x)$$

$$+ \sqrt{(n+1)(n+2)}\psi_{n+2}(x)].$$

The expectation values are

$$\langle p\rangle = \left\langle\psi_n\left|-i\hbar\frac{\mathrm{d}}{\mathrm{d}x}\right|\psi_n\right\rangle = 0,$$

$$\langle T\rangle = \left\langle\psi_n\left|-\frac{\hbar^2}{2m}\frac{\mathrm{d}^2}{\mathrm{d}x^2}\right|\psi_n\right\rangle = -\frac{\hbar^2}{2m}\frac{\beta^2}{2}[-(2n+1)] = \frac{1}{2}\left(n+\frac{1}{2}\right)\hbar\omega$$

$$= \frac{1}{2}E_n.$$

Solution 5.19. Recall that $[A, BC] = B[A, C] + [A, B]C$, thus

$$[q, f(q)\hat{p}^2] = f(q)[q, \hat{p}^2]$$

$$= f(q)(\hat{p}[q, \hat{p}] + [q, \hat{p}]\hat{p})$$

$$= f(q)(\hat{p}\cdot i\hbar + i\hbar\hat{p})$$

$$= 2i\hbar f(q)\hat{p}.$$

Consider the following commutator acting on a wavefunction ψ.

$$[\hat{p}, f(q)]\psi = \hat{p}(f(q)\psi) - f(q)\hat{p}\psi = -i\hbar\frac{\mathrm{d}f}{\mathrm{d}q}\psi.$$

Since ψ is arbitrary, we must have

$$[\hat{p}, f(q)] = -i\hbar\frac{\mathrm{d}f}{\mathrm{d}q}.$$

Thus

$$[\hat{p}, f(q)\hat{p}^2] = f(q)[\hat{p}, \hat{p}^2] + [\hat{p}, f(q)]\hat{p}^2$$
$$= [\hat{p}, f(q)]\hat{p}^2$$
$$= \frac{\hbar}{i}\left(\frac{\mathrm{d}f}{\mathrm{d}q}\right)\hat{p}^2.$$

Solution 5.20. Let $A = i(\hat{p}_x^2 x^2 - x^2\hat{p}_x^2)$. Since x^2 and $\hat{p}_x^2$ are Hermitian,

$$A^\dagger = [i(\hat{p}_x^2 x^2 - x^2\hat{p}_x^2)]^\dagger = -i(x^2\hat{p}_x^2 - \hat{p}_x^2 x^2) = A,$$

i.e. A is a Hermitian operator.

Solution 5.21. First consider $\hat{C}\hat{A}|\psi\rangle$.

$$\hat{C}\hat{A}|\psi\rangle = \hat{A}\hat{B}\hat{A}|\psi\rangle = \hat{A}(\hat{A}\hat{B} - 1)|\psi\rangle = \hat{A}\hat{C}|\psi\rangle - \hat{A}|\psi\rangle = (\lambda - 1)\hat{A}|\psi\rangle.$$

Therefore $\hat{A}|\psi\rangle$ is an eigenvector of $\hat{C}$ with eigenvalue $\lambda - 1$. Similarly for $\hat{C}\hat{B}|\psi\rangle$,

$$\hat{C}\hat{B}|\psi\rangle = \hat{A}\hat{B}\hat{B}|\psi\rangle = (\hat{B}\hat{A} + 1)\hat{B}|\psi\rangle = \hat{B}\hat{C}|\psi\rangle + \hat{B}|\psi\rangle = (\lambda + 1)\hat{B}|\psi\rangle.$$

Therefore $\hat{B}|\psi\rangle$ is an eigenvector of $\hat{C}$ with eigenvalue $\lambda + 1$.

Solution 5.22. For any linear combination of $|\psi_i\rangle$,

$$[\hat{A}, \hat{B}]\sum_i c_i|\psi_i\rangle = \hat{A}\hat{B}\sum_i c_i|\psi_i\rangle - \hat{B}\hat{A}\sum_i c_i|\psi_i\rangle$$
$$= \hat{A}\sum_i c_i b_i|\psi_i\rangle - \hat{B}\sum_i c_i a_i|\psi_i\rangle$$
$$= \sum_i c_i b_i a_i|\psi_i\rangle - \sum_i c_i a_i b_i|\psi_i\rangle$$
$$= 0.$$

Solution 5.23. If $\hat{U}^\dagger \hat{U} = 1$ and $\hat{U}|\psi\rangle = a_n|\psi\rangle$, then

$$\langle \hat{U}\psi|\hat{U}\psi\rangle = \langle \hat{U}^\dagger\hat{U}\psi|\psi\rangle = \langle\psi|\psi\rangle.$$

At the same time

$$\langle \hat{U}\psi|\hat{U}\psi\rangle = a_n^* \cdot a_n\langle\psi|\psi\rangle.$$

Therefore $|a_n| = 1$.

Solution 5.24. $\hat{U}^\dagger\hat{U} = 1$.

(1) $\langle \hat{U}\psi|\hat{U}\psi\rangle = \langle \hat{U}^\dagger\hat{U}\psi|\psi\rangle = \langle\psi|\psi\rangle = 1$.

(2) $\langle v_i|v_j\rangle = \langle \hat{U}u_i|\hat{U}u_j\rangle = \langle \hat{U}^\dagger\hat{U}u_i|u_j\rangle = \langle u_i|u_j\rangle = \delta_{ij}$.

Solution 5.25. $\langle E\rangle = \sum_i |c_i|^2 E_i$.

Solution 5.26. Since $\hat{A}$ is Hermition,

$$\langle\phi_j|\hat{A}|\psi_i\rangle^* = \langle\hat{A}\psi_i|\phi_j\rangle = \langle\psi_i|\hat{A}^\dagger\phi_j\rangle = \langle\psi_i|\hat{A}|\phi_j\rangle.$$

Thus

$$D^2 = \sum_{i,j}|\langle\phi_j|\hat{A}|\psi_i\rangle|^2 = \sum_{i,j}\langle\phi_j|\hat{A}|\psi_i\rangle\langle\psi_i|\hat{A}|\phi_j\rangle = \sum_j\langle\phi_j|\hat{A}^2|\phi_j\rangle$$

where we have used $\sum_i |\psi_i\rangle\langle\psi_i| = 1$. Similarly, $\sum_j |\phi_j\rangle\langle\phi_j| = 1$,

$$D^2 = \sum_{i,j}\langle\psi_i|\hat{A}|\phi_j\rangle\langle\phi_j|\hat{A}|\psi_i\rangle, = \sum_i\langle\psi_i|\hat{A}^2|\psi_i\rangle.$$

This shows that the value of D^2 is independent of the choice of basis. Thus we can choose the basis consist of the eigenvectors of $\hat{A}$, then D^2 obviously equals the sum of the squares of all eigenvalues of $\hat{A}$.

Solution 5.27. The plots may be found on the internet.

Solution 5.28. $F(\theta,\phi) = \frac{1}{\sqrt{12}}[(i-1)Y_1^1 + (i+1)Y_1^{-1} + 2\sqrt{2}Y_1^0]$.

(1) The probability of $L_z = -\hbar$ is given by the second term,

$$\frac{|i+1|^2}{12} = \frac{1}{6}.$$

(2) The expectation values are

$$\langle L^2\rangle = l(l+1)\hbar^2 = 2\hbar^2,$$

$$\langle L_z\rangle = 0.$$

(3) Recall that $Y_1^{\pm 1} = \mp\sqrt{\frac{3}{8\pi}}\sin\theta e^{\pm i\phi}$, $Y_1^0 = \sqrt{\frac{3}{4\pi}}\cos\theta$.

$$F(\theta,\phi) = \frac{1}{\sqrt{12}}[i(Y_1^1 + Y_1^{-1}) + (Y_1^{-1} - Y_1^1) + 2\sqrt{2}Y_1^0]$$

$$= \frac{1}{\sqrt{12}}\left[\sqrt{\frac{3}{2\pi}}\sin\theta\sin\phi + \sqrt{\frac{3}{2\pi}}\sin\theta\cos\phi + \sqrt{\frac{6}{\pi}}\cos\theta\right]$$

$$= \frac{1}{\sqrt{12}}\left[\sqrt{\frac{3}{2\pi}}\sin\theta(\sin\phi + \cos\phi) + \sqrt{\frac{6}{\pi}}\cos\theta\right].$$

So the probability per solid angle is

$$P(\theta,\phi) = |F(\theta,\phi)|^2$$

$$= \frac{1}{12}\left[\frac{3}{2\pi}\sin^2\theta(\sin\phi + \cos\phi)^2 + \frac{6}{\pi}\cos^2\theta\right.$$

$$\left. + 2\sqrt{\frac{3}{2\pi}}\sqrt{\frac{6}{\pi}}\sin\theta\cos\theta(\sin\phi + \cos\phi)\right]$$

$$= \frac{1}{12}\left[\frac{3}{2\pi}\sin^2\theta(1 + \sin 2\phi) + \frac{6}{\pi}\cos^2\theta\right.$$

$$\left. + \frac{6}{\pi}\sin\theta\cos\theta(\sin\phi + \cos\phi)\right].$$

5.7 Solutions of Quiz 1

(1) By de Broglie's relation $\lambda = \frac{h}{p}$, Bohr's condition $pr = n\hbar$ is equivalent to $2\pi r = n\lambda$, which corresponds to standing waves on a circle.

(2) The stationary-state Schrödinger equation is

$$-\frac{\hbar^2}{2m}\frac{\mathrm{d}^2}{\mathrm{d}x^2}\psi + \frac{1}{2}m\omega^2 x^2\psi = E\psi.$$

The solution ψ of this differential equation is the stationary-state wavefunction, and the corresponding E is the energy eigenvalue.

(3) The wavefunction is

$$\psi(x) = \begin{cases} e^{ikx} + Ae^{-ikx}, & x < 0, \quad k = \dfrac{\sqrt{2mE}}{\hbar}, \\[2ex] Be^{ik'x} + Ce^{-ik'x}, & 0 \leq x \leq a, \quad k' = \dfrac{\sqrt{2m(E - V_0)}}{\hbar}. \end{cases}$$

Using the continuity of ψ and ψ' at $x = 0$ and $x = a$, we obtain

$$1 + A = B + C,$$

$$k(1 - A) = k'(B - C),$$

$$Be^{ik'a} + Ce^{-ik'a} = 0.$$

The solution is

$$A = \frac{-e^{2ik'a}(k' + k) + k - k'}{e^{2ik'a}(k' - k) + k' + k},$$

$$B = \frac{2k}{e^{2ik'a}(k' - k) + k' + k},$$

$$C = \frac{-2ke^{2ik'a}}{e^{2ik'a}(k' - k) + k' + k}.$$

$|A|^2 = 1$, i.e. the reflection coefficient between region 1 and region 2 equals 1.

5.8   Solutions of Quiz 2

(1) First calculate ΔE.

$$E = \frac{p^2}{2m} \implies \Delta E = \frac{p}{m}\Delta p.$$

Thus

$$\Delta E \Delta t = \frac{p}{m}\Delta p \Delta t = \frac{p}{m}\Delta t \Delta p = \Delta x \Delta p > \hbar.$$

(2) (a) $\hat{O}_1 = \hat{A}\hat{B} - \hat{B}\hat{A}, \quad \hat{O}_1^\dagger = \hat{B}^\dagger\hat{A}^\dagger - \hat{A}^\dagger\hat{B}^\dagger.$
   (b) $\hat{O}_2 = \hat{A}\hat{B} + \hat{B}\hat{A}, \quad \hat{O}_2^\dagger = \hat{B}^\dagger\hat{A}^\dagger + \hat{A}^\dagger\hat{B}^\dagger.$
   (c) $\hat{O}_3 = i(\hat{A}\hat{B} - \hat{B}\hat{A}), \quad \hat{O}_3^\dagger = -i(\hat{B}^\dagger\hat{A}^\dagger - \hat{A}^\dagger\hat{B}^\dagger).$
   (d) $\hat{O}_4 = \hat{A}^\dagger, \quad \hat{O}_4^\dagger = \hat{A}.$
   (e) $\hat{O}_5 = \hat{A}^\dagger\hat{A}, \quad \hat{O}_5^\dagger = \hat{A}^\dagger\hat{A}.$
   If $\hat{A}$ and $\hat{B}$ are both Hermitian, $\hat{O}_2, \hat{O}_3, \hat{O}_4, \hat{O}_5$ are Hermitian.

(3) Combine the results of Problems 5.17 and 5.18, we obtain that

$$\langle V \rangle = \langle T \rangle = \frac{1}{2} E_n,$$

$$\frac{1}{2} m\omega^2 \langle x^2 \rangle = \frac{1}{2m} \langle p^2 \rangle = \frac{1}{2} E_n,$$

$$\implies \langle x^2 \rangle = \frac{E_n}{m\omega^2},$$

$$\langle p^2 \rangle = m E_n.$$

Since $\langle x \rangle = \langle p \rangle = 0$, we get

$$\Delta x \Delta p = \sqrt{\langle x^2 \rangle \langle p^2 \rangle} = \frac{E_n}{\omega}.$$

In the lowest energy state, $E_n = \frac{1}{2}\hbar\omega$, and $\Delta x \Delta p = \frac{\hbar}{2}$.

5.9 Solution of final examination

(1) Find the eigenfunction and eigenvalue of $\hat{p}$.

$$\hat{p}|p_0\rangle = p_0|p_0\rangle$$

$$\frac{\hbar}{i} \frac{d}{dx}|p_0\rangle = p_0|p_0\rangle$$

$$\frac{d|p_0\rangle}{|p_0\rangle} = \frac{i}{\hbar} p_0 \, dx$$

$$|p_0\rangle = A e^{\frac{i}{\hbar} p_0 x}.$$

To determine the normalizing coefficient,

$$\langle p'|p_0\rangle = \delta(p' - p_0) \implies A - \frac{1}{\sqrt{2\pi\hbar}}$$

$$|p_0\rangle = \frac{1}{\sqrt{2\pi\hbar}} e^{\frac{i}{\hbar} p_0 x}.$$

If ψ_{p_0} is an eigenstate,

$$\psi_{p_0} = \frac{1}{\sqrt{2\pi\hbar}} e^{\frac{i}{\hbar} p_0 x}$$

$$e^{icx} \psi_{p_0} = \frac{1}{\sqrt{2\pi\hbar}} e^{\frac{i}{\hbar} (p_0 + \hbar c) x}$$

$$\hat{p}(e^{icx} \psi_{p_0}) = (p_0 + \hbar c)(e^{icx} \psi_{p_0}).$$

Therefore $e^{icx}\psi_{p_0}$ is also an eigenstate of $\hat{p}$, and the corresponding eigenvalue is $p_0 + \hbar c$.

(2) Given a Hermitian operator $\hat{A}$. Let $\hat{A}'$ be the operator after the transformation.

$$\hat{A}' = \hat{U}\hat{A}\hat{U}^{-1},$$

$$\hat{A}'^\dagger = (\hat{U}\hat{A}\hat{U}^{-1})^\dagger$$

$$= (\hat{U}\hat{A}\hat{U}^\dagger)^\dagger$$

$$= \hat{U}\hat{A}^\dagger\hat{U}^\dagger$$

$$= \hat{U}\hat{A}\hat{U}^{-1}$$

$$= \hat{A}'.$$

Thus $\hat{A}'$ is Hermitian. If ψ is an eigenstate of $\hat{A}$ with eigenvalue a,

$$\hat{A}\psi = a\psi.$$

Let ψ' be the function after the transformation.

$$\psi' = \hat{U}\psi$$

$$\hat{A}'\psi' = \hat{U}\hat{A}\hat{U}^{-1}\hat{U}\psi$$

$$= \hat{U}\hat{A}\psi$$

$$= a\hat{U}\psi$$

$$= a\psi'.$$

Thus the eigenvalues are invariant.

(3) (a) $\frac{1}{3}$.

(b) The expectation values are

$$\langle E \rangle = \frac{2}{3}E_1 + \frac{1}{3}E_2 = \frac{2}{3}E_1 + \frac{1}{3}\frac{E_1}{4} = \frac{3}{4}E_1,$$

$$\langle L^2 \rangle = \frac{2}{3}\hbar^2,$$

$$\langle L_z \rangle = \frac{1}{3}\hbar.$$

(4) The matrix elements are

$$\langle \psi_2 | x | \psi_1 \rangle = \frac{1}{\beta} = \sqrt{\frac{\hbar}{m\omega}},$$

$$\langle \psi_2 | \hat{p}^2 | \psi_2 \rangle = mE_2 = m\frac{3}{2}\hbar\omega.$$

To calculate the uncertainty,

$$(\Delta x)^2 = \sqrt{\langle x^2 \rangle - \langle x \rangle^2} = \frac{E_2}{k},$$

$$(\Delta p)^2 = \sqrt{\langle p^2 \rangle - \langle p \rangle^2} = mE_2,$$

$$\Delta x \Delta p = \sqrt{\frac{m}{k}} E_2 = \sqrt{\frac{m}{k}} \frac{5}{2}\hbar\omega = \frac{5}{2}\hbar.$$

(5) For the notation, we separate the region into zone 1 and zone 2, see Fig. 5.7.

(a) The wavefunctions in zones 1 and 2 are given by

$$\psi_1 = Ae^{ik_1 x} + Be^{-ik_1 x},$$

$$\psi_2 = Ce^{k_2 x} + De^{-k_2 x}.$$

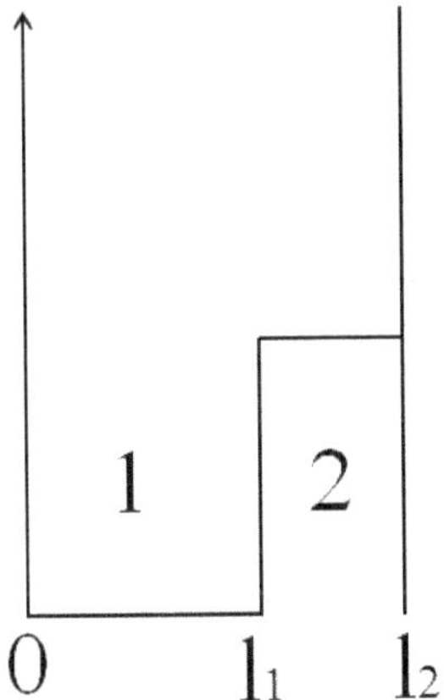

Fig. 5.7 Zones 1 and 2.

Using the continuity at 0, l_1 and l_2,

$$A + B = 0$$

$$C e^{k_2 l_1} + D e^{-k_2 l_1} = A e^{i k_1 l_1} + B e^{-i k_1 l_1} = 2 i A \sin k_1 l_1$$

$$k_2 C e^{k_2 l_1} - k_2 D e^{-k_2 l_1} = 2 i A \cos k_1 l_1$$

$$C e^{k_2 l_2} + D e^{-k_2 l_2} = 0 \implies \frac{D}{C} = -e^{2 k_2 l_2}.$$

$$\frac{e^{k_2 l_1} + \frac{D}{C} e^{-k_2 l_1}}{k_2 \left(e^{k_2 l_1} - \frac{D}{C} e^{-k_2 l_1} \right)} = \frac{1}{k_1} \tan k_1 l_1$$

$$k_2 \tan k_1 l_1 = -k_1 \frac{e^{-k_2 l_1 + 2 k_2 l_2} - e^{k_2 l_1}}{e^{-k_2 l_1 + 2 k_2 l_2} + e^{k_2 l_1}}$$

$$= -k_1 \frac{e^{-k_2 l_1 + k_2 l_2} - e^{k_2 l_1 - k_2 l_2}}{e^{-k_2 l_1 + k_2 l_2} + e^{k_2 l_1 - k_2 l_2}}$$

$$= -k_1 \tanh k_2 (l_2 - l_1).$$

(b) When $V_0 \gg E$, $k_2 \gg k_1$,

$$\tan k_1 l_1 \approx 0, \quad k_1 l_1 \approx n\pi$$

$$\implies E \approx \frac{\hbar^2}{2m} \left(\frac{n\pi}{l} \right)^2.$$

PART 2
Relativity

Chapter 6

Preliminaries

6.1 Introduction

Firstly we want to share with the readers about the aims of writing this second part the "essentials of relativity". We have the following aims.

(1) Strengthen and deepen physics background knowledge for students;
(2) Provide necessary and fundamental knowledge about relativity, in particular special relativity.

The main references we have used are:

(1) Robert Resnick, *Introduction to Special Relativity* (1968);
(2) R.D. Sard, *Relativistic Mechanics* (1970);
(3) J.D. Jackson, *Classical Electrodynamics* (1970);
(4) E.F. Taylor and J.A. Wheeler, *Spacetime Physics* (1966);
(5) H.C. Ohanian, *Gravitation and Spacetime* (1976);
(6) W. Rindler, *Introduction to Special Relativity* (1982);
(7) P.A.M. Dirac, *General Theory of Relativity* (1975).

Secondly, we want to reply three basic questions: 1) What is the meaning of relativity in physics? 2) How relativity governs the form of all physical laws? 3) Why do we say quantum mechanics and special relativity are the two pillars of modern physics? That is to say, we want to reply the three W's (What? How? Why?).

6.1.1 *What is the meaning of relativity in physics?*

Relativity is the *dynamics of spacetime*, it is the governing principle of spacetime and spacetime transformations for all physical laws.

Firstly we should know that we live in a very complex four-dimensional universe which we call spacetime. Up till now, we have found many physical laws because every event occurrs in spacetime, hence we must have one underlying principle to choose the required physical law. From vast number of observations and experiments, physicists' present answer is the relativity principle.

Relativity dictates that physical laws are *covariant* (*form invariant*) under certain spacetime transformations. Hence, relativity contains two related components: *physical laws* and *spacetime transformation*. For *Galilean relativity* and *special relativity* the spacetime transformation is limited to the transformation between two *inertial frames*, which is defined as follows.

Definition 6.1 (Inertial frame). A frame that is at rest or moves uniformly (i.e. not accelerated) relative to the average motion of the matter in the universe in a sufficiently large extent.

Definition 6.2 (Galilean relativity). For *Galilean relativity*, the law involved is Newton's law $\boldsymbol{F} = m\boldsymbol{a}$. The spacetime transformation is Galilean transformation:

$$x' = x - vt, \quad y' = y, \quad z' = z, \quad t' = t. \tag{6.1}$$

Definition 6.3 (Special relativity). For *special relativity* (SR), the laws involved are all physical laws. The spacetime transformation is Lorentz transformation:

$$x' = \frac{x - vt}{\sqrt{1 - \frac{v^2}{c^2}}}, \quad y' = y, \quad z' = z, \quad t' = \frac{t - \frac{v}{c^2}x}{\sqrt{1 - \frac{v^2}{c^2}}}. \tag{6.2}$$

Hence, according to SR, all physical laws have the same form when their spacetime are transformed by Lorentz transformation between any two different inertial reference frames. In other words, all inertial systems are equivalent, there is no preferred inertial system.

Definition 6.4 (General relativity). For *general relativity* (GR), the laws involved are all physical laws. The spacetime transformations are general coordinate transformations.

GR is only applied to curved spacetime, i.e. the spacetime with gravitation; whereas SR is only applied to flat spacetime, i.e. the spacetime without gravitation.

6.1.2 *How relativity governs the form of all physical laws?*

From SR we have already concluded that all laws must be Lorentz covariant.

Example 6.1. Newton's law is not covariant with Lorentz transformation (nonrelativistic). Hence when we acknowledge SR, we must modify Newton's law such that the modified law is covariant under Lorentz transformation, finally we obtain *relativistic mechanics*. The process of generalization from the nonrelativistic form to relativistic form is called *relativistization*.

Example 6.2. Coulomb's law is nonrelativistic. By the process of relativistization, we can obtain *relativistic electrodynamics*, i.e. the Maxwell equations.

Therefore, according to SR, every nonrelativistic law must be generalized to become a relativistic law.

6.1.3 *Why do we say quantum mechanics and special relativity are the two pillars of modern physics?*

Firstly, SR is very important. Since every event occurs in spacetime, every physical law must obey SR.

Secondly, it is well tested. SR is verified by vast number of observations and experiments [3].

For GR, there are only few observations verifying GR [4].

6.2 Vectors and tensors

Now we shall introduce two important mathematical tools to simplify our calculations.

6.2.1 *Vectors*

First we have the *vectors*: they have magnitudes, directions and obey vector addition law.

Example 6.3. Example of vectors.

(1) Radius vector $\boldsymbol{r} = r\boldsymbol{e}_r$, where r is the magniture of $\boldsymbol{r}$ and $\boldsymbol{e}_r$ is the unit vector along the direction of $\boldsymbol{r}$;
(2) Displacement vector $\mathrm{d}\boldsymbol{r}$;
(3) Velocity $\frac{\mathrm{d}\boldsymbol{r}}{\mathrm{d}t}$;
(4) Force $\boldsymbol{F}$.

Not every physical quantity with direction is a vector, the important thing is that it must obey vector addition law. How to define a vector mathematically?

In a coordinate system S a vector $\boldsymbol{A}$ is represented as $\boldsymbol{A} = A_j \boldsymbol{e}_j$; in another system S', the same vector is represented as $\boldsymbol{A}' = A'_i \boldsymbol{e}'_i$. Here we use again the *Einstein summation convention* (see Sec. 3.2). For a vector, there must be certain relation between its components A_j and A'_i. Recall the discussion about the transformation between representations in Sec. 3.4. The same also applies here, just in different notations. Now we shall give the following mathematical definition of a vector.

Definition 6.5 (Vectors). $\boldsymbol{A}$ is a vector if and only if $A'_i = a_{ij} A_j$, where $a_{ij} = \boldsymbol{e}'_i \cdot \boldsymbol{e}_j$.

Example 6.4. Consider $\frac{\partial \varphi}{\partial x_i} \equiv \varphi_{,i}$. We can check the relation $A'_i = a_{ij} A_j$. Since

$$A'_i = \frac{\partial \varphi}{\partial x'_i} = \frac{\partial \varphi}{\partial x_j} \frac{\partial x_j}{\partial x'_i} = a_{ij} A_j,$$

the definition of vector is satisfied.

Example 6.5. Consider the operator $\frac{\partial}{\partial x_i}$. It is evident that it satisfies the same relation $A'_i = a_{ij} A_j$, so it is a vector. It can also be denoted as ∂_i.

In other words, a geometrical object operator can also be defined as a vector.

6.2.2   *Tensors*

Definition 6.6. A geometrical object $\boldsymbol{B}$ with 9 components B_{ij} is a second-rank tensor if and only if

$$B'_{ij} = a_{ik} a_{jm} B_{km}. \tag{6.3}$$

Example 6.6 (Dyad). Consider the object $\boldsymbol{AB}$, $A_i B_j$. Let $C_{ij} = A_i B_j$.

$$C'_{ij} = A'_i B'_j = (a_{ik} A_k)(a_{jm} B_m) = a_{ik} a_{jm} C_{km}.$$

Hence we conclude that Dyad is a second-rank tensor.

Example 6.7 (Vector gradient). Consider the vector gradient $\nabla A = \frac{\partial A_j}{\partial x_i}$.

$$\frac{\partial A'_j}{\partial x'_i} = \frac{\partial}{\partial x'_i}(a_{jm}A_m)$$

$$= a_{jm}\frac{\partial}{\partial x'_i}A_m$$

$$= a_{jm}\frac{\partial x_k}{\partial x'_i}\left(\frac{\partial}{\partial x_k}A_m\right)$$

$$= a_{ik}a_{jm}\frac{\partial A_m}{\partial x_k}.$$

Hence we conclude that the vector gradient ∇A is a second-rank tensor.

Example 6.8 (Identity operator). Consider the identity operator $I = \delta_{ij}$ which is a second-rank tensor.

$$\delta'_{ij} = a_{ik}a_{jm}\delta_{km} = a_{ik}a_{jk} = a_{ik}\widetilde{a}_{kj} = \delta_{ij}.$$

This means that δ_{ij} is covariant under the transformation.

Similarly we can define nth-rank tensor.

Definition 6.7. $C_{i_1,i_2,\ldots,i_n}$ is an nth-rank tensor if and only if

$$C'_{i_1,i_2,\ldots,i_n} = a_{i_1,j_1}a_{i_2,j_2}\ldots a_{i_n,j_n}C_{j_1,j_2,\ldots,j_n}. \tag{6.4}$$

By the above definition, a scalar is a zeroth-rank tensor and a vector is a first-rank tensor.

Noted that the tensors defined above are *Cartesian tensors*, later we shall talk about Lorentz tensors.

6.2.3 *Permutation symbol e_{rst}*

Now we shall introduce a very useful symbol e_{rst}, the so-called permutation symbol.

Definition 6.8. The permutation symbol is defined as

$$e_{rst} = \begin{cases} 1, & \text{even permutation of } (123), \\ -1, & \text{odd permutation of } (123), \\ 0, & \text{otherwise.} \end{cases} \tag{6.5}$$

For example, $e_{123} = e_{231} = e_{312} = 1$, $e_{213} = e_{321} = e_{132} = -1$.

Table 6.1 Summary table.

Vector notation	Index notation	Rank of tensor
$\boldsymbol{v}$	v_i	1
$\lambda = \boldsymbol{u} \cdot \boldsymbol{v}$	$\lambda = u_i v_i$	0
$\boldsymbol{w} = \boldsymbol{u} \times \boldsymbol{v}$	$w_i = e_{ijk} u_j v_k$	1
$\operatorname{grad} \varphi = \nabla \varphi$	$\dfrac{\partial \varphi}{\partial x_i}$	1
$\operatorname{div} \boldsymbol{v} = \nabla \cdot \boldsymbol{v}$	$\dfrac{\partial v_i}{\partial x_i}$	0
$\operatorname{curl} \boldsymbol{v} = \nabla \times \boldsymbol{v}$	$e_{ijk} \dfrac{\partial}{\partial x_j} v_k$	1
$\operatorname{grad} \boldsymbol{v} = \nabla \boldsymbol{v}$	$v_{i,j} = \dfrac{\partial v_i}{\partial x_j}$	2
$\nabla^2 \boldsymbol{v} = \nabla \cdot \nabla \boldsymbol{v} = \Delta \boldsymbol{v}$	$\dfrac{\partial}{\partial x_j}\left(\dfrac{\partial v_i}{\partial x_j}\right) = \dfrac{\partial^2 v_i}{\partial x_j \partial x_j}$	1
$\boldsymbol{uv}$	$u_i v_j$	2

Example 6.9. When we use this symbol, many notations can be simplified.

(1) Determinant: $\|a_{ij}\| = e_{rst} a_{r1} a_{s2} a_{t3}$.
(2) Vector product, $\boldsymbol{w} = \boldsymbol{u} \times \boldsymbol{v}$: $w_i = e_{ijk} u_j v_k$.
(3) Curl: $(\nabla \times \boldsymbol{v})_i = e_{ijk} \dfrac{\partial}{\partial x_j} v_k$.

All the above results and some other results can be summarized in Table 6.1.

6.2.4   *The e–δ identity*

Now we shall introduce a very useful identity, the so-called e–δ identity, which is given by

$$e_{ijk} e_{ist} = \delta_{js}\delta_{kt} - \delta_{jt}\delta_{ks}. \tag{6.6}$$

When we use this identity, we can easily prove some seemingly complex mathematical results.

Example 6.10. Prove $\nabla \times (\varphi \boldsymbol{f}) = \nabla \varphi \times \boldsymbol{f} + \varphi \nabla \times \boldsymbol{f}$.

Proof.

$$[\nabla \times (\varphi \boldsymbol{f})]_i = e_{ijk} \frac{\partial}{\partial x_j}(\varphi f_k)$$

$$= e_{ijk} \frac{\partial \varphi}{\partial x_j} f_k + e_{ijk} \varphi \frac{\partial f_k}{\partial x_j}$$

$$= (\nabla \varphi \times \boldsymbol{f})_i + \varphi (\nabla \times \boldsymbol{f})_i. \qquad \square$$

Example 6.11. Prove $\nabla \times (\nabla \times \boldsymbol{f}) = \nabla(\nabla \cdot \boldsymbol{f}) - \nabla^2 \boldsymbol{f}$.

Proof.

$$[\nabla \times (\nabla \times \boldsymbol{f})]_i = e_{ijk}\frac{\partial}{\partial x_j}(\nabla \times \boldsymbol{f})_k$$

$$= e_{ijk}\frac{\partial}{\partial x_j}\left(e_{klm}\frac{\partial}{\partial x_l}f_m\right)$$

$$= e_{ijk}e_{klm}\frac{\partial^2 f_m}{\partial x_j \partial x_l}$$

$$= e_{kij}e_{klm}\frac{\partial^2 f_m}{\partial x_j \partial x_l}$$

$$= (\delta_{il}\delta_{jm} - \delta_{im}\delta_{jl})\frac{\partial^2 f_m}{\partial x_j \partial x_l}$$

$$= \frac{\partial^2 f_j}{\partial x_j \partial x_i} - \frac{\partial^2 f_i}{\partial x_j \partial x_j}$$

$$= \frac{\partial}{\partial x_i}\left(\frac{\partial f_j}{\partial x_j}\right) - (\nabla^2 \boldsymbol{f})_i$$

$$= [\nabla(\nabla \cdot \boldsymbol{f})]_i - (\nabla^2 \boldsymbol{f})_i. \qquad \square$$

Problem 6.1. *Prove* $\nabla \times \nabla\varphi = 0$ *and* $\nabla \times (\nabla \times \boldsymbol{f}) = 0$.

Problem 6.2. *Prove*

(1) $\nabla \cdot (\boldsymbol{a} \times \boldsymbol{b}) = \boldsymbol{b} \cdot (\nabla \times \boldsymbol{a}) - \boldsymbol{a} \cdot (\nabla \times \boldsymbol{b});$
(2) $\nabla \times (\boldsymbol{a} \times \boldsymbol{b}) = \boldsymbol{a} \cdot (\nabla \cdot \boldsymbol{b}) - \boldsymbol{b} \cdot (\nabla \cdot \boldsymbol{a}) - (\boldsymbol{a} \cdot \nabla)\boldsymbol{b} + (\boldsymbol{b} \cdot \nabla)\boldsymbol{a}.$

6.3 Principle of Galilean relativity

For Galilean relativity, the corresponding spacetime transformation is the Galilean transformation:

$$x' = x - vt, \tag{6.7}$$

$$y' = y, \tag{6.8}$$

$$z' = z, \tag{6.9}$$

$$t' = t. \tag{6.10}$$

The transformation is between two inertial frames. From this we can get some important results.

Firstly, because $t' = t$ and v is constant, we can measure the space interval between two neighboring points at the same time, i.e. simultaneously. The space interval is

$$dl'^2 = dx'^2 + dy'^2 + dz'^2$$

$$= dl^2 = dx^2 + dy^2 + dz^2. \tag{6.11}$$

Simultaneity is a crucial part of our definition of the length of the moving rod.

Secondly, we can see that space and time are separated. L (length), M (mass) and T (time) are the three basic quantities in mechanics, which here are all independent of the relative motion of the observer.

On the other hand, the corresponding physical law of Galilean relativity is Newton's law $\boldsymbol{F} = m\boldsymbol{a}$. With Galilean transformation, the physical quantities mass m and force $\boldsymbol{F}$ in inertial frame S remain unchanged in inertial frame S', i.e. $m' = m$ and $\boldsymbol{F}' = \boldsymbol{F}$. For the acceleration $\boldsymbol{a}$, it becomes $\boldsymbol{a}' = \frac{\mathrm{d}^2 \boldsymbol{r}'}{\mathrm{d}t'^2} = \frac{\mathrm{d}^2 \boldsymbol{r}}{\mathrm{d}t^2} = \boldsymbol{a}$, i.e. also unchanged. Hence Newton's law is covariant under the Galilean transformation.

In summary, we have the following equivalent statements of principle of Galilean relativity:

(1) Newton's law of motion is of the same form in all inertial systems;
(2) All inertial systems are equivalent in classical mechanics;
(3) There is no way at all (classically) of determining the absolute velocity of an inertial frame from our mechanical experiments;
(4) In classical mechanics, we can only speak of the relative velocity of one inertial frame with respect to another, and not of an absolute velocity.

6.4 Essentials of electrodynamics

Another important example of relativistic physics is the electrodynamics. Let us recall some basic notions here.

6.4.1 *Electrostatics and magnetostatics*

6.4.1.1 *Electric charge and electric field*

The electric field produced by a point charge q_i is given by $\boldsymbol{E}_i = \frac{1}{4\pi\epsilon_0} \frac{q_i \boldsymbol{r}}{r^3}$. By principle of superposition of field, we obtain the electric field produced

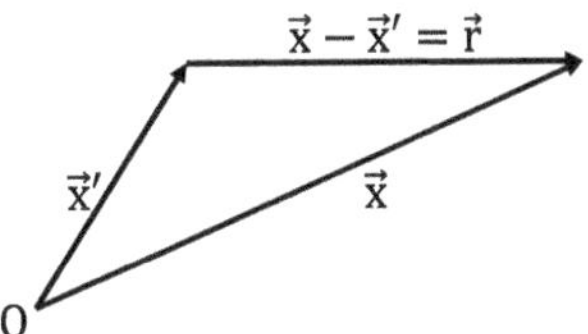

Fig. 6.1 Radius vectors.

by a charge distribution:

$$\boldsymbol{E} = \sum_i \boldsymbol{E}_i = \frac{1}{4\pi\epsilon_0} \int \frac{\rho \boldsymbol{r}}{r^3}\, \mathrm{d}V', \tag{6.12}$$

where ρ is the charge density in the volume element $\mathrm{d}V'$. The total force applied on the charge system by the electric fields can be calculated by means of the law

$$\boldsymbol{F} = q\boldsymbol{E}. \tag{6.13}$$

Now we shall use Fig. 6.1 to show the relationship between three radius vectors: radius vector of field point $\boldsymbol{x}$, radius vector of source point $\boldsymbol{x}'$, and radius vector from source point to field point $\boldsymbol{r} = \boldsymbol{x} - \boldsymbol{x}'$. Also we use two different operators to differentiate two different operation:

- ∇: operate on field point only, $\nabla = \boldsymbol{i}\frac{\partial}{\partial x} + \boldsymbol{j}\frac{\partial}{\partial y} + \boldsymbol{k}\frac{\partial}{\partial z}$, $\boldsymbol{x} = x\boldsymbol{i} + y\boldsymbol{j} + z\boldsymbol{k}$;
- ∇': operate on source point only, $\nabla' = \boldsymbol{i}\frac{\partial}{\partial x'} + \boldsymbol{j}\frac{\partial}{\partial y'} + \boldsymbol{k}\frac{\partial}{\partial z'}$, $\boldsymbol{x}' = x'\boldsymbol{i} + y'\boldsymbol{j} + z'\boldsymbol{k}$.

We first calculate $\boldsymbol{E}$. Since

$$\nabla\left(\frac{1}{r}\right) = -\frac{\boldsymbol{r}}{r^3},$$

$$\boldsymbol{E} = \frac{1}{4\pi\epsilon_0} \int \frac{\boldsymbol{r}\rho(\boldsymbol{x}')}{r^3}\, \mathrm{d}V'$$

$$= -\nabla \int \frac{\rho(\boldsymbol{x}')}{4\pi\epsilon_0 r}\, \mathrm{d}V'$$

$$= -\nabla\varphi, \tag{6.14}$$

where $\varphi(\boldsymbol{x})$ is the *scalar potential*. This means

$$\nabla \times \boldsymbol{E} = 0. \tag{6.15}$$

We can also calculate the divergence of $\boldsymbol{E}$,

$$\nabla \cdot \boldsymbol{E} = -\nabla^2 \varphi$$

$$= -\int \frac{\rho(\boldsymbol{x}')}{4\pi\epsilon_0} \nabla^2 \left(\frac{1}{r}\right) \, \mathrm{d}V'$$

$$= -\int \frac{\rho(\boldsymbol{x}')}{4\pi\epsilon_0} [-4\pi\delta(\boldsymbol{x} - \boldsymbol{x}')] \, \mathrm{d}V'$$

$$= \frac{\rho}{\epsilon_0}.$$

Hence we obtain the Gauss law:

$$\nabla \cdot \boldsymbol{E} = \frac{\rho}{\epsilon_0}. \tag{6.16}$$

Furthermore, we obtain the Poisson equation:

$$\nabla^2 \varphi = -\frac{\rho}{\epsilon_0}. \tag{6.17}$$

6.4.1.2 *Electric current*

Firstly we introduce a new vector $\boldsymbol{j}$ called current density, with direction same as the direction of motion of positive charges and magnitude the quantity of electricity crossing unit area (in the direction $\boldsymbol{j}$) per unit time. Then the current through a surface Σ is given by:

$$I = \int_\Sigma \boldsymbol{j} \cdot \mathrm{d}\boldsymbol{\Sigma}. \tag{6.18}$$

Secondly we introduce the law of conservation of charge. Since the current going out of a closed surface of a volume V is equal to the decrease of charge per unit time in the volume, we have:

$$\oint_{\partial V} \boldsymbol{j} \cdot \mathrm{d}\boldsymbol{\Sigma} = -\frac{\mathrm{d}}{\mathrm{d}t} \int_V \rho \, \mathrm{d}V, \tag{6.19}$$

where ∂V is the surface of V.

From Eq. (6.19) we obtain

$$\int_V \nabla \cdot \boldsymbol{j} \, \mathrm{d}V = -\int_V \frac{\partial \rho}{\partial t} \, \mathrm{d}V$$

$$\implies \nabla \cdot \boldsymbol{j} + \frac{\partial \rho}{\partial t} = 0. \tag{6.20}$$

This is the differential form of the law of conservation of charge.

6.4.1.3 *Biot–Savart law and Ampere's law*

Biot–Savart law gives the magnetic field produced by the current element $I\mathrm{d}\boldsymbol{l}$, then by the principle of superposition of field, the magnetic field of the whole current is given by integration:

$$\boldsymbol{B} = \frac{\mu_0}{4\pi} \int \frac{I\mathrm{d}\boldsymbol{l} \times \boldsymbol{r}}{r^3}. \tag{6.21}$$

Since

$$I\mathrm{d}\boldsymbol{l} = (\boldsymbol{j} \cdot \mathrm{d}\boldsymbol{\Sigma})\mathrm{d}\boldsymbol{l} = \boldsymbol{j}(\mathrm{d}\boldsymbol{\Sigma} \cdot \mathrm{d}\boldsymbol{l}) = \boldsymbol{j}\mathrm{d}V',$$

we have

$$\boldsymbol{B}(\boldsymbol{x}) = \frac{\mu_0}{4\pi} \int \frac{\boldsymbol{j}(\boldsymbol{x}) \times \boldsymbol{r}}{r^3}\mathrm{d}V'.$$

Since $\nabla(\frac{1}{r}) = \frac{\boldsymbol{r}}{r^3}$, $\nabla \times (\varphi\boldsymbol{a}) = \nabla\varphi \times \boldsymbol{a} + \varphi\nabla \times \boldsymbol{a}$,

$$\boldsymbol{B}(\boldsymbol{x}) = \frac{\mu_0}{4\pi}\nabla \times \int \frac{\boldsymbol{j}(\boldsymbol{x})}{r}\mathrm{d}V' = \nabla \times \boldsymbol{A}, \tag{6.22}$$

where

$$A = \frac{\mu_0}{4\pi} \int \frac{\boldsymbol{j}(\boldsymbol{x})}{r}\mathrm{d}V' \tag{6.23}$$

is the *vector potential* of the current distribution. From the above equation, we readily get

$$\nabla \cdot \boldsymbol{B} = 0. \tag{6.24}$$

Using $\nabla\cdot(\varphi\boldsymbol{a}) = \boldsymbol{a}\cdot\nabla\varphi + \varphi\nabla\cdot\boldsymbol{a}$ and $\nabla^2(\frac{1}{r}) = -4\pi\delta(\boldsymbol{x} - \boldsymbol{x}')$, we calculate

$$\begin{aligned}
\nabla \times \boldsymbol{B} &= \nabla \times (\nabla \times \boldsymbol{A})\\
&= \nabla(\nabla \cdot \boldsymbol{A}) - \nabla^2\boldsymbol{A}\\
&= \frac{\mu_0}{4\pi}\nabla \int \boldsymbol{j}(\boldsymbol{x}) \cdot \nabla\left(\frac{1}{r}\right)\mathrm{d}V' - \frac{\mu_0}{4\pi}\int \boldsymbol{j}(\boldsymbol{x})\nabla^2\left(\frac{1}{r}\right)\mathrm{d}V'\\
&= -\frac{\mu_0}{4\pi}\nabla \int \boldsymbol{j}(\boldsymbol{x}) \cdot \nabla'\left(\frac{1}{r}\right)\mathrm{d}V' + \mu_0\boldsymbol{j}\\
&= \frac{\mu_0}{4\pi}\nabla \int \frac{\nabla' \cdot \boldsymbol{j}(\boldsymbol{x})}{r}\mathrm{d}V' + \mu_0\boldsymbol{j},
\end{aligned}$$

Table 6.2 Comparison between electrostatics and magnetostatics.

Electrostatics	Magnetostatics
$\nabla \times \boldsymbol{E} = 0$	$\nabla \times \boldsymbol{B} = \mu_0 \boldsymbol{j}$
$\boldsymbol{E} = -\nabla \varphi$	$\boldsymbol{B} = \nabla \times \boldsymbol{A}$
$\nabla \cdot \boldsymbol{E} = \frac{\rho}{\epsilon_0}$	$\nabla \cdot \boldsymbol{B} = 0$
$\nabla^2 \varphi = -\frac{\rho}{\epsilon_0}$	$\nabla^2 \boldsymbol{A} = -\mu_0 \boldsymbol{j}$ (choose $\nabla \cdot \boldsymbol{A} = 0$)
$\varphi = \frac{1}{4\pi\epsilon_0} \int \frac{\rho}{r} \, \mathrm{d}V'$	$A = \frac{\mu_0}{4\pi} \int \frac{j}{r} \mathrm{d}V'$

where we have used integration by part in the last step. Since for steady current $\nabla' \cdot \boldsymbol{j}(\boldsymbol{x}) = 0$, we have

$$\nabla \times \boldsymbol{B} = \mu_0 \boldsymbol{j}. \tag{6.25}$$

This is Ampere's law.

As a summary, Table 6.2 shows the corresponding laws for electrostatics and magnetostatics.

6.4.2 *Maxwell equations*

6.4.2.1 *Faraday's law*

Faraday's law is expressed as

$$\mathcal{E} = -\frac{\mathrm{d}\psi}{\mathrm{d}t} = -\frac{\mathrm{d}}{\mathrm{d}t} \int \boldsymbol{B} \cdot \mathrm{d}\boldsymbol{\Sigma}, \tag{6.26}$$

where $\mathcal{E}$ is the electromotive force (emf) induced by the variable magnetic field in the loop we want to study,

$$\mathcal{E} = \oint \boldsymbol{E} \cdot \mathrm{d}\boldsymbol{l} = \int \nabla \times \boldsymbol{E} \cdot \mathrm{d}\boldsymbol{\Sigma},$$

and ψ is the magnetic flux through the surface bounded by the loop.

From the above two equations, we get

$$\nabla \times \boldsymbol{E} = -\frac{\partial \boldsymbol{B}}{\partial t}. \tag{6.27}$$

This equation shows that varying magnetic field produces electric field.

6.4.2.2 *Displacement current*

Ampere's law $\nabla \times \boldsymbol{B} = \mu_0 \boldsymbol{j}$ is correct only for steady-state current, i.e. $\nabla \cdot \boldsymbol{j} = 0$, or equivalently $\frac{\partial \rho}{\partial t} = 0$ from the conservation of charge (6.20).

In general, $\frac{\partial \rho}{\partial t} \neq 0$, to revise Ampere's law, Maxwell replaced $\boldsymbol{j}$ with $\boldsymbol{j} + \boldsymbol{j}_D$:

$$\nabla \times \boldsymbol{B} = \mu_0 (\boldsymbol{j} + \boldsymbol{j}_D) \implies \nabla \cdot (\boldsymbol{j} + \boldsymbol{j}_D) = 0.$$

Combine with the conservation of charge (6.20), we readily obtain the so-called *displacement current*,

$$\boldsymbol{j}_D = \epsilon_0 \frac{\partial \boldsymbol{E}}{\partial t}. \tag{6.28}$$

This shows the very important fact that *varying electric field can produce magnetic field, in the same way as conduction current* $\boldsymbol{j}$.

6.4.2.3 *Maxwell equation in vacuum*

Combining all that we have discussed thus far, we obtain the Maxwell equation in vacuum:

$$\begin{cases} \nabla \times \boldsymbol{E} = -\dfrac{\partial \boldsymbol{B}}{\partial t}, & \text{from generalization of slowly varying field case;} \\[2ex] \nabla \times \boldsymbol{B} = \mu_0 \boldsymbol{j} + \mu_0 \epsilon_0 \dfrac{\partial \boldsymbol{E}}{\partial t}, & \text{from Maxwell's suggestion;} \\[2ex] \nabla \cdot \boldsymbol{B} = 0, & \text{from generalization of static case;} \\[2ex] \nabla \cdot \boldsymbol{E} = \dfrac{\rho}{\epsilon_0}, & \text{from generalization of static case.} \end{cases} \tag{6.29}$$

The above equations have been verified by vast number of experiments and observations, they are always true. Note that however the following equations are false in general:

$$\boldsymbol{F} = \frac{1}{4\pi\epsilon_0} \frac{q_1 q_2 \boldsymbol{r}}{r^3},$$

$$\nabla \times \boldsymbol{E} = 0,$$

$$\nabla \times \boldsymbol{B} = \mu_0 \boldsymbol{j}.$$

6.4.2.4 *Energy density and energy flux, momentum density and momentum flux*

In a similar way as we introduce the charge density ρ and the current density j, we can introduce the energy density w and the energy flux S, which is called the Poynting vector, i.e.

$$w \longleftrightarrow \rho, \quad S \longleftrightarrow j.$$

Furthermore we introduce the force density f, then we have the Lorentz force law:

$$f \equiv \frac{F}{\Delta V} = \frac{qE + q(v \times B)}{\Delta V} = \rho E + j \times B. \quad (j = \rho v)$$

For a single charge q, the rate of doing work by external fields is $qv \cdot E$, where v is the velocity of the charge. The magnetic field does no work because the magnetic force is perpendicular to the velocity. For a continuous distribution of charge and current, the total rate of doing work by the fields in a finite volume is

$$\int_V f \cdot v \, dV = \int_V j \cdot E \, dV.$$

By the law of conservation of energy,

$$\int_V j \cdot E \, dV + \int_\Sigma S \cdot d\Sigma = -\frac{d}{dt} \int w \, dV$$

$$\Longrightarrow \nabla \cdot S + \frac{\partial w}{\partial t} = -j \cdot E.$$

For the vacuum case, from the Maxwell equation and Lorentz force law we can obtain

$$w = \frac{1}{2} \left(\epsilon_0 E^2 + \frac{1}{\mu_0} B^2 \right), \quad S = \frac{1}{\mu_0} E \times B.$$

Similar to how we introduce the energy density and energy flux, we can introduce the momentum density g and momentum flux $\mathcal{J}$, which is a second-rank tensor. By the law of conservation of momentum, we have

$$\nabla \cdot \mathcal{J} + \frac{\partial g}{\partial t} = -f.$$

From the Maxwell equation and Lorentz force law we obtain

$$\boldsymbol{g} = \epsilon_0 \boldsymbol{E} \times \boldsymbol{B}, \quad \boldsymbol{\mathcal{J}} = -\epsilon_0 \boldsymbol{E}\boldsymbol{E} - \frac{1}{\mu_0}\boldsymbol{B}\boldsymbol{B} + \frac{1}{2}\boldsymbol{I}\left(\epsilon_0 E^2 + \frac{1}{\mu_0}B^2\right).$$

6.4.2.5 *Wave equation*

From the Maxwell equation, we can see that there exist $\boldsymbol{E}(\boldsymbol{x},t)$ and $\boldsymbol{B}(\boldsymbol{x},t)$ even when $\rho = \boldsymbol{j} = 0$. This shows that the electromagnetic field can separate itself from charge and current; it can exist independently. The varying electric field and the varying magnetic field stimulate each other, thus there exists electromagnetic wave propagating in space.

In the vacuum $\rho = \boldsymbol{j} = 0$. Since

$$\nabla \times (\nabla \times \boldsymbol{E}) = \nabla(\nabla \cdot \boldsymbol{E}) - \nabla^2 \boldsymbol{E} = -\nabla^2 \boldsymbol{E},$$

$$-\nabla \times \frac{\partial \boldsymbol{B}}{\partial t} = -\frac{\partial}{\partial t}(\nabla \times \boldsymbol{B}) = -\mu_0\epsilon_0 \frac{\partial^2 \boldsymbol{E}}{\partial t^2},$$

we have

$$\nabla^2 \boldsymbol{E} - \frac{1}{c^2}\frac{\partial^2 \boldsymbol{E}}{\partial t^2} \equiv \Box^2 \boldsymbol{E} = 0, \quad c = \frac{1}{\sqrt{\mu_0 \epsilon_0}}. \tag{6.30}$$

The operator $\Box^2 = \nabla^2 - \frac{1}{c^2}\frac{\partial^2}{\partial t^2}$ is called the D'Alembertian.

Similarly,

$$\nabla \times (\nabla \times \boldsymbol{B}) = \nabla(\nabla \cdot \boldsymbol{B}) - \nabla^2 \boldsymbol{B} = -\nabla^2 \boldsymbol{B},$$

$$\nabla \times \left(\mu_0\epsilon_0 \frac{\partial \boldsymbol{E}}{\partial t}\right) = \mu_0\epsilon_0 \frac{\partial}{\partial t}(\nabla \times \boldsymbol{E}) = -\mu_0\epsilon_0 \frac{\partial^2 \boldsymbol{B}}{\partial t^2}$$

$$\implies \nabla^2 \boldsymbol{B} - \frac{1}{c^2}\frac{\partial^2 \boldsymbol{B}}{\partial t^2} = \Box^2 \boldsymbol{B} = 0.$$

This means that in the vacuum the propagation velocity of electromagnetic waves (light) along any direction is $c = \frac{1}{\sqrt{\mu_0\epsilon_0}}$.

From the Galilean transformation, we have the velocity addition law $\frac{d\boldsymbol{r}'}{dt} = \frac{d\boldsymbol{r}}{dt} - \boldsymbol{v}$, i.e. $\boldsymbol{u}' = \boldsymbol{u} - \boldsymbol{v}$. So if in one coordinate system S, the velocity of light is isotropic, then in another coordinate system S' with velocity $\boldsymbol{v}$ relative to S, the velocity of light cannot be isotropic, i.e. the Maxwell equation can only be true in a specific reference frame. The form of the wave equation $\Box^2 \boldsymbol{E} = \Box^2 \boldsymbol{B} = 0$ is not covariant under Galilean transformation.

Therefore if we use Galilean transformation as the spacetime transformation for Maxwell equation, then we must give up the form invariance between different inertial frames, i.e. there exists only one privileged reference frame where we can use Maxwell equation. In other words, there is no relativity for Maxwell equation.

From the mechanistic view of nature, we should ask the question: the electromagnetic waves (light) propagate in what medium? It seemed imperative to postulate the existence of a medium — the aether — which would serve as a carrier for these waves. So, from the mechanistic viewpoint, aether is the privileged reference frame. This led to the most urgent physical problem at the end of the 19th century: finding the aether and detecting the Earth's motion through the aether.

Chapter 7

Special Relativity: Kinematics

7.1 Attempts to locate the absolute frame: Michelson–Morley experiment

Chapter 6 ended with the conclusion of the existence of aether, i.e. the Maxwell equation under the Galilean transformation leads to the existence of a privileged reference frame, the aether frame. This is the mechanistic view of nature.

The aether was assumed to fill up all space and to be the medium with respect to which the light speed c applies. It followed then that an observer moving through the aether with velocity v_0 would measure a velocity c' for a light beam, where $c' = c - v_0$. This is the result that the Michelson–Morley experiment was designed to test. A.A. Michelson (1852–1931) invented the optical interferometer whose remarkable sensitivity made such an experiment possible. Michelson first performed the experiment in 1881, and then in 1887, in collaboration with E.W. Morley, carried out the more precise version of the investigation that was destined to lay the experimental foundations of special relativity.

A simplified version of the experiment is shown in Fig. 7.1. The beam of light from the laboratory source is split by the partially silvered mirror into two coherent beams, with beam 1 being transmitted through and beam 2 being reflected off. They are reflected by two mirrors back to the partially silvered mirror. Then the returning beam 1 is partially reflected and beam 2 is partially transmitted to a detector where they interfere. The interference is constructive or destructive depending on the phase difference of the beams.

Let us compute the phase difference between the beams. This difference can arise from two causes: the different path lengths traveled, l_1 and l_2,

135

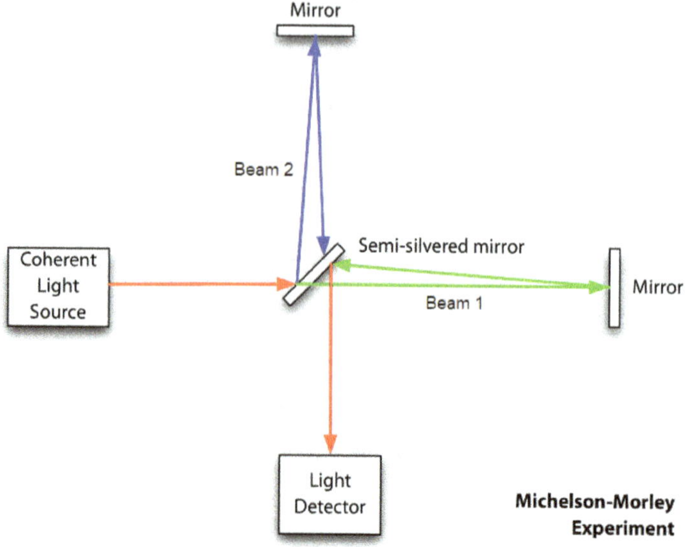

Fig. 7.1 A simplified version of the Michelson–Morley experiment showing how the beam from the source is split into two beams by the partially silvered mirror. The beams are reflected by the mirrors, returning to the partially silvered mirror, and then sent to the detector to form the interference pattern.

and the different speeds of travel with respect to the instrument because of the aether wind speed v. The latter is the crucial one. The different speeds are much like the difference between cross-stream and up-and-down-stream speeds of a swimmer in a moving stream with respect to shore. For beam 1, whose speed is c in the aether, it has an "upstream" speed of $c - v$ with respect to the apparatus and a "downstream" speed of $c + v$. Thus the time for beam 1 to travel to the mirror and back is

$$t_1 = \frac{l_1}{c - v} + \frac{l_1}{c + v} = l_1 \left(\frac{2c}{c^2 - v^2} \right) = \frac{2l_1}{c} \left(\frac{1}{1 - v^2/c^2} \right).$$

The path of beam 2, traveling up to the mirror and back, is a cross-stream path through the aether. By Galilean transformation, the light velocity in this direction is $\sqrt{c^2 - v^2}$, see Fig. 7.2. Thus the time for beam 2 to travel to the mirror and back is

$$t_2 = 2\frac{l_2}{\sqrt{c^2 - v^2}} = \frac{2l_2}{c} \left(\frac{1}{\sqrt{1 - v^2/c^2}} \right).$$

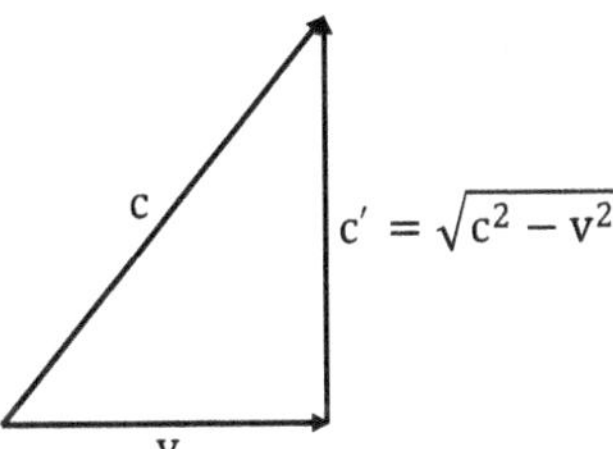

Fig. 7.2   Speed of light in the aether wind. $\boldsymbol{v}$ is the velocity of aether relative to Earth, $\boldsymbol{c}$ is the velocity of light relative to aether, and $\boldsymbol{c}'$ is velocity of light relative to Earth.

The difference in transit times is

$$\Delta t = t_2 - t_1 = \frac{2}{c}\left(\frac{l_2}{\sqrt{1 - v^2/c^2}} - \frac{l_1}{1 - v^2/c^2}\right).$$

Suppose that the instrument is rotated through $90°$, thereby making l_1 the cross-stream length and l_2 the downstream length. If the corresponding times are now designated by primes, the same analysis as above gives the transit time difference as

$$\Delta t' = t_2' - t_1' = \frac{2}{c}\left(\frac{l_2}{1 - v^2/c^2} - \frac{l_1}{\sqrt{1 - v^2/c^2}}\right).$$

Hence, the rotation changes the difference by

$$\Delta t' - \Delta t = \frac{2}{c}\left(\frac{l_2 + l_1}{1 - v^2/c^2} - \frac{l_2 + l_1}{\sqrt{1 - v^2/c^2}}\right)$$

$$\approx \frac{2}{c}(l_2 + l_1)\left(1 + \frac{v^2}{c^2} - 1 - \frac{1}{2}\frac{v^2}{c^2}\right)$$

$$= \left(\frac{l_1 + l_2}{c}\right)\frac{v^2}{c^2}.$$

Thus the rotation should cause a shift in the fringe pattern. There should be ΔN number of fringes moving past the crosshairs, where

$$\Delta N = \frac{\Delta t' - \Delta t}{T} \approx \frac{l_1 + l_2}{cT}\frac{v^2}{c^2} = \frac{l_1 + l_2}{\lambda}\frac{v^2}{c^2} \quad \left(\text{period } T = \frac{\lambda}{c}\right).$$

For the Michelson–Morley experiment, $l_1 = l_2 = 22$ m, $\lambda = 5.5\times10^{-7}$ m, and $\frac{v}{c} = 10^{-4}$, so we should have $\Delta N = 0.4$. However, the experiment concluded with no fringe shift at all.

The null result was such a blow to the aether hypothesis that the experiment was repeated by many scientists over a 50-year period. All the attempts to determine any motion relative to the hypothetical aether are proved to be in vain.

Before 1905 there were two choices for physicists:

(1) Retain the Maxwell equation and the Galilean transformation, abandon the relativity principle for electromagnetic phenomena;

(2) Retain the Maxwell equation and the relativity principle, abandon the Galilean transformation.

With the null result of the Michelson–Morley experiment and the failures to present the privileged aether frame, it appears more natural to choose (2) than (1), i.e. to extend the relativity principle from dynamics to the whole physics, this is what Einstein did in 1905. Einstein persisted in relativity principle and abandon the Newtonian spacetime, i.e. the Galilean transformation.

7.2   Postulates of special relativity and the Lorentz transformation

7.2.1 *Postulates of special relativity*

In 1905, Einstein abandoned Newtonian spacetime and found the true structure of spacetime. As he wrote in his autobiographical notes: "After ten years of reflection such a principle [the principle of special relativity] resulted from a paradox upon which I had already hit at the age of sixteen: If I pursue a beam of light with the velocity c (velocity of light in a vacuum), I should observe such a beam as an electromagnetic field constant in time, periodic in space. However, there seems to exist no such thing, neither on the basis of experience, nor according to Maxwell equations [...]"

He put forward two postulates:

(1) The laws of nature, whether mechanical or electromagnetic, are of the same form in all inertial systems.

(2) In all inertial systems the speed of propagation of electromagnetic waves is the same and is independent of direction.

7.2.2 *Lorentz transformation*

We consider two inertial reference systems Σ and Σ', with Σ' moving at a constant speed v relative to Σ, along the direction of the positive x_1 axis, see Fig. 7.3. This is sometimes called a boost in the direction of x_1.

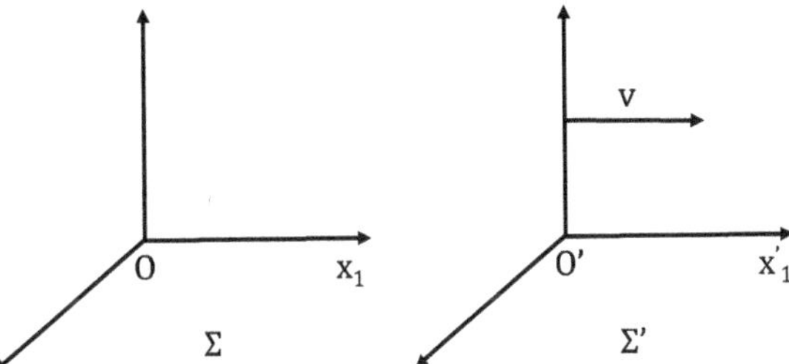

Fig. 7.3   Two inertial reference frames with one moving relative to the other at a constant speed in the x_1 direction.

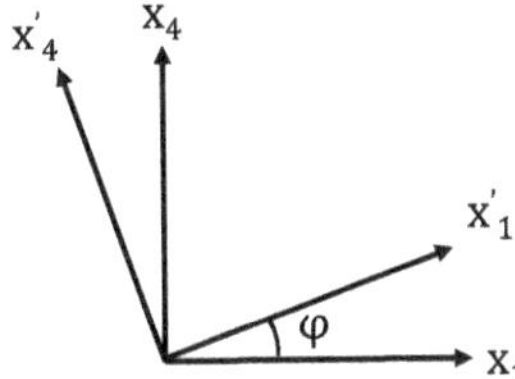

Fig. 7.4   Rotation of two-dimensional complex plane.

These two inertial reference systems coincide at $t = t' = 0$, at that time a pulse of light is emitted at the origin. By postulate (2), light speed equals c in both systems. Hence, in Σ, the pulse of light lies on the surface of sphere $x_1^2 + x_2^2 + x_3^2 = c^2 t^2$; whereas in Σ', the pulse of light is on the surface of sphere $x_1'^2 + x_2'^2 + x_3'^2 = c^2 t'^2$. Let $x_4 = ict$, $x_4' = ict'$, we obtain $x_1^2 + x_2^2 + x_3^2 + x_4^2 = x_1'^2 + x_2'^2 + x_3'^2 + x_4'^2 = 0$, i.e. $x_\mu x_\mu = x_\nu' x_\nu' = 0$. Therefore, we want to find a four-dimensional orthogonal transformation such that $x_\mu x_\mu = x_\nu' x_\nu'$.

Since in the current case $x_2' = x_2$, $x_3' = x_3$, we want $x_1'^2 + x_4'^2 = x_1^2 + x_4^2$, which corresponds to a rotation of two-dimensional complex plane, see Fig. 7.4.

$$\begin{cases} x_1' = x_1 \cos\varphi + x_4 \sin\varphi \\ x_4' = -x_1 \sin\varphi + x_4 \cos\varphi \end{cases}.$$

If we change back to (x_1, t),

$$\begin{cases} x_1' = x_1 \cos\varphi + ict \sin\varphi \\ t' = \frac{ix_1}{c} \sin\varphi + t \cos\varphi \end{cases}.$$

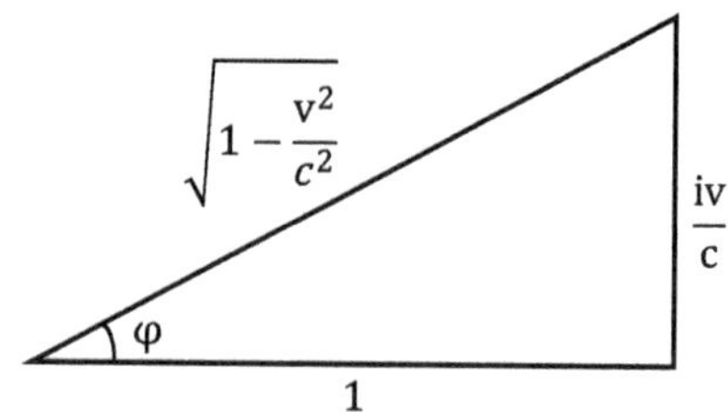

Fig. 7.5   Trigonometry of φ.

When $x_1' = 0$, $x_1 = vt$, hence $\tan \varphi = \frac{iv}{c}$. From Fig. 7.5, we have

$$\sin \varphi = \frac{\frac{iv}{c}}{\sqrt{1 - \frac{v^2}{c^2}}}, \quad \cos \varphi = \frac{1}{\sqrt{1 - \frac{v^2}{c^2}}}.$$

Finally, we obtain the Lorentz transformation:

$$x_1' = \frac{x_1 - vt}{\sqrt{1 - \frac{v^2}{c^2}}}, \tag{7.1}$$

$$x_2' = x_2, \tag{7.2}$$

$$x_3' = x_3, \tag{7.3}$$

$$t' = \frac{t - \frac{v}{c^2} x_1}{\sqrt{1 - \frac{v^2}{c^2}}}. \tag{7.4}$$

When $v \ll c$, the Lorentz transformation reduces to the Galilean transformation:

$$x_1' = x_1 - vt,$$

$$x_2' = x_2,$$

$$x_3' = x_3,$$

$$t' = t.$$

7.3   The structure of spacetime

7.3.1   *Comparison*

From the previous discussions we can compare the Newtonian spacetime and the relativistic spacetime by means of Table 7.1.

Table 7.1 Comparison between the Newtonian space-time and the relativistic spacetime.

Newtonian	Relativistic
Galilean transformation	Lorentz transformation
$\begin{cases} t' = t, \\ x' = x - vt, \\ y' = y, \\ z' = z. \end{cases}$	$\begin{cases} t' = \dfrac{t - \frac{v}{c^2} x_1}{\sqrt{1 - \frac{v^2}{c^2}}}, \\ x' = \dfrac{x - vt}{\sqrt{1 - \frac{v^2}{c^2}}}, \\ y' = y, \\ z' = z. \end{cases}$

In the case of Newtonian spacetime, space and time are separated. There are two separate geometries: a three-dimensional Euclidean geometry for space and a one-dimensional geometry for time. The space interval $\mathrm{d}l$ is given by $\mathrm{d}l^2 = \mathrm{d}x^2 + \mathrm{d}y^2 + \mathrm{d}z^2 \geq 0$. It can also be written as $\mathrm{d}l^2 = \delta_{ij}\mathrm{d}x_i\mathrm{d}x_j$.

For relativistic spacetime, space and time cannot be separated. There is one combined geometry that applies to both space and time, which is the four-dimensional pseudo-Euclidean geometry. In four-dimensional spacetime, every event is represented by the so-called *world point*, and a process of event is represented by a *world line*.

In four-dimensional spacetime, the elementary interval is given by

$$\mathrm{d}s^2 = \eta_{\mu\nu}\mathrm{d}x_\mu\mathrm{d}x_\nu, \tag{7.5}$$

where $\mu, \nu = 0, 1, 2, 3$, $x_0 = ct, x_1 = x, x_2 = y, x_3 = z$, and

$$(\eta_{\mu\nu}) = \begin{pmatrix} -1 & 0 & 0 & 0 \\ 0 & 1 & 0 & 0 \\ 0 & 0 & 1 & 0 \\ 0 & 0 & 0 & 1 \end{pmatrix}$$

is the Minkowski tensor. This spacetime is also called the Minkowski spacetime.

7.3.2 *Spacetime interval*

In four-dimensional spacetime, there are three kinds of spacetime interval: $\mathrm{d}s^2 < 0$, $\mathrm{d}s^2 = 0$, $\mathrm{d}s^2 > 0$.

7.3.2.1 *Time-like interval*

In a time-like interval, $ds^2 < 0$, the time part of the interval dominates over the space part. For example, we can have two events occur at the same place but different times, i.e. $ds^2 = -c^2 d\tau^2$. In another frame moving at speed v, the coordinates are connected to τ by Lorentz transformation:

$$d\tau = \sqrt{dt^2 - (dx^2 + dy^2 + dz^2)/c^2} = dt\sqrt{1 - \frac{v^2}{c^2}}.$$

In the particle's rest frame, $d\tau = dt$, thus the *proper time* $d\tau$ is the time interval measured by a clock that moves with the particle. As it is linked to the spacetime interval, $d\tau$ is a Lorentz scalar.

As seen in the calculation, two events separated by a time-like interval can be connected by a world line of a massive particle. Furthermore, since $ds^2 < 0$, we cannot find a frame where $dt = 0$, i.e. we cannot let the two events occur at the same time by Lorentz transformation. The time order of two such events is absolute: the earlier event is absolutely earlier, and the later event is absolutely later. In the particle rest frame, the time interval between the two events has its minimum value.

7.3.2.2 *Light-like interval*

In a light-like interval, $ds^2 = 0$, the time part of the interval equals the space part. For such intervals, we cannot talk about two events that occur at the same time or the same place, since if they have the same time, they necessarily have the same place, thus it is the same event.

On the other hand, two events can be connected by a world line of a light signal. The time order of two events is absolute, the spatial separation of two events is also absolute.

7.3.2.3 *Space-like interval*

In a space-like interval, $ds^2 > 0$, the space part of the interval dominates over the time part. For such interval, we can have two events occur at the same time but different places, i.e. $ds^2 = d\lambda^2$. If we apply Lorentz transformation, we have

$$d\lambda = \sqrt{dx^2 + dy^2 + dz^2 - c^2 dt^2}.$$

This means that $d\lambda$ is the distance between two events when measured simultaneously ($dt = 0$). As it is linked to the spacetime interval, $d\lambda$ is a Lorentz scalar.

For such intervals, two events cannot be connected by a world line of a massive particle or a world line of a light signal. Since $ds^2 > 0$, we cannot find a frame where $d\lambda = 0$, i.e. we cannot let the two events occur at the same place by Lorentz transformation. The spatial separation is absolute. In the frame where two events occur at the same time, the spatial distance between the two events has its maximum value.

7.3.3 *Proper time and coordinate time*

Recall that the proper time interval $d\tau$ is invariant under Lorentz transformation, i.e. it is a Lorentz scalar. However, it is not an exact differential, i.e. its value between two world points depends on the shape of the world line connecting these two points.

On the other hand, the coordinate time interval dt is not invariant under Lorentz transformation, i.e. it has different values in different inertial frame. However, dt is an exact differential. Once we have chosen the reference frame, every world point has its own definite coordinate time, so the coordinate time interval between two events is independent of the shape of the world line, i.e. independent of the processes of event.

7.3.4 *Light cone*

Along the world line of a light signal, the interval is of course light-like. The set of all world lines of light signals leaving or arriving at the origin forms a three-dimensional "surface" in the four-dimensional spacetime, called the light cone. This surface has the equation

$$-c^2t^2 + x^2 + y^2 + z^2 = 0,$$

Every world point has its own light cone. Since we cannot make a drawing of a four-dimensional space, let us omit the z-coordinate, then the light cone is simply the ordinary cone $-c^2t^2 + x^2 + y^2 = 0$, shown in Fig. 7.6.

The upper $(t > 0)$ and lower $(t < 0)$ parts of this cone are called the *forward (future) light cone* and the *backward (past) light cone* respectively. Particles which leave the origin must have world lines inside the future light cone; particles which arrive at the origin must have world lines inside the past light cone. Obviously, all world lines drawn from the origin to the interior of the cone has time-like interval, whereas all world lines drawn from the origin to the exterior point of the cone has space-like interval, and

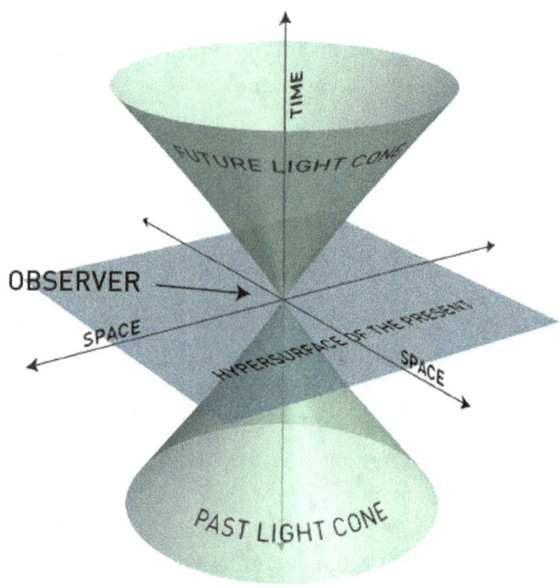

Fig. 7.6   Light cone. Credit: en.wikipedia/Stib.

all the exterior points can be transformed to have the same time as the origin by Lorentz transformation.

Problem 7.1. *Two events occur at the same place in the laboratory frame of reference and are separated in time by 3 seconds.*

(1) *What is the spatial distance between these two events in a rocket frame in which the events are separated in time by 5 seconds?*
(2) *What is the relative speed of the rocket and the laboratory frame?*

Answer. (1) 12×10^8 meters; (2) $\frac{4}{5}c$.

Problem 7.2. *Two events whose coordinates (x, y, z, t) in the inertial frame S are $(0, 0, 0, 0)$ and $(2c, 0, 0, 1)$. Find the speeds (relative to S) of frames in which*

(1) *the events are simultaneous;*
(2) *the second event precedes the first by one unit of time.*

Is there a frame in which the events occur at the same point?

Answer. (1) $\frac{1}{2}c$; (2) $\frac{4}{5}c$.

7.4 Law of causation and the relative character of simultaneity

For time-like and light-like intervals, since the time order of two events cannot be changed by Lorentz transformation, the causality will not be destroyed.

In a reference frame Σ,

$$\left. \begin{array}{l} \text{cause } (x_1, t_1) \\ \text{effect } (x_2, t_2) \end{array} \right\} \implies t_2 > t_1.$$

In reference frame Σ',

$$\left. \begin{array}{l} \text{cause } (x'_1, t'_1) \\ \text{effect } (x'_2, t'_2) \end{array} \right\} \implies t'_2 - t'_1 = \frac{(t_2 - t_1) - \frac{v}{c^2}(x_2 - x_1)}{\sqrt{1 - v^2/c^2}}.$$

The law of causation requires that

$$t'_2 > t'_1 \implies \left| \frac{x_2 - x_1}{t_2 - t_1} \right| < \frac{c^2}{v}.$$

Let $u = \frac{|x_2 - x_1|}{t_2 - t_1}$, then

$$uv < c^2 \implies u < c, \quad v < c.$$

Hence, law of causation requires the propagation speed u of all interactions to be less than c.

As for space-like interval, since there is no causality between two events in this case, time order can be changed by Lorentz transformation and simultaneity is relative: two events can occur at same time in one frame and not at the same time in other frames.

7.5 Time dilation and length contraction

7.5.1 *Time dilation*

Consider two inertial frames Σ and Σ'. Let a standard clock be fixed in Σ' and consider the two events at that clock when it indicates times t'_1 and t'_2, differing by $\Delta t'$. If Σ' is moving at a velocity v in the x-direction, we inquire what time interval Δt is ascribed to these events in Σ.

Let $\gamma = 1/\sqrt{1 - \frac{v^2}{c^2}}$, then the Lorentz transformation is given by

$$t' = \gamma\left(t - \frac{vx}{c^2}\right), \quad x' = \gamma(x - vt), \quad y' = y, \quad z' = z. \qquad (7.6)$$

The inverse Lorentz transformation is given by

$$t = \gamma\left(t' + \frac{vx'}{c^2}\right), \quad x = \gamma(x' + vt'), \quad y = y', \quad z = z'. \qquad (7.7)$$

Since the clock is fixed in Σ', $\Delta x' = 0$, so

$$\Delta t = \gamma\left(\Delta t' + \frac{v}{c^2}\Delta x'\right) = \gamma\Delta t'.$$

$\Delta t'$ is the proper time of the moving clock and $\Delta t' < \Delta t$. Therefore a clock moving uniformly with velocity v with respect to an inertial frame Σ goes slower by a factor of $\sqrt{1 - \frac{v^2}{c^2}}$ relative to the standard clocks at rest in Σ.

7.5.2 *Length contraction*

Consider two inertial frames Σ and Σ' where Σ' is moving at a velocity v relative to Σ. By Lorentz transformation,

$$\Delta x' = \gamma(\Delta x - v\Delta t).$$

If we measure the length $\Delta x = L$ at the same time in Σ, i.e. $\Delta t = 0$, then

$$L_0 = \Delta x' = \gamma\Delta x = \gamma L.$$

Hence the proper length of a body in the direction of its motion, L_0, is reduced by a factor of $\sqrt{1 - \frac{v^2}{c^2}}$ when measured by an observer at rest.

Problem 7.3. *A clock gives correct time in its rest frame. With what speed must the clock move relative to an observer so that it may seem to lose 1 minute in 24 hours?*

Answer. 1.1×10^7 m/s.

Problem 7.4. *Calculate the percentage contraction of a rod moving with a velocity 0.8c in a direction inclined at 60° to its own length.*

Answer. 8.4%.

Problem 7.5. *A particle with a mean lifetime of 1 μs moves through the laboratory at 2.7×10^8 m/s.*

(1) *What is its lifetime as measured by observers in the laboratory?*
(2) *If it was manufactured in the target of an accelerator, how far does it go, on the average, in the laboratory before disintegrating?*

Answer. (1) 2.3 μs; (2) 6.2×10^2 m.

Problem 7.6. *A rocket of proper length 10 m is moving away from the Earth at speed $\frac{4}{5}c$. A light signal is sent after it that arrived at the rocket's tail at time zero according to rocket and Earth clocks. Calculate in both frames the time at which the light signal reaches the head of the rocket. It is there reflected back by a mirror. Calculate in both frames the time at which the light signal again reaches the tail of the rocket. Compare the elapsed time in the two frames for the round-trip tail–head–tail.*

Answer. 6.7×10^{-8} s; 11.1×10^{-8} s.

7.6 Velocity transformation

The velocity components in a frame Σ are:

$$\left(\begin{array}{l} u_x = \dfrac{\mathrm{d}x}{\mathrm{d}t} \\[2mm] u_y = \dfrac{\mathrm{d}y}{\mathrm{d}t} \\[2mm] u_z = \dfrac{\mathrm{d}z}{\mathrm{d}t} \end{array} \right).$$

To calculate the velocity in another frame Σ', we apply the Lorentz transformation:

$$\mathrm{d}x' = \frac{\mathrm{d}x - v\mathrm{d}t}{\sqrt{1 - \frac{v^2}{c^2}}} = \frac{u_x - v}{\sqrt{1 - \frac{v^2}{c^2}}}\mathrm{d}t,$$

$$\mathrm{d}y' = \mathrm{d}y = u_y\mathrm{d}t,$$

$$\mathrm{d}z' = \mathrm{d}z = u_z\mathrm{d}t,$$

$$\mathrm{d}x' = \frac{\mathrm{d}t - \frac{v}{c^2}\mathrm{d}x}{\sqrt{1 - \frac{v^2}{c^2}}} = \frac{1 - \frac{v}{c^2}u_x}{\sqrt{1 - \frac{v^2}{c^2}}}\mathrm{d}t.$$

Thus the velocity in Σ' is given by

$$u'_x = \frac{dx'}{dt'} = \frac{u_x - v}{1 - \frac{v}{c^2}u_x}, \tag{7.8}$$

$$u'_y = \frac{dy'}{dt'} = \frac{u_y\sqrt{1 - \frac{v^2}{c^2}}}{1 - \frac{v}{c^2}u_x}, \tag{7.9}$$

$$u'_z = \frac{dz'}{dt'} = \frac{u_z\sqrt{1 - \frac{v^2}{c^2}}}{1 - \frac{v}{c^2}u_x}. \tag{7.10}$$

To get the inverse transformation, we can substitute v with $-v$.

$$u_x = \frac{u'_x + v}{1 + \frac{v}{c^2}u'_x}, \tag{7.11}$$

$$u_y = \frac{u'_y\sqrt{1 - \frac{v^2}{c^2}}}{1 + \frac{v}{c^2}u'_x}, \tag{7.12}$$

$$u_z = \frac{u'_z\sqrt{1 - \frac{v^2}{c^2}}}{1 + \frac{v}{c^2}u'_x}. \tag{7.13}$$

These equations can be regarded as giving the resultant u of the two velocities $v = (v, 0, 0)$ and u' and are therefore occasionally referred to as the relativistic velocity addition formula.

Problem 7.7. *In a given inertial frame, two particles are shot out simultaneously from a given point, with equal speeds v, in orthogonal directions. What is the speed of each particle relative to the other?*

Answer. $v\sqrt{2 - \frac{v^2}{c^2}}$.

Problem 7.8. *Two electrons are ejected in opposite directions from radioactive atoms in a sample of radioactive material at rest in the laboratory. Each electron has a speed $0.67c$ as measured by a laboratory observer. What is the speed of each electron as measured from the other?*

Answer. $0.92c$.

Problem 7.9. *Derive the relativistic acceleration transformation:*

$$a'_x = \frac{a_x \left(1 - \frac{v^2}{c^2}\right)^{3/2}}{\left(1 - \frac{v}{c^2} u_x\right)^3},$$
(7.14)

where $a_x = \frac{\mathrm{d}u_x}{\mathrm{d}t}$, $a'_x = \frac{\mathrm{d}u'_x}{\mathrm{d}t'}$.

Chapter 8

Relativistic Electrodynamics

8.1 Four-dimensional vectors and tensors

To understand the 4-dimensional spacetime, we can refer back to the familiar 3-dimensional space. In relativistic dynamics, we generally denote the indices of 3-dimensional variables with Roman alphabets, e.g. $i, j = 1, 2, 3$; for indices of 4-dimensional variables we use Greek alphabets, e.g. $\mu, \nu = 0, 1, 2, 3$.

Recall that for a 3-dimensional orthogonal transformation a_{ij},

$$x'_i = a_{ij} x_j, \quad i = 1, 2, 3,$$

the norm squared of a vector is invariant,

$$x'_i x'_i = x_j x_j,$$

and we have

$$a_{ij} a_{ik} = \delta_{jk}, \quad \widetilde{\underline{\underline{a}}}\, \underline{\underline{a}} = \underline{\underline{I}}.$$

Similarly for a 4-dimensional orthogonal transformation $a_{\mu\nu}$,

$$x'_\mu = a_{\mu\nu} x_\nu, \quad \mu = 0, 1, 2, 3,$$

$$a_{\mu\nu} a_{\mu\lambda} = \delta_{\nu\lambda}, \quad \widetilde{\underline{\underline{a}}}\, \underline{\underline{a}} = \underline{\underline{I}}.$$

The norm squared of a vector is still invariant, but recall from Sec. 7.3 that the norm squared in Minkowski spacetime is calculated as

$$\eta'_{\mu\nu} x'_\mu x'_\nu = \eta_{\mu\nu} x_\mu x_\nu.$$

In 3-dimensional space, we have scalars which are invariant under the coordinate transformation. In 4-dimensional spacetime, some invariant

151

scalars that we have already seen are

$$ds^2 = \eta_{\mu\nu}dx_\mu dx_\mu,$$

$$\frac{\partial}{\partial x_\mu}\frac{\partial}{\partial x_\mu} = \Box^2 = \nabla^2 - \frac{1}{c^2}\frac{\partial^2}{\partial t^2}.$$

For vectors, recall that a proper vector must follow the same orthogonal transformation as the basis when we change the coordinate system,

$$v'_i = a_{ij}v_j.$$

Similarly in 4-dimensional spacetime, when we change coordinate system, the vectors must change accordingly:

$$v'_\mu = a_{\mu\nu}v_\nu.$$

Some examples of 4-dimensional vectors are

$$x_\mu, \quad \frac{\partial}{\partial x_\mu}, \quad u_\mu = \frac{dx_\mu}{d\tau},$$

where τ is the proper time which is a scalar. The 4-dimensional vectors are also called four-vectors.

Take a closer look at the velocity four-vector, u_μ, also called the four-velocity. Recall the relation between coordinate time t and proper time τ: $dt = \gamma d\tau$. Thus we have

$$\begin{aligned} u_\mu &= \frac{dx_\mu}{d\tau} \\ &= \frac{d}{d\tau}(ct, \boldsymbol{x}) \\ &= \gamma\frac{d}{dt}(ct, \boldsymbol{x}) \\ &= \gamma(c, \boldsymbol{v}). \end{aligned} \tag{8.1}$$

Then the norm squared of the four-velocity is

$$\eta_{\mu\nu}u_\mu u_\nu = \gamma^2(-c^2 + v^2) = \frac{1}{1 - \frac{v^2}{c^2}}(-c^2 + v^2) = -c^2. \tag{8.2}$$

Recall that in 3-dimensional space, the second-rank tensors must transform as

$$T'_{ij} = a_{ik}a_{jl}T_{kl}.$$

Similarly for a 4-dimensional second-rank tensor, we have

$$T'_{\mu\nu} = a_{\mu\lambda}a_{\nu\rho}T_{\lambda\rho}.$$

8.2 Four-dimensional current density vector

Consider some charged particle that is moving at a velocity $\boldsymbol{v}$ relative to the laboratory frame Σ. In the frame where the particle is at rest, Σ', the charge density is ρ_0 while the current density is 0. To an observer in Σ, the charge density is ρ which is different from ρ_0 because of length contraction, i.e.

$$\rho = \gamma\rho_0.$$

For the current density in Σ,

$$\boldsymbol{j} = \rho\boldsymbol{v} = \gamma\rho_0\boldsymbol{v}.$$

Thus we can combine the charge density and current density to form a new variable j_μ,

$$j_\mu = (\rho c, \boldsymbol{j}) = \rho_0\gamma(c, \boldsymbol{v}) = \rho_0 u_\mu. \tag{8.3}$$

Since ρ_0 is a scalar and u_μ is a four-vector, j_μ is also a four-vector.

Recall the law of conservation of charge,

$$\nabla \cdot \boldsymbol{j} + \frac{\partial \rho}{\partial t} = 0.$$

In the new notation, it becomes

$$\frac{\partial j_\mu}{\partial x_\mu} = 0,$$

or simply

$$j_{\mu,\mu} = 0. \tag{8.4}$$

8.3 Four-dimensional potential vector

8.3.1 *Scalar potential and vector potential*

Recall that in Sec. 6.4.1 we introduced the vector potential $\boldsymbol{A}$ and scalar potential φ such that

$$\boldsymbol{B} = \nabla \times \boldsymbol{A}, \quad \boldsymbol{E} = -\nabla\varphi - \frac{\partial \boldsymbol{A}}{\partial t}.$$

Then two Maxwell equations

$$\nabla \cdot \boldsymbol{B} = 0, \quad \nabla \times \boldsymbol{E} = -\frac{\partial \boldsymbol{B}}{\partial t}$$

are automatically satisfied. We can use $(\varphi, \boldsymbol{A})$ to describe the electromagnetic field instead of $(\boldsymbol{E}, \boldsymbol{B})$.

Given an arbitrary scalar function ψ, if we have $(\varphi', \boldsymbol{A}')$ such that

$$\boldsymbol{A}' = \boldsymbol{A} + \nabla \psi, \quad \varphi' = \varphi - \frac{\partial \psi}{\partial t},$$

then $(\varphi', \boldsymbol{A}')$ gives the same $(\boldsymbol{E}, \boldsymbol{B})$. That means $(\varphi', \boldsymbol{A}')$ and $(\varphi, \boldsymbol{A})$ describe the same electromagnetic field. Each set $(\varphi, \boldsymbol{A})$ is called a gauge. The transformation $(\varphi, \boldsymbol{A}) \to (\varphi', \boldsymbol{A}')$ is called a gauge transformation. Observable quantities like $\boldsymbol{E}$ and $\boldsymbol{B}$ are unchanged by a gauge transformation and are said to be gauge invariant.

The Maxwell equations are gauge invariant. The gauge freedom in the choice of φ and $\boldsymbol{A}$ is often exploited in solving electromagnetic problems. The two most popular gauges are given below.

Coulomb gauge. $\nabla \cdot \boldsymbol{A} = 0$. Then we can solve for φ and $\boldsymbol{A}$ using the Maxwell equations.

$$\nabla \cdot \boldsymbol{E} = \frac{\rho}{\epsilon_0}$$

$$\nabla^2 \varphi + \frac{\partial}{\partial t}(\nabla \cdot \boldsymbol{A}) = -\frac{\rho}{\epsilon_0}$$

$$\nabla^2 \varphi = -\frac{\rho}{\epsilon_0}.$$

On the other hand,

$$\nabla \times \boldsymbol{B} = \mu_0 \boldsymbol{j} + \mu_0 \epsilon_0 \frac{\partial \boldsymbol{E}}{\partial t}$$

$$\Box^2 \boldsymbol{A} = -\mu_0 \boldsymbol{j} + \nabla \left(\nabla \cdot \boldsymbol{A} + \frac{1}{c^2} \frac{\partial \varphi}{\partial t} \right)$$

$$\Box^2 \boldsymbol{A} = -\mu_0 \boldsymbol{j} + \frac{1}{c^2} \nabla \left(\frac{\partial \varphi}{\partial t} \right).$$

Lorentz gauge. $\nabla\cdot\boldsymbol{A}+\frac{1}{c^2}\frac{\partial\varphi}{\partial t}=0$. Similarly, we use the Maxwell equations to find φ and $\boldsymbol{A}$.

$$\nabla\cdot\boldsymbol{E}=\frac{\rho}{\epsilon_0}$$

$$\nabla^2\varphi+\frac{\partial}{\partial t}(\nabla\cdot\boldsymbol{A})=-\frac{\rho}{\epsilon_0}$$

$$\Box^2\varphi=-\frac{\rho}{\epsilon_0}$$

$$\implies \varphi(\boldsymbol{x},t)=\int\frac{\rho(\boldsymbol{x},t-\frac{r}{c})}{4\pi\epsilon_0 r}\,\mathrm{d}V'.$$

On the other hand,

$$\nabla\times\boldsymbol{B}=\mu_0\boldsymbol{j}+\mu_0\epsilon_0\frac{\partial\boldsymbol{E}}{\partial t}$$

$$\Box^2\boldsymbol{A}=-\mu_0\boldsymbol{j}$$

$$\implies \boldsymbol{A}(\boldsymbol{x},t)=\int\frac{\epsilon_0\boldsymbol{j}(\boldsymbol{x},t-\frac{r}{c})}{4\pi r}\,\mathrm{d}V'.$$

These $\varphi(\boldsymbol{x},t)$ and $\boldsymbol{A}(\boldsymbol{x},t)$ are called the retarded potential solutions.

8.3.2 *Four-dimensional potential vector*

We can define a 4-vector by combining the scalar potential and vector potential,

$$A_\mu=\left(\frac{\varphi}{c},\boldsymbol{A}\right). \tag{8.5}$$

In this case, the Lorentz gauge

$$\nabla\cdot\boldsymbol{A}+\frac{1}{c^2}\frac{\partial\varphi}{\partial t}=0$$

can be simply written as

$$\frac{\partial A_\mu}{\partial x_\mu}=0 \quad\text{or}\quad A_{\mu,\mu}=0. \tag{8.6}$$

Recall the corresponding differential equation for φ,

$$\Box^2\varphi=-\frac{\rho}{\epsilon_0}=-\mu_0 c^2\rho$$

$$\implies \Box^2\left(\frac{\varphi}{c}\right)=-\mu_0\rho c.$$

Combine with the differential equation for $\boldsymbol{A}$,

$$\Box^2 \boldsymbol{A} = -\mu_0 \boldsymbol{j},$$

we have the differential equation for the four-potential

$$\Box^2 A_\mu = -\mu_0 j_\mu. \tag{8.7}$$

Since A_μ is a 4-vector, it transforms according to the following rule:

$$A'_\mu = a_{\mu\nu} A_\nu.$$

For the particular case of a boost along x-axis, we have

$$\varphi' = \gamma(\varphi - vA_x),$$
$$A'_x = \gamma\left(A_x - \frac{v}{c^2}\varphi\right),$$
$$A'_y = A_y,$$
$$A'_z = A_z.$$

8.4 Field tensor

From A_μ we can introduce a second-rank antisymmetric tensor:

$$F_{\mu\nu} \equiv A_{\nu,\mu} - A_{\mu,\nu}. \tag{8.8}$$

Using $\boldsymbol{E} = -\nabla\varphi - \frac{\partial \boldsymbol{A}}{\partial t}$ and $\boldsymbol{B} = \nabla \times \boldsymbol{A}$, $F_{\mu\nu}$ can be expressed as

$$F_{\mu\nu} = \begin{pmatrix} 0 & -\frac{E_x}{c} & -\frac{E_y}{c} & -\frac{E_z}{c} \\ \frac{E_x}{c} & 0 & -B_z & B_y \\ \frac{E_y}{c} & B_z & 0 & -B_x \\ \frac{E_z}{c} & -B_y & B_x & 0 \end{pmatrix}. \tag{8.9}$$

Problem 8.1. *Using the Maxwell equations,*

$$\nabla \cdot \boldsymbol{E} = \frac{\rho}{\epsilon_0},$$
$$\nabla \times \boldsymbol{B} = \mu_0 \boldsymbol{j} + \mu_0 \epsilon_0 \frac{\partial \boldsymbol{E}}{\partial t},$$

show that $F_{\mu\nu,\nu} = \mu_0 j_\mu.$

Problem 8.2. *Using the Maxwell equations,*

$$\nabla \cdot \boldsymbol{B} = 0,$$

$$\nabla \times \boldsymbol{E} = -\frac{\partial \boldsymbol{B}}{\partial t},$$

show that

$$F_{\mu\nu,\lambda} + F_{\nu\lambda,\mu} + F_{\lambda\mu,\nu} = 0,$$

i.e. $F_{\{\mu\nu,\lambda\}} = 0$.

Problem 8.3. *From* $F'_{\mu\nu} = a_{\mu\lambda} a_{\nu\rho} F_{\lambda\rho}$, *prove that*

$$\boldsymbol{E}'_{\parallel} = \boldsymbol{E}_{\parallel}$$

$$\boldsymbol{B}'_{\parallel} = \boldsymbol{B}_{\parallel}$$

$$\boldsymbol{E}'_{\perp} = \gamma(\boldsymbol{E} + \boldsymbol{v} \times \boldsymbol{B})_{\perp},$$

$$\boldsymbol{B}'_{\perp} = \gamma\left(\boldsymbol{B} - \frac{1}{c^2}\boldsymbol{v} \times \boldsymbol{E}\right)_{\perp}.$$

Assuming $\boldsymbol{v}$ *is in the positive* x *direction, the above is same as*

$$E'_x = E_x, \quad E'_y = \gamma(E_y - vB_z), \quad E'_z = \gamma(E_z + vB_y),$$

$$B'_x = B_x, \quad B'_y = \gamma\left(B_y + \frac{v}{c^2}E_z\right), \quad B'_z = \gamma\left(B_z - \frac{v}{c^2}E_y\right).$$

When $v \ll c$, $\gamma \to 1$,

$$\boldsymbol{E}' = \boldsymbol{E} + \boldsymbol{v} \times \boldsymbol{B}, \quad \boldsymbol{B}' = \boldsymbol{B} - \frac{\boldsymbol{v}}{c^2} \times \boldsymbol{E}.$$

8.5 Electric field of a point charge in uniform motion

Consider a point charge moving with velocity $\boldsymbol{v}$ along the x-axis. Let Σ denote the laboratory frame and Σ' denote the rest frame of the point charge. In Σ',

$$\boldsymbol{E}' = \frac{q\boldsymbol{x}'}{4\pi\epsilon_0 r'^3},$$

$$\boldsymbol{B}' = 0.$$

By the result of Problem 8.3, we can apply the inverse transformation to obtain the quantities in Σ.

$$E_x = E'_x = \frac{qx'}{4\pi\epsilon_0 r'^3},$$

$$E_y = \gamma(E'_y + vB'_z) = \gamma\frac{qy'}{4\pi\epsilon_0 r'^3},$$

$$E_z = \gamma(E'_z - vB'_y) = \gamma\frac{qz'}{4\pi\epsilon_0 r'^3},$$

$$B_x = B'_x = 0,$$

$$B_y = \gamma\left(B'_y - \frac{v}{c^2}E'_z\right) = -\frac{\gamma v}{c^2}\frac{qz'}{4\pi\epsilon_0 r'^3},$$

$$B_z = \gamma\left(B'_z + \frac{v}{c^2}E'_y\right) = \frac{\gamma v}{c^2}\frac{qy'}{4\pi\epsilon_0 r'^3}.$$

Now we want to express $\boldsymbol{E}$, $\boldsymbol{B}$ in terms of coordinates of Σ. Suppose the time when the point charge pass through the origin of Σ is $t = 0$. From Lorentz transformation we obtain

$$x' = \gamma x, \quad y' = y, \quad z' = z,$$

assuming all distances are determined simultaneously for Σ. With these we can obtain the expressions in (x, y, z).

Problem 8.4. *Consider the following equations:*

$$r'^2 = \gamma^2 x^2 + y^2 + z^2 = r^2 + \frac{\left(\frac{\boldsymbol{v}\cdot\boldsymbol{x}}{c}\right)^2}{1 - \frac{v^2}{c^2}},$$

$$\boldsymbol{E} = \left(1 - \frac{v^2}{c^2}\right)\frac{q\boldsymbol{x}}{4\pi\epsilon_0\left[\left(1 - \frac{v^2}{c^2}\right)\gamma^2 + \left(\frac{\boldsymbol{v}\cdot\boldsymbol{x}}{c}\right)^2\right]^{3/2}},$$

$$\boldsymbol{B} = \frac{\boldsymbol{v}}{c^2} \times \boldsymbol{E}.$$

Discuss the two limiting cases:

(1) $v \ll c$;
(2) $v \approx c$.

Answer. (1) Substituting into the equations, we have

$$\boldsymbol{E} = \boldsymbol{E}_0, \quad \boldsymbol{B} = \frac{\boldsymbol{v}}{c^2} \times \boldsymbol{E}_0 = \frac{\mu_0 q \boldsymbol{v} \times \boldsymbol{x}}{4\pi r^3}.$$

(2) In the direction perpendicular to $\boldsymbol{v}$,

$$\boldsymbol{E} = \gamma \frac{q\boldsymbol{x}}{4\pi\epsilon_0 r^3} \gg \boldsymbol{E}_0.$$

In the direction parallel to $\boldsymbol{v}$,

$$\boldsymbol{E} = \left(1 - \frac{v^2}{c^2}\right) \frac{q\boldsymbol{x}}{4\pi\epsilon_0 r^3} \ll \boldsymbol{E}_0.$$

Problem 8.5.

(1) *If $\boldsymbol{B} = 0$ in a frame Σ, in another frame Σ' moving with velocity $\boldsymbol{v}$ along the x direction, prove that*

$$\boldsymbol{B}' = -\frac{1}{c^2}(\boldsymbol{v} \times \boldsymbol{E}').$$

(2) *If $\boldsymbol{E} = 0$ in Σ, prove that in Σ'*

$$\boldsymbol{E}' = \boldsymbol{v} \times \boldsymbol{B}'.$$

Problem 8.6. *Find the magnetic field of a point charge in uniform motion and obtain the non-relativistic limit ($v \ll c$).*

Problem 8.7.

(1) *Show that $\boldsymbol{E} \times \boldsymbol{B}$ and $E^2 - c^2 B^2$ are relativistically invariant.*
(2) *Suppose the magnetic field at some point is zero in one frame. Is it possible to find another frame in which the electric field at that point is zero?*

Problem 8.8. *For perpendicular $\boldsymbol{E}$ and $\boldsymbol{B}$ fields (i.e. $\boldsymbol{E} \cdot \boldsymbol{B} = 0$), show that you can transform to a moving frame having either a pure electric or magnetic field except when $E = cB$.*

Problem 8.9. *A point charge e moves with constant velocity $\boldsymbol{v}$ in the z direction so that at time t it is at the point Q with coordinates $x = 0$, $y = 0$ and $z = vt$. Now consider the time t and the point P with coordinates $x = b$, $y = 0$ and $z = 0$. Find the scalar potential, vector potential and the electric field in the x direction E_x.*

8.6 Generalization of Coulomb's law to Maxwell equations using special relativity

Coulomb's law is non-relativistic. In the charge rest frame it is correct. If we introduce a potential φ, we obtain

$$\frac{\partial \varphi}{\partial t} = 0,$$

$$\nabla^2 \varphi = -\frac{\rho}{\epsilon_0}.$$

We can see the correspondence with the 4-vector:

$$\frac{\partial \varphi}{\partial t} = 0 \quad \longleftrightarrow \quad \frac{\partial A_0}{\partial x_0} = 0$$

$$\nabla^2 \varphi = -\frac{\rho}{\epsilon_0} \quad \longleftrightarrow \quad \Box^2 A_0 = -\mu_0 j_0.$$

Thus to relativistize a non-relativistic equation is nothing but changing every physical quantity to a Lorentz tensor, i.e. to let every equation become a Lorentz tensor equation.

$$\frac{\partial A_0}{\partial x_0} = 0 \quad \longrightarrow \quad \frac{\partial A_\mu}{\partial x_\mu} = 0$$

$$\Box^2 A_0 = -\mu_0 j_0 \quad \longrightarrow \quad \Box^2 A_\mu = -\mu_0 j_\mu.$$

Hence, by relativistization of Coulomb's law we can obtain the Maxwell equations.

Chapter 9

Relativistic Dynamics

9.1 Equation of motion

To discuss about relativistic dynamics, first we need the relativistic equation of motion. Recall in Sec. 8.6 we have introduced the three steps to obtain a relativistic equation.

(1) Find a specific inertial frame in which the non-relativistic law is correct. In this case, in the particle rest frame Σ', we have classical mechanics, which is governed by Newton's law. Newton's law of motion is given by

$$m_0 \frac{\mathrm{d}^2 x_i}{\mathrm{d}t^2} = F_i,$$

where F_i is the component of force $\boldsymbol{F}$ in classical mechanics. Since m_0 is the mass of the particle in its rest frame, it is called the rest mass.

(2) Write the non-relativistic equation in the form of Lorentz tensor equation. In Σ', $\mathrm{d}t = \mathrm{d}\tau$. We can write the classical law of motion in the form:

$$m_0 \frac{\mathrm{d}^2 x_\mu}{\mathrm{d}\tau^2} = K_\mu,$$

where the component of K_μ in Σ' are

$$K_\mu = (0, \boldsymbol{F}).$$

(3) Generalize the obtained tensor equation to all inertial systems, completing the procedure of relativistization.

$$m_0 \frac{\mathrm{d}^2 x_\mu}{\mathrm{d}\tau^2} = K_\mu. \tag{9.1}$$

161

In summary, following the general procedure of relativistization, we have obtained the relativistic law and the transformation law of the four-vector K_μ.

9.2   Relativistic force K_μ: Force four-vector

With the expression of K_μ in the rest frame, we want to find its expression in the laboratory frame. In the laboratory frame Σ, consider a particle moving with velocity $\boldsymbol{v}$ along the x-axis, i.e. Σ has a velocity $-\boldsymbol{v}$ relative to the particle rest frame Σ'.

In Σ',

$$K'_\mu = (0, \boldsymbol{F}).$$

In Σ,

$$K_\mu = b_{\mu\nu} K'_\nu,$$

where $b_{\mu\nu}$ is the inverse Lorentz transformation of the boost along x-axis:

$$(b_{\mu\nu}) = \begin{pmatrix} \gamma & \beta\gamma & 0 & 0 \\ \beta\gamma & \gamma & 0 & 0 \\ 0 & 0 & 1 & 0 \\ 0 & 0 & 0 & 1 \end{pmatrix}, \quad \beta = \frac{v}{c}.$$

Hence,

$$K_\mu = (\beta\gamma F_1, \gamma F_1, F_2, F_3)$$
$$= \left(\gamma \frac{\boldsymbol{v}}{c} \cdot \boldsymbol{F}, \gamma F_1, F_2, F_3\right).$$

We can generalize the above calculation to a velocity $\boldsymbol{v}$ in any direction, then the true expression of the four-force should be have γ applied to all components of $\boldsymbol{F}$,

$$K_\mu = \left(\gamma \frac{\boldsymbol{v}}{c} \cdot \boldsymbol{F}, \gamma \boldsymbol{F}\right). \tag{9.2}$$

Let the spatial components of K_μ be $\boldsymbol{K}$, then $\boldsymbol{K} = \gamma \boldsymbol{F}$.

$$K_\mu = \left(\boldsymbol{K} \cdot \frac{\boldsymbol{v}}{c}, \boldsymbol{K}\right).$$

9.3 Momentum four-vector

Now that we have the relativistic equation of motion, the force four-vector, the next important physical quantity in mechanics is the momentum. Recall that

$$K_\mu = m_0 \frac{\mathrm{d}^2 x_\mu}{\mathrm{d}\tau^2} = m_0 \frac{\mathrm{d}u_\mu}{\mathrm{d}\tau},$$

where u_μ is the four-velocity. This suggests that we define the momentum four-vector as

$$p_\mu \equiv m_0 u_\mu = m_0 \gamma(c, \boldsymbol{v}). \tag{9.3}$$

Then the spatial component of p_μ is

$$\boldsymbol{p} = m_0 \gamma \boldsymbol{v} = m \boldsymbol{v}, \tag{9.4}$$

where m is the mass of the particle in the laboratory frame,

$$m = \gamma m_0 = \frac{m_0}{\sqrt{1 - \frac{v^2}{c^2}}}. \tag{9.5}$$

Note that $\boldsymbol{p}$ is the same as the momentum in classical mechanics.

To understand the physical meaning of $p_0 = \gamma m_0 c$, we can look at K_μ again.

$$\frac{\mathrm{d}p_0}{\mathrm{d}\tau} = K_0 = \frac{\gamma}{c} \boldsymbol{F} \cdot \boldsymbol{v}$$

$$\implies \frac{\mathrm{d}p_0}{\mathrm{d}t} = \frac{1}{c} \boldsymbol{F} \cdot \boldsymbol{v}.$$

Since the energy E of the particle satisfies $\frac{\mathrm{d}E}{\mathrm{d}t} = \boldsymbol{F} \cdot \boldsymbol{v}$, from the above equation we deduce

$$p_0 = \frac{E}{c}. \tag{9.6}$$

In other words,

$$E = mc^2 = \gamma m_0 c^2. \tag{9.7}$$

We can calculate the norm squared of p_μ,

$$\eta_{\mu\nu} p_\mu p_\nu = m_0^2 \eta_{\mu\nu} u_\mu u_\nu = -m_0^2 c^2. \tag{9.8}$$

This means the four components of p_μ are not independent of each other. In particular,

$$-\frac{E^2}{c^2} + p^2 = -m_0^2 c^2$$

$$\implies E = \sqrt{p^2 c^2 + m_0^2 c^4}. \tag{9.9}$$

In relativistic mechanics, the kinetic energy is given by

$$T = E - m_0 c^2 = (\gamma - 1) m_0 c^2. \tag{9.10}$$

In the non-relativistic limit ($v \ll c$), we have

$$T = \left(\frac{1}{\sqrt{1 - \frac{v^2}{c^2}}} - 1 \right) m_0 c^2$$

$$= \left[\left(1 + \frac{1}{2}\frac{v^2}{c^2} + \frac{3}{8}\frac{v^4}{c^4} + \cdots \right) - 1 \right] m_0 c^2$$

$$= \frac{1}{2} m_0 v^2 + \frac{3}{8} m_0 \frac{v^4}{c^2} + \cdots$$

$$\approx \frac{1}{2} m_0 v^2.$$

Thus we recover the classical equation.

9.4   Lorentz force

Recall the relativistic equation of motion,

$$K_\mu = \frac{\mathrm{d}p_\mu}{\mathrm{d}\tau}.$$

This can be broken down into two parts:

$$\text{temporal part: } \boldsymbol{K} \cdot \boldsymbol{v} = \frac{\mathrm{d}E}{\mathrm{d}t} \longrightarrow \boldsymbol{F} \cdot \boldsymbol{v} = \frac{\mathrm{d}E}{\mathrm{d}t},$$

$$\text{spatial part: } \boldsymbol{K} = \frac{\mathrm{d}\boldsymbol{p}}{\mathrm{d}\tau} \longrightarrow \boldsymbol{F} = \frac{\mathrm{d}\boldsymbol{p}}{\mathrm{d}t},$$

where $\boldsymbol{F} = \boldsymbol{K}/\gamma$ is the three-dimensional force vector.

Problem 9.1. *Consider the relativistic dynamics involving the electromagnetic field, i.e.*

$$K_\mu = qF_{\mu\nu}u_\nu,$$

where $F_{\mu\nu}$ is the field tensor defined in Sec. 8.4. Show that

$$\boldsymbol{K} = \gamma q(\boldsymbol{E} + \boldsymbol{v} \times \boldsymbol{B}).$$

The result of the previous problem tells us that the Lorentz force can be obtained from relativistic dynamics with the field tensor,

$$\boldsymbol{F} = \frac{\boldsymbol{K}}{\gamma} = q(\boldsymbol{E} + \boldsymbol{v} \times \boldsymbol{B}).$$

9.5 Acceleration in relativistic dynamics

Now we shall discuss the acceleration of a particle under the influence of a force.

$$
\begin{aligned}
\boldsymbol{F} &= \frac{d\boldsymbol{p}}{dt} = \frac{d}{dt}(m\boldsymbol{v}) \\
&= m\frac{d\boldsymbol{v}}{dt} + \boldsymbol{v}\frac{dm}{dt} \\
&= m\frac{d\boldsymbol{v}}{dt} + \frac{\boldsymbol{v}}{c^2}\frac{dE}{dt} \\
&= m\frac{d\boldsymbol{v}}{dt} + \frac{\boldsymbol{v}}{c^2}(\boldsymbol{F} \cdot \boldsymbol{v}).
\end{aligned}
$$

In general, the acceleration $\boldsymbol{a} = \frac{d\boldsymbol{v}}{dt}$ in relativistic dynamics is not parallel to the force.

In the following two special cases, we have the acceleration in the same direction as the force.

Case. $\boldsymbol{F} \parallel \boldsymbol{v}$. In this case,

$$
\begin{aligned}
m\boldsymbol{a} &= \boldsymbol{F} - \frac{\boldsymbol{v}}{c^2}(\boldsymbol{F} \cdot \boldsymbol{v}) \\
&= \boldsymbol{F} - \frac{v^2}{c^2}\boldsymbol{F} \\
&= \left(1 - \frac{v^2}{c^2}\right)\boldsymbol{F}.
\end{aligned}
$$

Thus $\boldsymbol{a}$ is also parallel to $\boldsymbol{F}$ and $\boldsymbol{v}$ and the particle moves in a straight line. An example of this is a charged particle that starts from rest in a uniform electric field.

If we rearrange the equation,

$$\boldsymbol{F} = \frac{m}{\left(1 - \frac{v^2}{c^2}\right)}\boldsymbol{a}$$

$$= \frac{m_0}{\left(1 - \frac{v^2}{c^2}\right)^{3/2}}\boldsymbol{a}.$$

Thus we can define the longitudinal mass m_L,

$$m_L = \frac{m_0}{\left(1 - \frac{v^2}{c^2}\right)^{3/2}}.$$

Case. $\boldsymbol{F} \perp \boldsymbol{v}$. In this case, we have simply

$$m\boldsymbol{a} = \boldsymbol{F}.$$

Thus $\boldsymbol{a}$ is also parallel to $\boldsymbol{F}$ and is perpendicular to $\boldsymbol{v}$, so the particle moves in a circle. An example of this is a charged particle that moves in a uniform magnetic field.

If we look at the equation again,

$$\boldsymbol{F} = m\boldsymbol{a}$$

$$= \frac{m_0}{\sqrt{1 - \frac{v^2}{c^2}}}\boldsymbol{a}.$$

Thus we can define the transverse mass m_T,

$$m_T = \frac{m_0}{\sqrt{1 - \frac{v^2}{c^2}}}.$$

Problem 9.2. *What is the kinetic energy acquired by an electron starting from rest when it falls through an electrostatic potential difference of $V_0 = 10^4$ V? Find its speed and its mass at the end of the acceleration.*

Note. Charge on the electron $e = 1.602 \times 10^{-19}$ C, rest mass $m_0 = 9.109 \times 10^{-31}$ kg.

Answer. 1.602×10^{-5} J, 9.287×10^{-31} kg, 5.85×10^7 m/s.

Problem 9.3. *In a region in which there is a uniform magnetic field $B = 2$ Wb/m^2, a 10 MeV electron entering at right angles to the field moves in a circle of radius $r = \frac{mv}{qB}$, classically $m = m_0$ and $T = \frac{1}{2}m_0 v^2$. Compute the radius both classically and relativistically.*

Answer. 0.53 cm, 1.8 cm.

Problem 9.4. *A spaceman wishes to travel to a star at a distance 10^3 light years away and is willing to spend the rest of his life, say 50 years, on the journey. Is this possible? If so, how fast must he go? If the mass of the spaceship is 10^5 kg, what is the kinetic energy of the ship?*

Answer. $v = 0.9988c$, 1.71×10^{23} J.

Problem 9.5. *Consider two identical bodies of rest mass m_0, each with kinetic energy T as seen by a particular observer Σ'. The two bodies collide and stick together forming a single body of rest mass M_0. In the frame Σ', before collision, bodies A and B have a speed u', with velocities oppositely directed and along the x'-axis. The combined body C, formed by the collision, is at rest in Σ'. In another frame Σ, moving with respect to Σ' with a speed v $(= u')$ to the left along the common x, x'-axis, so the combined body C has a velocity of magnitude v directed to the right along x-axis. Using the conservation law of momentum in Σ, prove that*

$$M_0 = \frac{2m_0}{\sqrt{1 - \frac{u'^2}{c^2}}}.$$

Problem 9.6. *Show that total energy is conserved in both frames Σ and Σ' for Problem 9.5.*

Problem 9.7. *Show that the velocity of an object is given by $\boldsymbol{v} = \frac{c^2}{E}\boldsymbol{p}$ and that the magnitude of the velocity can be expressed as $v = \frac{\mathrm{d}E}{\mathrm{d}p}$.*

Problem 9.8. *One atomic mass unit (1 amu) is equal to 1.66×10^{-27} kg. The rest mass of the proton is 1.00731 amu and that of the neutron is 1.00867 amu. The rest mass of the deuteron is found to be 2.01360 amu, find the binding energy of the deuteron.*

Answer. 2.22 MeV.

Problem 9.9. *The Earth receives radiant energy from the Sun at the rate of 1.34×10^3 W/m^2, the mean Earth–Sun separation is $r = 1.49 \times 10^{11}$ m.*

Find the fractional decrease of solar rest mass. The Sun's rest mass is now about 2.0×10^{30} *kg.*

Answer. 6.8×10^{-14}/yr.

Problem 9.10. *An electron is accelerated in a synchrotron to an energy of 1.0 BeV, find the effective mass and speed of the electron.*

Problem 9.11. *Show that if momentum and energy are conserved in frame* Σ *(i.e.* $\Delta p_x = \Delta p_y = \Delta p_z = \Delta E = 0$*), they are also conserved in any frame* Σ'.

9.6   Quiz

(1) Prove

$$\nabla(a \cdot b) = (a \cdot \nabla)b + (b \cdot \nabla)a + a \times (\nabla \times b) + b \times (\nabla \times a).$$

(2) From the Maxwell equations and Lorentz force law deduce

$$w = \frac{1}{2}\left(\epsilon_0 E^2 + \frac{1}{\mu_0}B^2\right), \quad S = \frac{1}{\mu_0}E \times B.$$

(3) A pair of twins is 20 years old when one of them leaves for a space voyage to a planet 20 light years away. The round trip is accomplished with no delays and the traveling twin comes home to find his brother twice as old as himself. What was the traveling twin's speed?

(4) The quantity $F_{\mu\nu}F_{\mu\nu}$ is Lorentz invariant. What is this invariant in terms of electric and magnetic fields?

(5) Write out the covariant equation of motion of a charged particle in an electromagnetic field. A charged particle moves at right angles to a uniform magnetic field. Find the expression for rB in terms of the energy of the particle, where r is the radius of the orbit, B is the magnetic field.

Chapter 10

More about Tensors

10.1 Contravariant and covariant vectors

10.1.1 *Vectors in Euclidean space*

There are 2 kinds of components of a vector in Euclidean space.

(1) Components along the direction of basis vectors.

$$\boldsymbol{v} = v^i \boldsymbol{e}_i.$$

v^i are called the contravariant components of vector $\boldsymbol{v}$.

(2) Projection on the basis vectors.

$$v_i = \boldsymbol{v} \cdot \boldsymbol{e}_i.$$

v_i are called the covariant components of vector $\boldsymbol{v}$.

In an orthogonal coordinate system, since $\boldsymbol{e}_i \cdot \boldsymbol{e}_j = \delta_{ij}$,

$$v_i = \boldsymbol{v} \cdot \boldsymbol{e}_i = (v^j \boldsymbol{e}_j) \cdot \boldsymbol{e}_i = v^i.$$

Hence, in orthogonal coordinate systems there is no difference between contravariant components and covariant components. For non-orthogonal coordinate systems, contravariant components are different from covariant components.

Consider the scalar product between two vectors.

$$\boldsymbol{u} \cdot \boldsymbol{v} = \boldsymbol{u} \cdot (v^i \boldsymbol{e}_i) = (\boldsymbol{u} \cdot \boldsymbol{e}_i)v^i = u_i v^i.$$

Hence the scalar product is given by the sum of the product of the contravariant and covariant components. On the other hand,

$$\boldsymbol{u} \cdot \boldsymbol{v} = (u^i (\boldsymbol{e})_i) \cdot (v^j \boldsymbol{e}_j) = (\boldsymbol{e}_i \cdot \boldsymbol{e}_j)u^i v^i = g_{ij} u^i v^i,$$

169

where g_{ij} is the *metric*, which is a second-rank symmetric tensor:

$$g_{ij} \equiv \boldsymbol{e}_i \cdot \boldsymbol{e}_j = \boldsymbol{e}_j \cdot \boldsymbol{e}_i = g_{ji}. \tag{10.1}$$

For the infinitesimal displacement vector $\mathrm{d}\boldsymbol{x}$, its contravariant components are $\mathrm{d}x^i$ and its covariant components are $\mathrm{d}x_i = g_{ij}\mathrm{d}x^j$. Thus the norm squared is given by

$$\mathrm{d}s^2 = \mathrm{d}x^i \mathrm{d}x_i = g_{ij}\mathrm{d}x^i \mathrm{d}x^j. \tag{10.2}$$

10.1.2   *Vectors in four-dimensional Minkowski spacetime*

For Euclidean space, we can always choose an orthogonal coordinate system, such that $g_{ji} = \delta_{ij}$,

$$\mathrm{d}s^2 = \mathrm{d}x^i \mathrm{d}x_i = \delta_{ij}\mathrm{d}x^i \mathrm{d}x^j.$$

However, recall that for four-dimensional Minkowski spacetime ($x^0 = ct, x^1 = x, x^2 = y, x^3 = z$), we have

$$\mathrm{d}s^2 = \eta_{\mu\nu}\mathrm{d}x^\mu \mathrm{d}x^\nu, \quad \eta_{\mu\nu} = \begin{pmatrix} -1 & 0 & 0 & 0 \\ 0 & 1 & 0 & 0 \\ 0 & 0 & 1 & 0 \\ 0 & 0 & 0 & 1 \end{pmatrix}. \tag{10.3}$$

Here we must differentiate *contravariant components* and *covariant components*. In this case,

$$x_0 = -ct, \tag{10.4}$$

$$x_1 = x, \tag{10.5}$$

$$x_2 = y, \tag{10.6}$$

$$x_3 = z, \tag{10.7}$$

$$\mathrm{d}x_\mu = \eta_{\mu\nu}\mathrm{d}x^\nu. \tag{10.8}$$

The proper time interval is given by

$$-c^2\mathrm{d}\tau^2 = \mathrm{d}x^\mu \mathrm{d}x_\mu.$$

The inverse of $\eta_{\mu\nu}$ is $\eta^{\mu\nu}$, i.e.

$$\eta^{\mu\nu}\eta_{\nu\lambda} = \delta^\mu{}_\lambda, \tag{10.9}$$

where δ^μ_ν is the Kronecker delta. Hence we have

$$x^\mu = \eta^{\mu\nu} x_\nu. \tag{10.10}$$

Problem 10.1. *Prove that*

$$\eta^{\mu\nu} = \begin{pmatrix} -1 & 0 & 0 & 0 \\ 0 & 1 & 0 & 0 \\ 0 & 0 & 1 & 0 \\ 0 & 0 & 0 & 1 \end{pmatrix}.$$

In this notation, the Lorentz transformation can be written as

$$x'^\mu = a^\mu_{\ \nu} x^\nu.$$

If we adopt the convention $c = 1$, then the boost transformation along x-axis can be written as

$$a^\mu_{\ \nu} = \begin{pmatrix} \frac{1}{\sqrt{1-v^2}} & \frac{-v}{\sqrt{1-v^2}} & 0 & 0 \\ \frac{-v}{\sqrt{1-v^2}} & \frac{1}{\sqrt{1-v^2}} & 0 & 0 \\ 0 & 0 & 1 & 0 \\ 0 & 0 & 0 & 1 \end{pmatrix}.$$

The matrix $a^\mu_{\ \nu}$ for a general Lorentz transformation must satisfy the equation

$$a^\alpha_{\ \mu} \eta_{\alpha\beta} a^\beta_{\ \nu} = \eta_{\mu\nu}.$$

Problem 10.2. *From the invariance of the norm of vector x^μ,*

$$x'^\alpha \eta_{\alpha\beta} x'^\beta = x^\mu \eta_{\mu\nu} x^\nu,$$

prove the above equation for a^μ_ν, i.e.

$$a^\alpha_{\ \mu} \eta_{\alpha\beta} a^\beta_{\ \nu} = \eta_{\mu\nu}.$$

10.2 Lorentz tensors

Definition 10.1. A tensor A of rank r is an object with 4^r components which transforms according to Lorentz transformation:

$$A'^{\alpha\beta\ldots\gamma} = a^\alpha_{\ \mu} a^\beta_{\ \nu} \cdots a^\gamma_{\ \rho} A^{\mu\nu\ldots\rho}. \tag{10.11}$$

10.2.1 *Lowering and raising an index*

Definition 10.2. Any index of a tensor A can be lowered or raised by the metric:

$$\text{Lowering: } A^{\alpha}{}_{\beta}{}^{\gamma\ldots\rho} = \eta_{\beta\mu}A^{\alpha\mu\gamma\ldots\rho}, \tag{10.12}$$

$$\text{Raising: } A^{\alpha\beta\gamma\ldots\rho} = \eta^{\beta\mu}A^{\alpha}{}_{\mu}{}^{\gamma\ldots\rho}. \tag{10.13}$$

The same procedure can be applied to one or more indices of a tensor of arbitrary rank. For example,

$$A_{\mu}{}^{\nu} = \eta_{\mu\alpha}A^{\alpha\nu},$$

$$A^{\mu}{}_{\nu} = \eta_{\nu\alpha}A^{\mu\alpha},$$

$$A_{\mu\nu} = \eta_{\mu\alpha}\eta_{\nu\beta}A^{\alpha\beta},$$

$$A^{\mu\nu} = \eta^{\mu\alpha}\eta^{\nu\beta}A_{\alpha\beta}.$$

Our rules for raising and lowering indices can also be applied to the Lorentz transformation matrix a^{μ}_{ν},

$$a^{\alpha}{}_{\mu}\eta_{\alpha\beta}a^{\beta}{}_{\nu} = \eta_{\mu\nu}$$

$$\implies a_{\beta\mu}a^{\beta}{}_{\nu} = \eta_{\mu\nu}.$$

If we raise the index μ, we get

$$a_{\beta}{}^{\mu}a^{\beta}{}_{\nu} = \eta^{\mu}{}_{\nu} = \delta^{\mu}{}_{\nu}.$$

This shows that $a_{\beta}{}^{\mu}$ is the inverse of $a^{\beta}{}_{\nu}$,

$$(a^{-1})^{\mu}{}_{\beta} = a_{\beta}{}^{\mu}.$$

Hence the $a_{\beta}{}^{\mu}$ can be used to write the transformation law for covariant tensors in the form:

$$A'_{\alpha\beta\ldots\gamma} = a_{\alpha}{}^{\mu}a_{\beta}{}^{\nu}\cdots a_{\gamma}{}^{\rho}A_{\mu\nu\ldots\rho}. \tag{10.14}$$

Problem 10.3. *Prove the above equation*

$$A'_{\alpha\beta\ldots\gamma} = a_{\alpha}{}^{\mu}a_{\beta}{}^{\nu}\cdots a_{\gamma}{}^{\rho}A_{\mu\nu\ldots\rho}.$$

Problem 10.4. *Prove*

$$x_{\nu} = a^{\mu}{}_{\nu}x'_{\mu}.$$

Problem 10.5. *Prove*

$$x^{\nu} = a_{\mu}{}^{\nu}x'^{\mu}.$$

Problem 10.6. *Given a second-rank tensor $A^{\mu\nu}$, the quantity $A^\mu{}_\mu$ is called the trace of the tensor. Prove*

$$A^\mu{}_\mu = A_\mu{}^\mu.$$

10.2.2 *Contraction*

A tensor of rank higher than two has more than one trace. For example, $A^\mu{}_\mu{}^\alpha$, $A^{\mu\nu}{}_\mu$, $A^{\mu\nu}{}_\nu$ are three different traces of a third-rank tensor $A^{\mu\nu\alpha}$. We say $A^\mu{}_\mu{}^\alpha$ is the contraction of the first two indices, etc. Contraction reduces the rank of a tensor by two.

As a special case, the contraction of the dyad $B^\mu C^\nu$ (i.e. $B^\mu C_\mu$) is simply the scalar product of vectors B^μ and C^μ.

Problem 10.7. *Show that $u^\mu u_\mu = 1$, $p^\mu p_\mu = m_0^2$.*

Problem 10.8. *Show that $B^\mu C^\mu$ is not a scalar, $A^{\mu\mu\alpha}$ is not a vector.*

10.3 General tensors

In a general coordinate transformation from x^μ to x'^μ, the transformation can be a function of the coordinates.

$$x'^\mu = x'^\mu(x),$$

$$x^\mu = x^\mu(x'),$$

$$dx'^\mu = \frac{\partial x'^\mu}{\partial x^\nu} dx^\nu,$$

$$dx^\mu = \frac{\partial x^\mu}{\partial x'^\nu} dx'^\nu.$$

In the case of Lorentz transformation, $\frac{\partial x'^\mu}{\partial x^\nu}$ and $\frac{\partial x^\mu}{\partial x'^\nu}$ are constant. In the present case, these partial derivatives are functions of x^μ (or x'^μ). However, since they are inverse of each other, we still have

$$\frac{\partial x'^\mu}{\partial x^\alpha} \frac{\partial x^\alpha}{\partial x'^\nu} = \delta^\mu{}_\nu.$$

A four-dimensional object A^μ is a contravariant vector under general coordinate transformations if it transforms according to

$$A'^\mu = \frac{\partial x'^\mu}{\partial x^\nu} A^\nu. \tag{10.15}$$

Conversely a four-dimensional object B_μ is a covariant vector if

$$B'_\mu = \frac{\partial x^\nu}{\partial x'^\mu} B_\nu. \tag{10.16}$$

Contravariant vectors follow the transformation law of dx^μ, while covariant vectors follow the transformation law of $\frac{\partial}{\partial x^\mu}$, since

$$\frac{\partial}{\partial x'^\mu} = \frac{\partial x^\nu}{\partial x'^\mu} \frac{\partial}{\partial x^\nu}. \tag{10.17}$$

Note that x^μ is not a vector with respect to general coordinate transformations. In the exceptional case of a linear transformation,

$$x'^\mu = a^\mu{}_\nu x^\nu$$

$$\implies x'^\mu = \frac{\partial x'^\mu}{\partial x^\nu} x^\nu.$$

This means that x^μ is a vector with respect to all linear transformations.

Similar to Lorentz transformation, tensors of higher rank follow the following transformation laws:

$$A'^{\alpha\beta\ldots\gamma} = \frac{\partial x'^\alpha}{\partial x^\mu} \frac{\partial x'^\beta}{\partial x^\nu} \cdots \frac{\partial x'^\gamma}{\partial x^\rho} A^{\mu\nu\ldots\rho},$$

$$B'_{\alpha\beta\ldots\gamma} = \frac{\partial x^\mu}{\partial x'^\alpha} \frac{\partial x^\nu}{\partial x'^\beta} \cdots \frac{\partial x^\rho}{\partial x'^\gamma} B_{\mu\nu\ldots\rho}.$$

Problem 10.9. *Show that $\delta_\mu{}^\nu$ transforms as a tensor covariant in μ, contravariant in ν.*

Problem 10.10. *Show that $A^\mu B_\mu$ is a scalar and $A^{\mu\nu\alpha}_\alpha$ is a second-rank contravariant tensor, if A^μ and B_ν are vectors, and that $A^{\mu\nu\alpha}_\beta$ is a tensor contravariant in $\mu\nu\alpha$ and covariant in β.*

It is easy to show that the derivative of a scalar function φ, i.e. $\partial_\mu \varphi = \frac{\partial \varphi}{\partial x^\mu}$, is a vector. However, in general the derivative of a vector A_ν, i.e. $\partial_\mu A_\nu$, is not a tensor.

Problem 10.11. *Show that the transformation law of the derivative of a vector is*

$$\frac{\partial A'^\mu}{\partial x'^\nu} = \frac{\partial x'^\mu}{\partial x^\alpha} \frac{\partial x^\beta}{\partial x'^\nu} \frac{\partial A^\alpha}{\partial x^\beta} + A^\alpha \frac{\partial x^\beta}{\partial x'^\nu} \frac{\partial^2 x'^\mu}{\partial x^\beta \partial x^\alpha}.$$

This is not the transformation law of a tensor.

10.4 Metric tensor

The metric tensor $g_{\mu\nu}(x)$ was defined in terms of the spacetime interval ds^2 between two points separated by a coordinate interval dx^μ as

$$ds^2 = g_{\mu\nu}(x)dx^\mu dx^\nu. \tag{10.18}$$

The spacetime interval is a scalar, so it is invariant under coordinate transformation.

$$ds^2 = ds'^2 = g'_{\alpha\beta}dx'^\alpha dx'^\beta.$$

Thus,

$$g'_{\alpha\beta} = g_{\mu\nu}\frac{\partial x^\mu}{\partial x'^\alpha}\frac{\partial x^\nu}{\partial x'^\beta}.$$

Hence $g_{\mu\nu}$ is a second-rank tensor.

The inverse $g^{\mu\nu}$ of the metric tensor is defined by

$$g_{\mu\alpha}g^{\alpha\nu} = \delta_\mu^{\ \nu}.$$

We will use the tensors $g^{\mu\nu}$ and $g_{\mu\nu}$ to raise and lower indices:

$$dx_\mu = g_{\mu\nu}dx^\mu,$$

$$\partial^\mu = g^{\mu\nu}\partial_\nu,$$

$$A^{\mu\alpha\beta\dots} = g^{\mu\nu}A_\nu^{\ \alpha\beta\dots},$$

$$B_\mu^{\ \mu,\alpha} = g^{\alpha\beta}B_\mu^{\ \mu}_{\ ,\beta}.$$

Problem 10.12. *Find the tensor inverse to*

$$g_{\mu\nu} = \begin{pmatrix} -1 & 0 & 0 & 0 \\ 0 & 1 & 0 & 0 \\ 0 & 0 & r^2 & 0 \\ 0 & 0 & 0 & r^2\sin^2\theta \end{pmatrix}.$$

10.5 Christoffel symbol and the covariant derivative of a vector

The Christoffel symbol

$$\Gamma^\mu_{\ \alpha\beta} = \frac{1}{2}g^{\mu\nu}(g_{\nu\alpha,\beta} + g_{\beta\nu,\alpha} - g_{\alpha\beta,\nu}) \tag{10.19}$$

is a $4 \times 4 \times 4$-dimensional object, symmetric in α and β. It is not a tensor under general coordinate transformations. The transformation law of the Christoffel symbol is

$$\Gamma'^{\mu}_{\alpha\beta} = \frac{\partial x'^{\mu}}{\partial x^{\nu}} \frac{\partial x^{\sigma}}{\partial x'^{\alpha}} \frac{\partial x^{\tau}}{\partial x'^{\beta}} \Gamma^{\nu}_{\sigma\tau} - \frac{\partial x^{\nu}}{\partial x'^{\alpha}} \frac{\partial x^{\sigma}}{\partial x'^{\beta}} \frac{\partial^2 x'^{\mu}}{\partial x^{\sigma} \partial x^{\nu}}. \tag{10.20}$$

Problem 10.13. *Suppose*

$$\mathrm{d}s^2 = -e^{N(r)}\mathrm{d}t^2 + e^{L(r)}\mathrm{d}^2 + r^2\mathrm{d}\theta^2 + r^2\sin^2\theta\mathrm{d}\varphi^2,$$

where N, L are some functions of r, with $x^0 = t$, $x^1 = r$, $x^2 = \theta$, $x^3 = \varphi$. Derive the following expressions for the Christoffel symbols:

$$\Gamma^0_{10} = \Gamma^0_{01} = \frac{1}{2}N',$$

$$\Gamma^1_{00} = \frac{1}{2}N'e^{N-L}, \quad \Gamma^1_{11} = \frac{1}{2}L',$$

$$\Gamma^1_{22} = -re^{-L}, \quad \Gamma^1_{33} = -r\sin^2\theta\, e^{-L},$$

$$\Gamma^2_{12} = \Gamma^2_{21} = \frac{1}{r}, \quad \Gamma^2_{33} = -\sin\theta\cos\theta,$$

$$\Gamma^3_{13} = \Gamma^3_{31} = \frac{1}{r}, \quad \Gamma^3_{23} = \Gamma^3_{32} = \cot\theta.$$

All other Christoffel symbols are zero. ($L' = \frac{\partial L}{\partial r}$, $N' = \frac{\partial N}{\partial r}$.)

The importance of the objects $\Gamma^{\mu}_{\alpha\beta}$ lies in that they can be used for the construction of tensors by differentiation of lower-rank tensors. We can show that

$$A^{\mu}_{;\beta} \equiv A^{\mu}_{,\beta} + \Gamma^{\mu}_{\alpha\beta}A^{\alpha} \tag{10.21}$$

is a second-rank tensor if A^{μ} is a vector. The quantity $A^{\mu}_{;\beta}$ is called the covariant derivative of A^{μ}. The covariant derivative of a covariant vector is

$$B_{\mu;\beta} \equiv B_{\mu,\beta} - \Gamma^{\alpha}_{\mu\beta}B_{\alpha}. \tag{10.22}$$

10.6 Geodesics and parallel transport

We can gain a better understanding of the meaning of the covariant derivative if we introduce the concept of geodesics and parallel transport of a vector.

A geodesic is a curve representing the shortest path between two points. In Euclidean space, that is just the straight line connecting the two points, but that is not the case in a curved space. For example, the geodesics on a sphere is an arc.

In curved space we can parallel transport a vector such that the angle between the vector and the geodesics remains constant. Since in curved space the geodesics are not straight lines, the vector $A^\mu(x)$ will be changed to $A^\mu(x) + \delta A^\mu(x)$ when parallel transport from x^μ to $x^\mu + \mathrm{d}x^\mu$, where

$$\delta A^\mu(x) = -\Gamma^\mu{}_{\alpha\nu}\mathrm{d}x^\alpha A^\nu.$$

Note that $A^\mu(x)$ is a contravariant vector at x^μ, $A^\mu(x) + \delta A^\mu(x)$ is a contravariant vector at $x^\mu + \mathrm{d}x^\mu$, so the difference $\delta A^\mu(x)$ is in general not a vector, and hence $\Gamma^\mu{}_{\nu\lambda}$ is in general not a tensor.

If we want to define the derivative of a vector field A^μ in the usual way as

$$\lim_{\mathrm{d}x \to 0} \frac{A^\mu(x+\mathrm{d}x) - A^\mu(x)}{\mathrm{d}x^\beta},$$

then the obtained result is not a tensor, since the numerator is not a vector — it is the difference between two vectors evaluated at different points. To get the numerator to be a vector, we should parallel transport $A^\mu(x)$ from x to $x + \mathrm{d}x$ before carrying out the subtraction. We then have

$$\lim_{\mathrm{d}x \to 0} \frac{A^\mu(x+\mathrm{d}x) - A^\mu(x) - \delta A^\mu(x)}{\mathrm{d}x^\beta}$$

$$= \lim_{\mathrm{d}x \to 0} \frac{A^\mu(x+\mathrm{d}x) - A^\mu(x)}{\mathrm{d}x^\beta} + \Gamma^\mu{}_{\alpha\nu} A^\nu \frac{\mathrm{d}x^\alpha}{\mathrm{d}x^\beta}$$

$$= A^\mu{}_{,\beta} + \Gamma^\mu{}_{\beta\nu} A^\nu.$$

This is exactly our earlier expression for the covariant derivative.

10.7 Riemann curvature tensor

From the definition of covariant derivative we can get

$$A_{\beta;\mu;\nu} = A_{\beta;\mu,\nu} - \Gamma^\sigma{}_{\beta\nu} A_{\sigma;\mu} - \Gamma^\sigma{}_{\mu\nu} A_{\beta;\sigma},$$

thence we can obtain

$$A_{\beta;\mu;\nu} - A_{\beta;\nu;\mu} = R^\alpha{}_{\beta\mu\nu} A_\alpha,$$

where

$$R^{\alpha}{}_{\beta\mu\nu} = -\Gamma^{\alpha}{}_{\beta\mu,\nu} + \Gamma^{\alpha}{}_{\beta\nu,\mu} + \Gamma^{\sigma}{}_{\beta\nu}\Gamma^{\alpha}{}_{\sigma\mu} - \Gamma^{\sigma}{}_{\beta\mu}\Gamma^{\alpha}{}_{\sigma\nu}. \tag{10.23}$$

This is called the *Riemann curvature tensor*.

Problem 10.14. *Show that*

$$A_{\beta;\mu;\nu} - A_{\beta;\nu;\mu} = R^{\alpha}{}_{\beta\mu\nu}A_{\alpha}.$$

Problem 10.15. *Prove that*

$$R_{\alpha\beta\mu\nu} = -R_{\beta\alpha\mu\nu} \quad (R_{\alpha\beta\mu\nu} = g_{\alpha\sigma}R^{\sigma}{}_{\beta\mu\nu}),$$

$$R_{\alpha\beta\mu\nu} = -R_{\alpha\beta\nu\mu},$$

$$R_{\alpha\beta\mu\nu} = R_{\mu\nu\alpha\beta},$$

$$R_{\alpha\beta\mu\nu} + R_{\alpha\mu\nu\beta} + R_{\alpha\nu\beta\mu} = 0.$$

From the definition, we can easily verify the Bianchi relations:

$$R^{\alpha}{}_{\mu\rho\sigma;\tau} + R^{\alpha}{}_{\mu\sigma\tau;\rho} + R^{\alpha}{}_{\mu\tau\rho;\sigma} = 0. \tag{10.24}$$

From the Riemann curvature tensor, we can contract two indices to obtain the *Ricci tensor*

$$R_{\beta\mu} \equiv R^{\alpha}{}_{\beta\mu\alpha}. \tag{10.25}$$

From the result of Problem 10.15, we can readily see that the Ricci tensor is symmetric:

$$R_{\beta\mu} = R_{\mu\beta}.$$

If we take the trace of the Ricci tensor, we obtain the *curvature scalar*

$$R \equiv R^{\beta}{}_{\beta} = R^{\alpha\beta}{}_{\beta\alpha}. \tag{10.26}$$

From the definitions of Ricci tensor and curvature scalar, we can obtain the Bianchi identity:

$$\left(R^{\mu}{}_{\nu} - \frac{1}{2}\delta^{\mu}{}_{\nu}R \right)_{;\mu} = 0. \tag{10.27}$$

From the equations above, although the tensor $R_{\alpha\beta\mu\nu}$ has 256 components, only 20 of these are independent.

Chapter 11

General Relativity

11.1 Principle of equivalence

It is well known that there are two different definitions of mass: Newton's second law involves the so-called inertial mass m_I, whereas Newton's law of gravitation involves the so-called gravitational mass m_G. Thence questions about the equality arise. Einstein from the experimental results suggested the principle of equivalence.

11.1.1 *Weak and strong principle of equivalence*

Many experiments have been performed to test the equality of inertial mass m_I and gravitational mass m_G.

Scientist	Year	Accuracy
Galileo	~ 1610	$|m_I - m_G|/m_I < 2 \times 10^{-3}$
Newton	~ 1680	$|m_I - m_G|/m_I < 10^{-3}$
Eötvös	1890	$|m_I - m_G|/m_I < 5 \times 10^{-8}$
Braginsky *et al.*	1971	$|m_I - m_G|/m_I < 9 \times 10^{-13}$

In a system with gravitational potential φ,

$$m_I \frac{d^2 x^k}{dt^2} = -m_G \frac{\partial \varphi}{\partial x^k}.$$

If $m_I = m_G$,

$$\frac{d^2 x^k}{dt^2} = -\frac{\partial \varphi}{\partial x^k}.$$

179

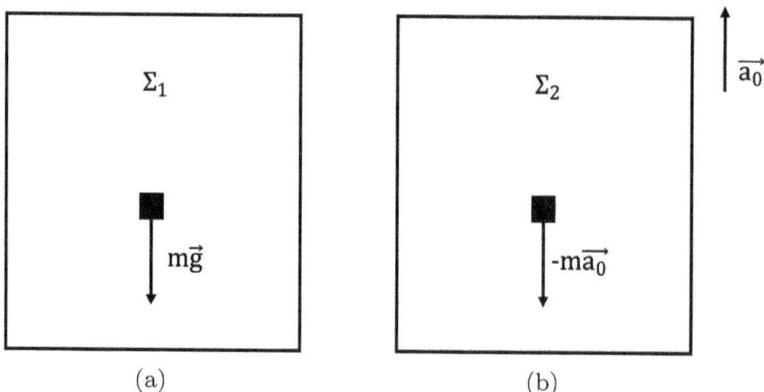

Fig. 11.1   Equivalence principle. (a) A system in a uniform gravitational field $\boldsymbol{g}$. (b) A system in constant acceleration $\boldsymbol{a}$. According to the principle of equivalence, if $\boldsymbol{g} = -\boldsymbol{a}$, the two frames are equivalent, i.e. any observer in the frame won't be able to tell which frame he is in.

Hence all particles in a given gravitational field have the same acceleration. Figure 11.1 shows the equivalence of a system with homogeneous gravitational field and an accelerated system.

Einstein then suggested the following weak principle of equivalence:

Weak principle of equivalence. All mechanical phenomena (i.e. all equations in mechanics) are the same in the system with homogeneous gravitational field and in the accelerated system.

Einstein further generalize the above weak principle of equivalence to the so-called strong principle of equivalence:

Strong principle of equivalence. All physical phenomena (i.e. all equations in physics) are the same in the system with homogeneous gravitational field and in the accelerated system.

Hereafter we shall simply refer to the strong principle of equivalence as the principle of equivalence.

According to the principle of equivalence, for any local region (for any spacetime point), we can always find a frame in which there is no action of gravitation (i.e. there exists a local inertial frame in which the gravitational action is totally canceled), and the physical law in such a frame is Lorentz covariant, i.e. the same as that in special relativity. A freely falling system is an inertial system.

11.1.2 *Description of gravitation*

In a local inertial system, according to the principle of equivalence, if there are no other forces than the gravitational force, then the equation of motion of a particle is

$$\frac{\mathrm{d}^2 \xi^\alpha}{\mathrm{d}\tau^2} = 0,$$

where ξ^α are coordinates in the local inertial system, $\mathrm{d}\tau$ is the proper time. If we choose $c = 1$, then $\mathrm{d}s = \mathrm{d}\tau$,

$$\mathrm{d}s^2 = \eta_{\mu\nu}\mathrm{d}\xi^\mu \mathrm{d}\xi^\nu.$$

In another frame,

$$x^\alpha = x^\alpha(\xi),$$

we have

$$\frac{\mathrm{d}^2 x^\lambda}{\mathrm{d}\tau^2} + \Gamma^\lambda{}_{\mu\nu}\frac{\mathrm{d}x^\mu}{\mathrm{d}\tau}\frac{\mathrm{d}x^\nu}{\mathrm{d}\tau} = 0,$$

where

$$\Gamma^\lambda{}_{\mu\nu} \equiv \frac{\partial x^\lambda}{\partial \xi^\alpha}\frac{\partial^2 \xi^\alpha}{\partial x^\mu \partial x^\nu}.$$

Combining the above formula with

$$\mathrm{d}s^2 = g_{\mu\nu}\mathrm{d}x^\mu \mathrm{d}x^\nu,$$

we get

$$\Gamma^\sigma{}_{\lambda\mu} = \frac{1}{2}g^{\nu\sigma}\left(g_{\mu\nu,\lambda} + g_{\lambda\nu,\mu} - g_{\mu\lambda,\nu}\right).$$

11.2 Principle of general covariance

The principle of general covariance is the principle to express the general laws in a way that holds for any reference frame, i.e. the expression should be covariant with respect to any change of coordinates.

The principle is given by Einstein as follows:

Principle of general covariance. All physical laws must be written in the form of general tensor equations, such that it can be valid in any reference frame.

11.3   Principle of minimal coupling

The principle of minimal coupling is the guiding principle to get the equations of motion of matter in the presence of gravitation.

Principle of minimal coupling. The equations of motion of matter in the presence of gravitation are obtained from those that hold in the absence of gravitation (equations of special relativity) by replacing $\eta^{\mu\nu}$ by $g^{\mu\nu}$ and ordinary derivatives by covariant derivatives; no other changes are to be made.

Example 11.1 (Particle dynamics). For a free particle in special relativity, we have

$$\frac{\mathrm{d}^2\xi^\alpha}{\mathrm{d}s^2} = 0 \quad \text{or} \quad \frac{\mathrm{d}u'^\alpha}{\mathrm{d}s} = 0,$$

where $u'^\alpha = \frac{\mathrm{d}\xi^\alpha}{\mathrm{d}s}$.

In a general coordinate frame x^μ,

$$u^\mu = \frac{\partial x^\mu}{\partial \xi^\alpha} u'^\alpha.$$

By principle of minimal coupling, the equation of motion is

$$\frac{\mathrm{D}u^\mu}{\mathrm{d}s} = 0,$$

$$u^\mu_{\;;\nu} \frac{\mathrm{d}x^\nu}{\mathrm{d}s} = 0,$$

$$\frac{\mathrm{d}u^\mu}{\mathrm{d}s} + \Gamma^\mu_{\;\nu\lambda} \frac{\mathrm{d}x^\nu}{\mathrm{d}s} u^\lambda = 0,$$

$$\frac{\mathrm{d}^2 x^\mu}{\mathrm{d}s^2} + \Gamma^\mu_{\;\nu\lambda} \frac{\mathrm{d}x^\nu}{\mathrm{d}s} \frac{\mathrm{d}x^\lambda}{\mathrm{d}s} = 0.$$

If there is a force f'^α in the local inertial system, in special relativity we have

$$\frac{\mathrm{d}u'^\alpha}{\mathrm{d}s} = \frac{f'^\alpha}{m}.$$

Again in a general coordinate frame x^μ,

$$f^\mu = \frac{\partial x^\mu}{\partial \xi^\alpha} f'^\alpha.$$

By principle of minimal coupling, the equation of motion is

$$m\frac{Du^\mu}{ds} = f^\mu,$$

$$m\frac{d^2x^\mu}{ds^2} = f^\mu - m\Gamma^\mu{}_{\nu\lambda}\frac{dx^\nu}{ds}\frac{dx^\lambda}{ds}.$$

Example 11.2 (Electrodynamics). In a local inertial frame, the Maxwell equations are

$$F'^{\alpha\beta}{}_{,\alpha} = \mu_0 j'^\beta,$$

$$F'^{\alpha\beta,\gamma} + F'^{\gamma\alpha,\beta} + F'^{\beta\gamma,\alpha} = 0.$$

In a general coordinate frame x^μ, we have

$$F^{\mu\nu} = \frac{\partial x^\mu}{\partial \xi^\alpha}\frac{\partial x^\nu}{\partial \xi^\beta} F'^{\alpha\beta},$$

$$j^\mu = \frac{\partial x^\mu}{\partial \xi^\alpha} j'^\alpha.$$

By principle of minimal coupling, we get the Einstein–Maxwell equations

$$F^{\mu\nu}{}_{;\mu} = \mu_0 j^\nu,$$

$$F^{\mu\nu;\lambda} + F^{\lambda\mu;\nu} + F^{\nu\lambda;\mu} = 0.$$

In a local inertial frame, the Lorentz force is

$$f'^\alpha = \frac{q}{m}p'_\beta F'^{\alpha\beta} = qu'_\beta F'^{\alpha\beta} = q\eta_{\beta\gamma}u'^\gamma F'^{\alpha\beta}.$$

In a general coordinate frame x^μ, we have

$$f^\mu = qg_{\nu\lambda}F^{\mu\nu}\frac{dx^\lambda}{ds} = qF^\mu{}_\lambda\frac{dx^\lambda}{ds}.$$

Combining with the result of Example 11.1, the equation of motion of the charged particle in electromagnetic and gravitational field is

$$m\frac{d^2x^\mu}{ds^2} = qF^\mu{}_\lambda\frac{dx^\lambda}{ds} - m\Gamma^\mu{}_{\nu\lambda}\frac{dx^\nu}{ds}\frac{dx^\lambda}{ds}.$$

11.4   Field equations: Relations between metric field and matter field

From the principle of equivalence we know that the gravitational field is equivalent to the metric field. From the principle of general covariance we know that all physical laws must be written in the form of general tensor equations. Finally from the principle of minimal coupling we know that the method of general relativistization of an equation that is only true for local inertial system is by charging $\eta_{\mu\nu}$ to $g_{\mu\nu}$ and ordinary derivative to covariant derivative. Now the problem is how to find the relation between metric field and matter field (distribution of matter and the motion of the matter, which is characterized by the so-called energy-momentum tensor $T^{\mu\nu}$).

For a system of non-interacting particles:

$$T^{\mu\nu} = \rho_0 u^\mu u^\nu \quad (\rho_0\text{: proper particle density}), \tag{11.1}$$

$$T^{00} = \text{energy density}, \tag{11.2}$$

$$T^{0k} = T^{k0} = \text{momentum density} = \text{energy flux density}, \tag{11.3}$$

$$T^{kj} = \text{flux density of } k - \text{momentum in } j \text{ direction}. \tag{11.4}$$

In this notation, the conservation of energy and momentum is expressed as

$$T^{\mu\nu}{}_{,\nu} = 0. \tag{11.5}$$

11.4.1   *Einstein tensor*

From the principle of general covariance, the field equations must be of the form:

$$G_{\mu\nu} = kT_{\mu\nu}. \tag{11.6}$$

In special relativity, we have

$$T^{\mu\nu}{}_{,\nu} = 0.$$

Hence in general relativity, we have

$$T^{\mu\nu}{}_{;\nu} = 0. \tag{11.7}$$

From Eq. (11.6), we obtain

$$G^{\mu\nu}{}_{;\nu} = 0. \tag{11.8}$$

Since the tensor $R_{\mu\nu} - \frac{1}{2}g_{\mu\nu}R$ satisfies the above requirement, Einstein chose that to be $G_{\mu\nu}$, i.e.

$$G_{\mu\nu} = R_{\mu\nu} - \frac{1}{2}g_{\mu\nu}R. \tag{11.9}$$

This is called the *Einstein tensor*.

11.4.2 *Einstein's field equations*

To find the equation that governs the dynamics in general relativity, we can refer to classical mechanics. For weak and static field, the field equations should reduce to Poisson's equation:

$$\nabla^2 \varphi = 4\pi G\rho.$$

By this requirement we obtain for Eq. (11.6)

$$k = -8\pi G.$$

Hence, *Einstein's field equations* are

$$R_{\mu\nu} - \frac{1}{2}g_{\mu\nu}R = -8\pi G T_{\mu\nu}. \tag{11.10}$$

Details about Einstein's field equations are given in Appendix G.

11.5 Final examination

(1) What you have learned from this course? Write down your own understanding about relativity (classical, special and general).
(2) Discuss simultaneity and causality in special relativity and compare the spacetime of classical mechanics and special relativity.
(3) How did we obtain the transformation law of electromagnetic field? Describe the electromagnetic field of a uniformly moving charge.
(4) State the method of obtaining relativistic equation of motion of a particle and discuss its important consequences.
(5) Discuss the relation between mathematics and physics in relativity and give a brief summary of general relativity.
(6) Find one or two problems to calculate with the knowledge of relativity.

Appendices of Essentials of Quantum Mechanics

Appendix A

Wave Mechanics or Wave Statistical Mechanics

Wave Mechanics or Wave Statistical Mechanics

QIAN Shang-Wu[1] and XU Lai-Zi[2]

[1] Physics Department, Peking University, Beijing 100871, China

[2] Institute of Technical Application of Rare Earth, Baotou 014000, China

(Received September 1, 2006)

Abstract *By comparison between equations of motion of geometrical optics and that of classical statistical mechanics, this paper finds that there should be an analogy between geometrical optics and classical statistical mechanics instead of geometrical mechanics and classical mechanics. Furthermore, by comparison between the classical limit of quantum mechanics and classical statistical mechanics, it finds that classical limit of quantum mechanics is classical statistical mechanics not classical mechanics, hence it demonstrates that quantum mechanics is a natural generalization of classical statistical mechanics instead of classical mechanics. Thence quantum mechanics in its true appearance is a wave statistical mechanics instead of a wave mechanics.*

PACS numbers: 03.65.Ta
Key words: geometrical optics, classical statistical mechanics, quantum mechanics, classical limit

1 Introduction

Quantum mechanics (QM) plays a fundamental role in the description and understanding of natural phenomena. It is the pillar of modern physics. There is a prevalent view that the relation between QM and classical mechanics is the same as that between wave optics and geometrical optics, and hence QM is often entitled wave mechanics. Recently, one of the present authors has proved that the classical limit of any stationary state wavefunction in QM describes not a single particle, but a certain kind of single particle ensemble (SPE) called "homogeneous ensemble" (HMES),[1,2] the classical limit of a quantum mechanical wave packet describes a mixture of HMES.[3] From these facts we should recognize that the classical limit of quantum mechanics is not the classical mechanics, but the classical statistical mechanics. Furthermore we should obtain the conclusion that QM is the natural generalization of classical statistical mechanics in the area of microscopic phenomena.

In Sec. 2 considering not only the path but also the intensity of the light ray, we find that there is an analogy between geometrical optics and classical statistical mechanics, instead of the analogy between geometrical optics and classical mechanics. From this point of view, since wave optics is the natural generalization of geometrical optics when we consider the wave character of light, i.e. wave length does not approach to zero, the new mechanics discovered by de Broglie and Schrödinger, i.e. quantum mechanics, should be considered as the natural generalization of classical statistical mechanics, and should be correctly entitled "wave statistical mechanics", not "wave mechanics". In Sec. 3 we shall further demonstrate the analogy between the classical limit of quantum mechanics and classical statistical mechanics, instead of the analogy between classical limit of quantum mechanics and classical mechanics, thus we go a step further to the judgment that QM is a natural generalization of classical statistical mechanics, hence it is undoubtedly logical, that classical statistical mechanics should be the classical limit of QM,

i.e. SPE should be the classical limit of wave function in QM, thence the another appropriate name of QM should be "wave statistical mechanics", not "wave mechanics".

2 Analogy Between Geometrical Optics and Classical Statistical Mechanics

In classical statistical mechanics, the equations of motion of SPE are

$$\dot{x}_i = \frac{\partial H}{\partial p_i}, \quad \dot{p}_i = \frac{\partial H}{\partial x_i},$$

$$\frac{\partial \rho}{\partial t} + \nabla \cdot (\boldsymbol{v}\rho) + \sum_{i=1}^{3} \frac{\partial}{\partial p_i}(\dot{p}_i\rho) = 0, \tag{1}$$

where x_i and p_i are generalized coordinates and generalized momentum respectively, $v_i = \mathrm{d}x_i/\mathrm{d}t$, $\rho(x_i, p_i, t)$ is the density of the representative points in phase space, and H is the Hamiltonian. In terms of the Hamilton's principal function S,[4] equation (1) can be rewritten as

$$\frac{\partial S}{\partial t} + \frac{1}{2m}(\nabla S)^2 + U = 0,$$

$$\frac{\partial \rho}{\partial t} + \nabla \cdot (\boldsymbol{v}\rho) + \sum_{i=1}^{3} \frac{\partial}{\partial p_i}(\dot{p}_i\rho) = 0. \tag{2}$$

For a conservative system equation (2) becomes

$$\frac{1}{2m}(\nabla S)^2 + U = E,$$

$$\frac{\partial \rho}{\partial t} + \nabla \cdot (\boldsymbol{v}\rho) + \sum_{i=1}^{3} \frac{\partial}{\partial p_i}(\dot{p}_i\rho) = 0, \tag{3}$$

where U is the potential energy, and E is the total energy.

For geometrical optics, the fundamental equations can be obtained from the scalar wave equation of optics,[4]

$$\nabla^2 \Phi - \frac{1}{v^2}\frac{\partial^2 \Phi}{\partial t^2} = 0, \tag{4}$$

where v is the velocity of light in the medium, Φ is a scalar quantity, e.g., the scalar electromagnetic potential. Let

$$\Phi = A \exp \mathrm{i}(\phi - \omega t), \tag{5}$$

where A is the amplitude, the phase ϕ is a function of space coordinates only, and ω is the angular frequency of the wave. Substituting Eq. (5) into Eq. (4), equating the real part and imaginary part of the resulting equation to zero separately, we obtain

$$(\nabla\phi)^2 = \frac{4\pi^2}{\lambda^2} + \frac{1}{A}\left(\nabla^2 A - \frac{1}{v^2}\frac{\partial^2 A}{\partial t^2}\right),$$

$$\frac{\partial A^2}{\partial t} + \nabla\cdot\left[\left(\frac{\lambda}{2\pi}\nabla\phi\right)vA^2\right] = 0, \qquad (6)$$

where the wave length $\lambda = 2\pi v/\omega$. If the amplitude A is a slowly varying function of space and time, then when $\lambda \to 0$, equation (6) reduces to

$$(\nabla\phi)^2 = \frac{4\pi^2}{\lambda^2}, \qquad \frac{\partial I}{\partial t} + \nabla\cdot(v\boldsymbol{n}I) = 0, \qquad (7)$$

where the intensity $I = A^2$ and $\boldsymbol{n} = (\lambda/2\pi)\nabla\phi$. The first equation of Eq. (7) is known as the eikonal equation of geometrical optics. The surfaces of constant ϕ are the surfaces of constant optical phase and thus define the wave fronts. The ray trajectories are everywhere perpendicular to the wave fronts. The second equation of Eq. (7) determines the intensity of light ray.

Since in general $\dot{p}_i$ is independent of p_i, hence in the case that ρ is independent of p_i we have $\sum_{i=1}^{3}\frac{\partial}{\partial p_i}(\dot{p}_i\rho) = 0$, then equation (3) becomes

$$(\nabla S)^2 = 2m(E - U), \qquad \frac{\partial\rho}{\partial t} + \nabla\cdot(v\rho) = 0. \qquad (8)$$

Comparing Eq. (7) with Eq. (8), we readily find that there is an analogy between geometrical optics and classical statistical mechanics, instead of the analogy between geometrical optics and classical mechanics. 170 years ago, classical statistical mechanics is not yet established, at that time (1834) Hamilton only compared the eikonal equation with the Hamilton–Jacobi equation and hence found inappropriately the analogy between geometrical optics and classical mechanics. Evidently, if we consider not only the path but also the intensity of the light ray, i.e. the second equation of Eq. (7), then we should correctly find the analogy between geometrical optics and classical statistical mechanics. Because of this historical misunderstanding, when de Broglie and Schrödinger discovered the new mechanics in the microscopic world, they gave it a name "wave mechanics" not "wave statistical mechanics".

3 Classical Limit of Schrödinger Equation Versus Classical Statistical Mechanics

If we write the wave function $\Psi(\boldsymbol{x}, t)$ in the following form:

$$\Psi(\boldsymbol{x}, t) = R\exp\frac{iS}{\hbar}, \qquad (9)$$

where R and S are real functions, then Schrödinger equation becomes[5]

$$\frac{\partial S}{\partial t} + \frac{1}{2m}(\nabla S)^2 + U = \frac{\hbar^2}{2m}\frac{\nabla^2 R}{R},$$

$$\frac{\partial\rho_s}{\partial t} + \nabla\cdot(v\rho_s) = 0, \qquad (10)$$

where $\rho_s = R^2$ and $v = \nabla S/m$. In the case

$$\sum_{i=1}^{3}\frac{\partial}{\partial p_i}(\dot{p}_i\rho) = 0,$$

equation (2) becomes

$$\frac{\partial S}{\partial t} + \frac{1}{2m}(\nabla S)^2 + U = 0, \qquad \frac{\partial\rho}{\partial t} + \nabla\cdot(v\rho) = 0. \quad (11)$$

In the classical limit, $\hbar \to 0$, equation (10) becomes

$$\frac{\partial S}{\partial t} + \frac{1}{2m}(\nabla S)^2 + U \cong 0, \qquad \frac{\partial\rho_s}{\partial t} + \nabla\cdot(v\rho_s) = 0. \quad (12)$$

Comparing Eq. (12) with Eq. (11) we should attribute the analogy of the classical limit of quantum mechanics to classical statistical mechanics, instead of the analogy of classical limit of quantum mechanics to classical mechanics. Since many authors, including the author of Ref. [5], David Bohm, neglected the analogy between the second equation of Eqs. (11) and (12), hence they drew the inappropriate conclusion, that the classical limit of quantum mechanics is classical mechanics.

4 Conclusions

From the above-mentioned analyses, we have drawn two conclusions:

(i) There is an analogy between geometrical optics and classical statistical mechanics, instead of the analogy between geometrical optics and classical mechanics, which Hamilton found inappropriately 170 years ago.

(ii) Classical limit of Schrödinger equation is not the equation of motion of a particle in classical mechanics, but the equation of motion of single particle ensemble in classical statistical mechanics.

These facts lead us naturally to such an understanding that quantum mechanics is a natural generalization of classical statistical mechanics in microscopic world. The misunderstanding that quantum mechanics is a natural generalization of classical mechanics roots in the research work of Hamilton 170 years ago. Now we have sufficient reasons to say, that it is the time to dispel such historical misunderstanding and reveal quantum mechanics in its true appearance as a wave statistical mechanics instead of a wave mechanics.

References

[1] S.W. Qian and X.Y. Huang, Phys. Lett. A **115** (1986) 319.

[2] S.W. Qian and X.Y. Huang, Phys. Lett. A **117** (1986) 166.

[3] S.W. Qian and X.Y. Huang, Phys. Lett. A **121** (1987) 211.

[4] Herbert Goldstein, *Classical Mechanics*, second edition, Addison-Wesley, Singapore (1980).

[5] David Bohm, *Quantum Theory*, Constable, London (1954).

Alternative Derivation of the Propagator in Polar Coordinates by Feynman's Physical Interpretation of the Characteristic Function

Alternative Derivation of the Propagator in Polar Coordinates by Feynman's Physical Interpretation of the Characteristic Function

QIAN Shang-Wu,[1] CHAN King-Man,[1] and GU Zhi-Yu[2]

[1]Physics Department, Peking University, Beijing 100871, China

[2]Physics Department, Capital Normal University, Beijing 100037, China

(Received May 29, 2002)

Abstract *This article revisits Feynman's characteristic function, and points out the insight and usefulness of his physical interpretation. As an example, the tedious and rather long derivation of the propagator in polar coordinates can be easily and clearly obtained by merely using Feynman's physical intepretation of the characteristic function and some well-known results of central force problem.*

PACS numbers: 03.65.Bz, 03.65.Dd
Key words: Feynman's characteristic function, propagator in polar coordinates

1 Introduction

R.P. Feynman was a famous theoretical physicist with deep physical insight. He relied heavily on his profound intuition for physics. The "Feynman diagrams" are widely used in theoretical physics today. He always obtained important physical results from physical consideration and physical interpretation. In his famous book "Quantum Mechanics and Path Integrals", he introduced the concept "characteristic function" and gave it a physical interpretation. If $\Phi_n(x)$ is the eigenfunction of the Hamiltonian H with energy E_n, according to Feynman, $\Phi_n(x)$ and $\Phi_n^*(x)$ have different physical meaning. He asserted that $\Phi_n(x)$ is the probability amplitude of the particle to be found at x if the particle has energy E_n, whereas $\Phi_n^*(x)$ is the probability amplitude of the particle to be found in the energy state E_n if the particle is at x, i.e., he emphasized that $\Phi_n(x)$ and $\Phi_n^*(x)$ have quite different physical meaning, for $\Phi_n(x)$, firstly we give the energy E_n, i.e., in the condition that particle has energy E_n, then we get the information of the particle at x, for $\Phi_n^*(x)$, firstly we give the position of the particle, x, i.e., in the condition that the position of the particle is x, then we get the information of the particle with energy E_n. Furthermore, for the eigenfunction $g_G(x)$ of an arbitrary observable $\hat{G}$, Feynman introduced the concept of characteristic function $g_G^*(x)$. He endowed different physical meaning of $g_G^*(x)$ from $g_G(x)$. This is a very important assertion of deep physical insight, regrettably, his assertion is often ignored by most physicists thus far. The aim of this article is to call attention to this very important physical interpretation and to show its powerfulness in solving physical problems. Using the above physical interpretation of energy eigenfunction, without any detailed mathematical derivation, Feynman readily obtained the famous Feynman–Kac expansion theorem:[1]

$$K(x_2, t_2; x_1, t_1) = \sum_n \Phi_n(x_2)\Phi_n^*(x_1)$$

$$\times \exp\left[-\frac{\mathrm{i}}{\hbar}E_n(t_2 - t_1)\right]. \qquad (1)$$

Though Feynman's physical interpretation on characteristic functions is very interesting, attractive and powerful in dealing physical problems, it has not yet attracted public attention and is totally forgotten by most physicists. In Sec. 2 we will revisit Feynman's characteristic function and his physical interpretation. Furthermore, we will point out that some seemingly complicated formulae will be easily understood and derived from the physical interpretation of Feynman's characteristic functions. As an example, in Sec. 3 we will introduce propagator in polar coordinates, which is derived by a very tedious and difficult method. In Sec. 4 we will show an alternative derivation of the propagator in polar coordinates by merely using Feynman's physical interpretation of the characteristic function and some well-known results of central force problem. Though this result is not new, the method is new enough, extraordinary physical, very simple and very enlightening, and would be of use in obtaining some important physical results.

2 Feynman's Characteristic Function and His Physical Interpretation

In subsection 5.2 of Ref. [1] Feynman introduced the concept about characteristic function. For an observable $\hat{G}$, the eigenvalue equation is

$$\hat{G}g_G(x) = Gg_G(x). \qquad (2)$$

The eigenfunction $g_G(x)$ has the physical meaning that it is the probability amplitude of the particle to be found at x in the state that the observable $\hat{G}$ takes the value

G. Feynman introduced the characteristic function $g_G^*(x)$, which is the complex conjugate of $g_G(x)$. He endowed the physical meaning of $g_G^*(x)$, which is the probability amplitude of the particle to be found with the observable $\hat{G}$ taking the value G if the particle is at x. Since $\Psi(x)\mathrm{d}x$ [$\Psi(x)$ is the wave function] is the probability amplitude of the particle to be found in the region $(x, x + \mathrm{d}x)$, hence we obtain the result: the probability amplitude that the system has G as the value of $\hat{G}$ is

$$\Psi(G) = \int g_G^*(x)\Psi(x)\mathrm{d}x, \tag{3}$$

e.g. for momentum $\hat{p}$, the eigenvalue equation is

$$\hat{p} = p\,\psi_p(x). \tag{4}$$

From Eq. (4) we get

$$\psi_p(x) = \frac{1}{\sqrt{2\pi\hbar}}\exp\left(\frac{\mathrm{i}}{\hbar}px\right), \tag{5}$$

the corresponding characteristic function is

$$\psi_p^*(x) = \frac{1}{\sqrt{2\pi\hbar}}\exp\left(-\frac{\mathrm{i}}{\hbar}px\right). \tag{6}$$

For Hamiltonian operator $\hat{H}$, the eigenvalue equation is

$$\hat{H}\Phi_n(x) = E_n\Phi_n(x), \tag{7}$$

the corresponding characteristic function is $\Phi_n^*(x)$.

3 Propagator in Polar Coordinates

Feynman's path integral formulation provided an approach to solve quantum mechanical problems which is alternative to the methods of Heisenberg and Schrödinger. The central concept in Feynman's approach is the propagator which, as a Green's function of the Schrödinger equation, contains all the information about the system. In previous literatures[2−4] the propagator in polar coordinates was treated in the standard path integration method by using polygonal approach. The essential steps of the derivation were first given by Edwards and Gulyaev.[2] Subsequently Peak and Inomata[3] rederived the propagator in polar coordinates. For 3-dimensional case, the resulting expression is

$$K_l(\vec{r}'', t''; \vec{r}', t') = \sum_{l=0}^{\infty} \sum_{m=-l}^{l} K_l(r'', t''; r', t')Y_{lm}^*(\theta', \phi')Y_{lm}(\theta'', \phi''), \tag{8}$$

where the radial propagator

$$K_l(r'', t''; r', t') = \frac{1}{r'r''}\int \exp\left[\frac{\mathrm{i}}{\hbar}\int_{t'}^{t''} L_{\text{eff}}\,\mathrm{d}t\right] D[r(t)]. \tag{9}$$

The effective Lagrangian is

$$L_{\text{eff}} = \frac{1}{2}\mu\dot{r}^2 - \frac{l(l+1)\hbar^2}{2\mu r^2} - V(r, t). \tag{10}$$

Though the derivation is rather tedious and makes use of some unfamiliar expansion formulae of the Bessel function, the resulting expressions are impressively simple. In fact, if we compare Eq. (8) with the Feynman–Kac expansion theorem and use Feynman's physical interpretation of characteristic function,[1] we can readily write out Eq. (8), and if we further use some well known results in central force problem,[5] we can readily obtain Eq. (9), and hence get an alternative derivation of the propagator in polar coordinates.

4 An Alternative Derivation of the Propagator in Polar Coordinates

We have already pointed out that Feynman introduced "characteristic function" and gave it a clear physical interpretation. Feynman used the physical meaning of $\phi_n^*(x)$ and $\phi_n(x)$ in subsection 5.3 of Ref. [1] to obtain Feynman-Kac expansion theorem. Similarly, we can endow $Y_{lm}^*(\theta, \phi)$ and $Y_{lm}(\theta, \phi)$ with the following physical interpretations and use them to obtain Eq. (8):

i) $Y_{lm}^*(\theta, \psi)$ is the probability amplitude of the particle to be found in the angular momentum's eigenstate state lm, if the particle is at θ, ϕ;

ii) $Y_{lm}(\theta, \phi)$ is the probability amplitude to be found at θ, ϕ, if the particle is in the angular momentum state lm.

Using these physical interpretations and introducing the radial propagator $K_l(r'', t''; r', t')$ which is the probability amplitude for a particle from r', t' to r'', t'', we can readily obtain Eq. (8). It remains to obtain the expression of radial propagator from some physical considerations.

The transition probability amplitude connects the wave function $\Psi(\vec{r}', t')$ at time t' to the wave function $\Psi(\vec{r}'', t'')$ at a later time t'' by means of the integral equation

$$\Psi(\vec{r}'', t'') = \int K(\vec{r}'', t''; \vec{r}', t')\Psi(\vec{r}', t')\mathrm{d}\vec{r}'. \tag{11}$$

For central potential $V(r, t)$, use polar coordinates and take

$$\Psi(r, \theta, \phi, t) = R(r, t)Y_{lm}(\theta, \phi) = \frac{1}{r}\Xi(r, t)Y_{lm}(\theta, \phi) \,. \tag{12}$$

Substituting Eqs. (8) and (12) into Eq. (11) and removing the angular part we obtain

$$\Xi(r'', t'') = \int [r''r'K_l(r'', t''; r', t')]\Xi(r', t')\mathrm{d}r' \,. \tag{13}$$

Since $\Xi(r, t)$ satisfies the equivalent radial equation

$$\hat{H}_{\text{eff}}\Xi(r, t) = \left[-\frac{\hbar^2}{2\mu}\frac{\mathrm{d}^2}{\mathrm{d}r^2} + V_{\text{eff}}\right]\Xi(r, t) = \mathrm{i}\hbar\frac{\partial}{\partial t}\Xi(r, t) \,, \tag{14}$$

where we use the effective potential

$$V_{\text{eff}} = \frac{l(l+1)\hbar^2}{2\mu r^2} + V(r, t) \tag{15}$$

to express the effective Hamiltonian $\hat{H}_{\text{eff}}$, the corresponding effective Lagrangian is

$$L_{\text{eff}} = \frac{1}{2}\mu\dot{r}^2 - \frac{l(l+1)\hbar^2}{2\mu r^2} - V(r, t) \,. \tag{16}$$

By Feynman's postulate, from Eq. (14) we can introduce an "effective radial propagator" $K_r(r'', t''; r', t')$ and let

$$K_r(r'', t''; r', t') = \int \exp\left[\frac{\mathrm{i}}{\hbar}\int_{t'}^{t''} L_{\text{eff}}\mathrm{d}t\right]D[r(t)] \,. \tag{17}$$

From Eqs. (13) and (17) we readily obtain the required result

$$K_l(r'', t''; r', t') = \frac{1}{r'r''}K_r(r'', t''; r', t') = \frac{1}{r'r''}\int \exp\left[\frac{\mathrm{i}}{\hbar}\int_{t'}^{t''} L_{\text{eff}}\mathrm{d}t\right]D[r(t)] \,.$$

Let us further consider the two-dimensional case. In that case, we can let $\Psi(r, \varphi, t) = R(r, t)\Phi_l(\varphi)$, then

$$K(\vec{r}'', t''; \vec{r}', t') = \sum_{l=0}^{\infty} K_l(r'', t''; r', t')\Phi_l^*(\varphi')\Phi_l(\varphi'') \,, \tag{18}$$

where

$$\Phi_l(\varphi) = \frac{1}{\sqrt{2\pi}}\exp(\mathrm{i}l\phi) \,, \tag{19}$$

and put

$$R(r, t) = \frac{1}{\sqrt{r}}\Xi(r, t) \,, \tag{20}$$

then the equation corresponding to Eq. (13) will be

$$\Xi(r'', t'') = \int \left[\sqrt{r''r'}K_l(r'', t''; r', t')\right]\Xi(r', t')\mathrm{d}r' \,. \tag{21}$$

The effective Lagrangian in this case is

$$L_{\text{eff}} = \frac{1}{2}\mu\dot{r}^2 - \frac{(l^2 - 1/4)\hbar^2}{2\mu r^2} - V(r, t) \,. \tag{22}$$

Finally we obtain

$$K_l(r'', t''; r', t') = \frac{1}{\sqrt{r'r''}}\int \exp\left[\frac{\mathrm{i}}{\hbar}\int_{t'}^{t''} L_{\text{eff}}\mathrm{d}t\right]D[r(t)] \,. \tag{23}$$

If we use the symbol d to denote the dimensionality of the problem, then we can combine Eqs. (9) and (23) into one equation

$$K_l(r'', t''; r', t') = (r'r'')^{1/2-d/2}\int \exp\left[\frac{\mathrm{i}}{\hbar}\int_{t'}^{t''} L_{\text{eff}}\mathrm{d}t\right]D[r(t)] \,, \tag{24}$$

where

$$L_{\text{eff}} = \frac{1}{2}\mu\dot{r}^2 - \frac{(\nu^2 - 1/4)\hbar^2}{2\mu r^2} - V(r, t) \,, \tag{25}$$

and

$$\begin{aligned} \nu &= l + \frac{1}{2} \,, \quad \text{when } d = 3 \,, \\ \nu &= l \,, \qquad \text{when } d = 2 \,. \end{aligned} \tag{26}$$

5 Conclusion

The above discussion shows that the propagator in polar coordinates can indeed be obtained by merely using Feynman's physical interpretation of the characteristic function and some well-known results of central force problem. This alternative derivation gives a vivid physical picture. Though mathematics is uttermost useful in treating physical problems, we think physical insight and physical interpretation in some cases should be more important than tedious mathematical derivation. Feynman introduced the physical interpretation of the characteristic function many years ago. But it has not yet aroused general attention. In this article, we have illustrated its powerful application and we believe that this idea has a wide field of applications, giving useful insights and methods to solve those seemingly complicated physical problems.

References

[1] R.P. Feynman and A.R. Hibbs, *Quantum Mechanics and Path Integrals*, McGraw-Hill, New York (1965).

[2] S.F. Edwards and Y.V. Gulyaev, Proc. R. Soc. London **A279** (1964) 229.

[3] D. Peak and A. Inomata, J. Math. Phys. **8** (1969) 1422l, New York (1965).

[4] D.C. Khandekar and S.V. Lawande, Phys. Rep. **137** (1986) 115.

[5] A. Messiah, *Quantum Mechanics*, North-Holland Publishing Company, Amsterdam (1972).

Separation of Variable Treatment for Solving Time-Dependent Potentials

Separation of Variable Treatment for Solving Time-Dependent Potentials*

QIAN Shang-Wu,[1] GU Zhi-Yu[2] and XIE Guo-Qiang[2]

[1]Department of Physics, Peking University, Beijing 100871, China

[2]Department of Physics, Capital Normal University, Beijing 100037, China

(Received December 26, 2000)

Abstract   We use the separation of variable treatment to treat some time-dependent systems, and point out that the condition of separability is the same as the condition of existence of invariant, and the separation of variable treatment is interrelated with the quantum-invariant method and the propagator method. We directly use the separation of variable treatment to obtain the wavefunctions of the time-dependent Coulomb potential and the time-dependent Hulthén potential.

PACS numbers: 03.65.Ge, 03.65.Bz

Key words: time-dependent system, separation of variable treatment, time-dependent Coulomb potential, time-dependent Hulthén potential, wavefunction

There is a great deal of interest in solving time-dependent problems in quantum mechanics. During the past several decades, there were several techniques to treat the time-dependent systems (TDSs).[1−6] Among these methods, the quantum-invariant operator method and the propagator method[7] are particularly well known. Recently a new method based on separation of variables and time-dependent redefinition of the spatial coordinate has been developed by Efthimiou and Spector (ES).[8] Using this method, we can easily obtain the exact solution of some TDSs. In Ref. [9] we have used this method to obtain the propagator of a complex TDS discussed in Ref. [6]. In Ref. [10] we have used this method to obtain the eigenfunctions of the invariant operator and the Lewis–Riesenfeld phase for any admissible potential. In order to show the powerfulness of this method, in this article we shall further directly use this method to solve exactly two famous TDSs, the time-dependent Coulomb potential and the time-dependent Hulthén potential. In principle, we can also use the method developed by us in Ref. [10] to obtain the same results for these two important time-dependent potentials, here we only want to emphasize that the condition of separability is the same as the condition of existence of invariant, and the separation of variable treatment is interrelated with the quantum-invariant method and the propagator method.

Firstly we shall briefly review the separation of variable treatment. For nonrelativistic quantum mechanics the Schrödinger equation is given by

$$i\hbar(\partial/\partial t)\Psi(x,t) = H\Psi(x,t), \tag{1}$$

where the Hamiltonian with time-dependent potential is

$$H = -(\hbar^2/2m)(\partial^2/\partial x^2) + V(x,t). \tag{2}$$

In Ref. [8], ES defined new variables

$$y = [x - \alpha(t)]/\rho, \tag{3}$$

$$\tau(t) = \int_0^t \frac{\mathrm{d}t}{\rho^2(t)}. \tag{4}$$

They wrote the potential in the form

$$V[x(y,t),t] = [1/\rho^2(t)]\tilde{V}(y) + U(y,t) + g(t), \tag{5}$$

then they showed that whenever

$$V(x,t) = \frac{1}{\rho^2(t)}\tilde{V}(y) - m\Big[\ddot{\alpha} - \alpha\frac{\ddot{\rho}}{\rho}\Big]x - \frac{m}{2}\frac{\ddot{\rho}}{\rho}x^2 + h(t), \tag{6}$$

the model is separable. Comparing Eq. (5) with the condition for admissible potential,[10,11] we readily know that the condition of separability is the same as the condition for the existence of invariant, i.e. the condition for admissible potential. Moreover, we know that the propagator method[7] is only applicable for admissible potential. Hence the separation of variable treatment is interrelated with the invariant operator method and the propagator method.

In addition, when the Schrödinger equation is exactly soluble for the potential $\tilde{V}(y)$, then the corresponding TDSs is also exactly soluble. The propagator $K(x,x';t,t')$ for $V(x,t)$ can be determined by the propagator $K_0(y,y';t,t')$ for $\tilde{V}(y)$,[8]

$$K(x,x';t,t') = [\rho(t)\rho(t')]^{-1/2}\exp\Big(-\frac{\mathrm{i}}{\hbar}\int_{t'}^t \mathrm{d}s h(s)\Big)$$

$$\times \exp\Big[\frac{-\mathrm{i}m}{2\hbar}\int_{t'}^t \mathrm{d}s\Big(\dot{\alpha}(s) - \frac{\dot{\rho}(s)}{\rho(s)}\alpha(s)\Big)^2\Big]$$

$$\times \exp\Big[\frac{\mathrm{i}m}{2\hbar}\Big(\frac{\dot{\rho}(t)}{\rho(t)}x^2 - \frac{\dot{\rho}(t')}{\rho(t')}x'^2\Big)\Big]$$

$$\times \exp\Big[\frac{\mathrm{i}m}{\hbar}\Big(\dot{\alpha}(t) - \frac{\dot{\rho}(t)}{\rho(t)}\alpha(t)\Big)x - \frac{\mathrm{i}m}{\hbar}\Big(\dot{\alpha}(t') - \frac{\dot{\rho}(t')}{\rho(t')}\alpha(t')\Big)x'\Big]$$

$$\times K_0\Big(\frac{x - \alpha(t)}{\rho(t)}, \frac{x' - \alpha(t')}{\rho(t')}; \tau(t), \tau(t')\Big). \tag{7}$$

The above result is quite effective in handling TDSs.

Now we use the above method suggested by ES to treat the time-dependent Coulomb potential. In this case, the Hamiltonian is

$$H = -\frac{\hbar^2}{2m}\frac{\mathrm{d}^2}{\mathrm{d}r^2} - \frac{Ze^2}{r}\exp\Big(-\frac{1}{2}\beta t\Big) - \frac{1}{8}m\beta^2 r^2, \tag{8}$$

where $r \in [0,\infty)$, $\beta > 0$. Applying the above method, we

*The project supported by National Natural Science Foundation of China (No. 19774007)

can easily obtain the exact propagator for Eq. (8) from the time-independent Coulomb potential propagator. According to Eq. (6), when we take

$$\alpha(t) = 0\,, \quad h(t) = 0\,, \quad \rho(t) = \exp(\tfrac{1}{2}\beta t)\,, \qquad (9)$$

the time-independent potential corresponding to Eq. (8) is exactly the Coulomb potential

$$\tilde{V}(y) = -Ze^2/y\,. \qquad (10)$$

The exact propagator for Coulomb potential $-Ze^2/y$ is given by Ref. [12], which is expressed in the confluent hypergeometric form,

$$K_0(y,y';t,t') = \sum_{n=0}^{\infty} \left(\frac{2k^{3/2}}{(n+1)!}\right)^2 \exp[-k(y+y')]yy'$$

$$\times \Phi(-n,2;2ky)\Phi^*(-n,2;2ky') \exp\left[-\frac{i}{\hbar}E_{n+1}(t-t')\right], (11)$$

where $\Phi(a,b;c)$ is the confluent hypergeometric function; or expressed in the Laguerre polynomial form,

$$K_0(y,y';t,t') = \sum_{n=0}^{\infty} \left(\frac{2k^{3/2}}{(n+1)!}\right)^2 \exp[-k(y+y')]yy'$$

$$\times \mathrm{L}_n^1(2ky)\mathrm{L}_n^{1*}(2ky') \exp\left[-\frac{i}{\hbar}E_{n+1}(t-t')\right], \qquad (12)$$

where

$$k = mZe^2/\hbar^2(n+1)\,, \qquad E_n = -mZ^2e^4/2\hbar^2n^2\,. \qquad (13)$$

Considering that from Eqs (4) and (9) we have $\tau = -1/\beta\rho^2$, we can use Eq. (7) to write out the propagator of time-dependent Coulomb potential,

$$K(r,r';t,t') = \sum_{n=0}^{\infty} \left(\frac{2k^{3/2}}{(n+1)!(n+1)}\right)^2 \exp\left[\frac{im\beta}{4\hbar}(r^2-r'^2)\right]$$

$$\times \exp\{-k[r\exp(-\tfrac{1}{2}\beta t) + r'\exp(-\tfrac{1}{2}\beta t')]\}rr'$$

$$\times \exp[-\tfrac{3}{4}\beta(t+t')]\,\mathrm{L}_n^1[2kr\exp(-\tfrac{1}{2}\beta t)]$$

$$\times \mathrm{L}_n^{1*}[2kr'\exp(-\tfrac{1}{2}\beta t')]$$

$$\times \exp\left\{-\frac{i}{\hbar\beta}E_{n+1}[-\exp(-\beta t) + \exp(-\beta t')]\right\}. \quad (14)$$

Substituting Eq. (13) into Eq. (14), considering that $K(r,r';t,t') = \sum_n \Psi_n(r,t)\Psi_n^*(r',t')$, we get the corresponding wavefunction,

$$\Psi_n(r,t) = \frac{1}{(n+1)!(n+1)^{5/2}} \left(\frac{mZe^2}{\hbar^2}\right)^{3/2} \exp\left[\frac{im\beta}{4\hbar}r^2\right]$$

$$\times \exp\left[-\frac{mZe^2}{\hbar^2(n+1)}r\exp\left(-\frac{1}{2}\beta t\right)\right] \exp\left(-\frac{3}{4}\beta t\right)$$

$$\times \mathrm{L}_n^1\left[\frac{2mZe^2}{\hbar^2(n+1)}r\exp\left(-\frac{1}{2}\beta t\right)\right]$$

$$\times \exp\left[-\frac{imZ^2e^4}{2\hbar^3\beta(n+1)^2}\exp(-\beta t)\right]. \qquad (15)$$

This is the same result as that given by S.N. Storchak.[13] He obtained it by a rheonomic homogeneous point transformation and reparametrization in the path integral, which is much more complex than the method we adopted here.

Now we turn to consider another important time-dependent Hulthén potential. In this case, the Hamiltonian is

$$H = -\frac{\hbar^2}{2m}\frac{\mathrm{d}^2}{\mathrm{d}r^2} - V_0\exp(-\beta t)$$

$$\times \frac{\exp[-\frac{r}{a}\exp(-\frac{1}{2}\beta t)]}{1 - \exp[-\frac{r}{a}\exp(-\frac{1}{2}\beta t)]} - \frac{1}{8}m\beta^2 r^2\,. \qquad (16)$$

According to Eq. (6), when we take

$$\alpha(t) = 0\,, \quad h(t) = 0\,, \quad \rho(t) = \exp(\tfrac{1}{2}\beta t)\,, \qquad (17)$$

the time-independent potential corresponding to Eq. (16) is exactly the Hulthén potential

$$\tilde{V}(y) = -V_0\frac{\exp(-y/a)}{1 - \exp(-y/a)}\,. \qquad (18)$$

So the exact propagator $K(r,r';t,t')$ and wavefunction $\Psi_n(r,t)$ can be obtained easily by using the propagator $K_0(y,y';t,t')$[12] for time-independent Hulthén potential. The wavefunctions and energy eigenvalues E_n are

$$\Psi_n(r,t) = [p_{n+1}(p_{n+1}+n+1)(2p_{n+1}+n+1)/a]^{1/2}$$

$$\times \exp\left[-\frac{i}{\hbar\beta}E_{n+1}\exp(-\beta t)\right]\left[1 - \exp\left(-\frac{r}{a}\exp\left(-\frac{1}{2}\beta t\right)\right)\right]$$

$$\times \exp\left[-\frac{1}{4}\beta t - p_{n+1}\frac{r}{a}\exp\left(-\frac{1}{2}\beta t\right)\right]\exp\left(\frac{im\beta}{4\hbar}r^2\right)$$

$$\times \mathrm{F}\left(2p_{n+1}+n+2, -n, 2; 1-\exp\left[-\frac{r}{a}\exp\left(-\frac{1}{2}\beta t\right)\right]\right), (19)$$

$$E_n = -(\hbar^2/8m^2a^2)b^2(b^2-2n^2)\,, \qquad (20)$$

where $\mathrm{F}(a,b,c;z)$ is the hypergeometric function, and

$$p_n = (-2ma^2E_n/\hbar^2)^{1/2}\,, \quad b = 2ma^2V_0/\hbar^2\,, \quad b^2 > 1\,. \quad (21)$$

From the above two interesting examples, it is evident that the separation of variable treatment is very elegant and effective in handling TDSs, this method is closely interrelated with the invariant operator method and the propagator method.

References

[1] J.R. Ray, Phys. Rev. **A26** (1982) 729.

[2] S. Takagi, Prog. Theor. Phys. **85** (1991) 463.

[3] G.J. Reid, SIAM J. Math. Anal. **17** (1986) 646.

[4] V.V. Dadonov, V.I. Man'ko and D.E. Nikonov, Phys. Lett. **A162** (1992) 359.

[5] R.S. Kaushal, Phys. Rev. **A46** (1992) 2941.

[6] Z.Y. Gu and S.W. Qian, J. Phys. **A27** (1994) 3989.

[7] D.C. Khandekar and S.V. Lawande, Phys. Rep. **137** (1986) 180.

[8] C.J. Efthimiou and D. Spector, Phys. Rev. **A49** (1994) 2301.

[9] G.Q. Xie, S.W. Qian and Z.Y. Gu, Phys. Lett. **A207** (1995) 11.

[10] S.W. Qian, G.Q. Xie and Z.Y. Gu, Ann. Phys. **266** (1998) 497.

[11] S.W. Qian and Z.Y. Gu, Ann. Phys. **250** (1996) 420.

[12] J.M. Cai, P.Y. Cai and A. Inomata, Phys. Rev. **A34** (1986) 4621.

[13] S.N. Storchak, Phys. Lett. **A161** (1992) 397.

Supersymmetry and Shape Invariance of the Effective Screened Potential

Supersymmetry and shape invariance of the effective screened potential

Shang-Wu Qian[1], Bo-Wen Huang[2] and Zhi-Yu Gu[3,2]

[1] Physics Department, Peking University, Beijing 100871,
People's Republic of China
[2] Physics Department, Capital Normal University, Beijing 100037,
People's Republic of China
[3] CCAST(World Laboratory), PO Box 8730, Beijing 100080,
People's Republic of China
E-mail: swqian@pku.edu.cn

New Journal of Physics **4** (2002) 13.1–13.6 (http://www.njp.org/)
Received 30 November 2001, in final form 18 February 2002
Published 7 March 2002

Abstract. This paper discusses the supersymmetry and shape invariance (SI) of the effective screened potential. It is shown that the effective screened potential (ESP) has SI and belongs to the first class of shape-invariant potentials, thence the energy levels of this potential are obtained. Furthermore, by using the method of point canonical transformation we find that the ESP belongs to the same subclass as Pöschl–Teller potential 1; the bound-state spectra and the eigenfunctions of this potential are obtained. The results obtained can readily be applied to the special case of the ESP, the famous Hulthén potential.

1. Introduction

Many exactly solvable potentials are exponential functions of the spatial coordinate. These exponential potentials are widely used in many branches of physics. The effective screened potential (ESP) [1] belongs to this group of exactly solvable exponential potentials. Though the solutions of the ESP have already been given by some authors [1], its principal features, such as supersymmetry (SUSY) and shape invariance (SI) [2, 3], have not been systematically discussed in the recent literature. Since the ESP is of considerable importance in nuclear physics and elementary particle physics, the famous Hulthén potential is just a special case of the ESP and also it is closely related to the famous Yukawa potential, we feel that it is necessary to systematically discuss the SUSY and SI of the ESP, such that one can easily grasp the principal features of this very important potential. First we shall discuss the SUSY and SI of the ESP from the Hamiltonian formulation of supersymmetric quantum mechanics (SUSYQM) [4, 5]. It is well known that SUSY relates bosonic and fermionic degrees of freedom; it is a necessary ingredient in any unifying approach. The algebra involved in SUSY is a graded Lie algebra, which closes under a combination of commutation and anti-commutation relations. SUSYQM is the simplest case. Once people started studying various aspects of SUSYQM, it was soon clear that this field was interesting in its own right, not just as a model for testing field theory methods. Gradually a whole technology was evolved based on SUSY to understand the solvable potential problem. Gendenstein [6] in 1983 pointed out that all analytically solvable potentials in quantum mechanics have the property of SI. Recently it has been shown that almost all of the second-order differential equations in mathematical physics have the convenient property of SUSY and SI [7]. Nine years ago it was found that there are two classes of shape-invariant potentials (SIPs) [8, 9]: for the first class of SIPs (SIP1), the parameters a_1 and a_2 of the two supersymmetric partners are related to each other by translation $a_2 = a_1 + \alpha$; for the second class of SIPs (SIP2), the parameters a_1 and a_2 of the two supersymmetric partners are related to each other by scaling $a_2 = q a_1$. For SIP1 and SIP2, we have shown that these two classes are interrelated with each other [10]; in general SIP2 with $0 < q < 1$ can be regarded as the multi-parameter deformation of SIP1 with q acting as the deformation parameter. In section 2 of this article we shall show that the ESP has SI and belongs to SIP1, and thence we obtain energy levels of the ESP. At present we know that all SIPs in SIP1 can be grouped into two subclasses in the sense that the potentials in any subclass can be mapped to a single potential of that class through point canonical transformation (PCT) [11]. In order to find out which subclass ESP belongs to, in section 3 we shall use the method of PCT to map ESP into the Pöschl–Teller-I potential (PT-I), thus we show that the ESP belongs to the same subclass of SIP1 as PT-I and its eigenfunctions correspond to hypergeometric functions; meanwhile we shall use this mapping to find the bound-state spectra and eigenfunctions of the ESP. In principle we can obtain completely identical bound-state eigenfunctions from the well known method suggested by SUSYQM [2, 3].

2. Shape invariance of the ESP

The ESP is of the form

$$V(r) = -\frac{\lambda - \mu}{\mathrm{e}^{r/a} - 1} + \frac{\mu}{(\mathrm{e}^{r/a} - 1)^2}, \qquad \mu = \frac{\hbar^2 l(l+1)}{2ma^2}. \tag{1}$$

The famous Hulthén potential is the particular case when $\mu = 0$. It should be pointed out that in SUSYQM we only consider the bound-state spectra and the corresponding bound state wavefunctions. The configuration space for the ESP is obviously the positive half-line $r > 0$. Since we only consider the bound-state problem, in that case the bound-state energy $E < 0$. In order to find the physically acceptable solutions for the bound-state problem, we have imposed a boundary condition concerning the energy eigenfunctions: they tend to zero when $r \to \infty$. As for the boundary condition at $r = 0$, it has already been discussed in detail in the book *Quantum Mechanics* written by F. Merzbacher for all central force problems, i.e. the wavefunction tends to zero when $r \to 0$. When we introduce the quantities

$$E_0 = -\frac{\hbar^2}{2ma^2}\left[\frac{b^2 - (l+1)^2}{2(l+1)}\right], \qquad b^2 = \frac{2ma^2\lambda}{\hbar^2}, \tag{2}$$

we can define a superpotential $W(r)$ of the ESP:

$$W(r) = \sqrt{-E_0} - \frac{\hbar}{a\sqrt{2m}}\frac{l+1}{\mathrm{e}^{r/a} - 1}. \tag{3}$$

From (3) we obtain the supersymmetric partner potentials:

$$V_- = W^2 - \frac{\hbar}{\sqrt{2m}}W' = -E_0 - \frac{\lambda}{e^{r/a}-1} + \frac{\hbar^2 l(l+1)}{2ma^2}\left[\frac{1}{e^{r/a}-1} + \frac{1}{(e^{r/a}-1)^2}\right] \tag{4}$$

$$V_+ = W^2 + \frac{\hbar}{\sqrt{2m}}W' = -E_0 - \frac{\lambda}{e^{r/a}-1} + \frac{\hbar^2 l(l+1)(l+2)}{2ma^2}\left[\frac{1}{e^{r/a}-1} + \frac{1}{(e^{r/a}-1)^2}\right]. \tag{5}$$

Obviously

$$V(r) = V_- + E_0. \tag{6}$$

If we take $a_0 = l$, $a_1 = l+1$, then from (4) and (5) we readily see

$$V_+(r, a_0) = V_-(r, a_1) + R(a_1) \tag{7}$$

where

$$R(a_1) = -E_0 + E_1, \qquad E_1 = -\frac{\hbar^2}{2ma^2}\left[\frac{b^2 - (l+2)^2}{2(l+2)}\right]. \tag{8}$$

Equation (7) is just the mathematical expression of SI, hence the ESP has the required property of SI; furthermore, from the relation $a_1 = l+1 = a_0 + 1$, we know that the ESP belongs to SIP1. The series of Hamiltonians $H^{(s)}(s = 0, 1, 2, \ldots)$ is defined as

$$H^{(s)} = -\frac{\hbar^2}{2m}\frac{\mathrm{d}^2}{\mathrm{d}r^2} + V_-(r, a_s) + \sum_{k=1}^{s} R(a_k) \tag{9}$$

where $H^{(0)} \equiv H_-$, $H^{(1)} = H_+$.

$$a_s = f^s(a_0) = a_0 + s = l + s \tag{10}$$

hence

$$V_-(r, a_s) = -E_s - \frac{\lambda}{e^{r/a}-1} + \frac{\hbar^2(l+s)(l+s+1)}{2ma^2}\left[\frac{1}{e^{r/a}-1} + \frac{1}{(e^{r/a}-1)^2}\right]. \tag{11}$$

In view of (7) and (9) we have

$$H^{(s+1)} = -\frac{\hbar^2}{2m}\frac{\mathrm{d}^2}{\mathrm{d}r^2} + V_-(r; a_{s+1}) + \sum_{k=1}^{s+1} R(a_k) = -\frac{\hbar^2}{2m}\frac{\mathrm{d}^2}{\mathrm{d}r^2} + V_+(r, a_s) + \sum_{k=1}^{s} R(a_k). \tag{12}$$

Hence the complete energy spectrum of H_- is given by

$$E_n^{(-)} = \sum_{k=1}^{n} R(a_k) = E_n - E_0, \qquad (n \geq 1),\ E_0^{(-)} = 0 \tag{13}$$

thence we obtain the energy levels of the ESP:

$$E_n^{(-)} + E_0 = E_n = -\frac{\hbar^2}{2ma^2}\left[\frac{b^2 - (n+l+1)^2}{2(n+l+1)}\right]^2 = -\frac{\hbar^2 p_{n+1}^2}{2ma^2}, \qquad n = 0, 1, 2, \ldots$$

$$p_{n+1} = \frac{b^2 - (n+l+1)^2}{2(n+l+1)}. \tag{14}$$

From equation (10), we already know the ESP belongs to class ESP1; in order to reveal the relationship between the ESP and other SIPs in class SIP1, in the next section we shall introduce the method of mapping of SIPs under the PCT. Incidentally, we can easily obtain the corresponding energy eigenfunctions, hence we can also use the standard method given by SUSYQM to obtain the bound-state wavefunctions [2, 3], but we think there is no need to perform such exercises in this paper; this method has already been systematically discussed in [3].

3. Mapping of the ESP into Pöschl–Teller potential 1 (PT-1)

First we briefly review the method of mapping of an SIP under PCT. For a given potential $V(\alpha_i; x)$ the Schrödinger equation is

$$\left[-\frac{\hbar^2}{2m}\frac{d^2}{dx^2} + V_-(\alpha_i; x) - E(\alpha_i)\right]\psi(\alpha_i; x) = 0 \tag{15}$$

where $[\alpha_i]$ represents set of parameters of the given potential. We invoke a transformation of both the independent and dependent variables of the form

$$x = f(z), \qquad \psi(\alpha_i; x) = v(z)\tilde{\psi}(\tilde{\alpha}_i; z) \tag{16}$$

then equation (15) becomes

$$-\frac{\hbar^2}{2m}\frac{d^2\tilde{\psi}}{dz^2} - \frac{\hbar^2}{m}\frac{d\tilde{\psi}}{dz}\left\{\frac{v'}{v} - \frac{f''}{2f'}\right\} + \left[f'^2\{V(\alpha_i; f(z)) - E(\alpha_i)\} + \frac{\hbar^2}{2m}\left\{\frac{f''v'}{f'v} - \frac{v''}{v}\right\}\right]\tilde{\psi} = 0. \tag{17}$$

We require (17) in the form

$$\left[-\frac{\hbar^2}{2m}\frac{d^2}{dz^2} + \tilde{V}(\tilde{\alpha}_i; z) - \tilde{E}(\tilde{\alpha}_i)\right]\tilde{\psi}(\tilde{\alpha}_i; z) = 0 \tag{18}$$

for which $\tilde{\psi}_n(\tilde{\alpha}_i; z)$ and $\tilde{E}_n(\tilde{\alpha}_i)$ are known for the SIP $\tilde{V}(\tilde{\alpha}_i; z)$ for each state labelled by the quantum number $n = 0, 1, 2, \ldots$. Here $[\tilde{\alpha}_i]$ represents set of parameters of the transformed potential. To remove the first-derivative term from (17) one requires

$$v(z) = C\sqrt{f'(z)}. \tag{19}$$

Using (19) and comparing (17) and (18) we obtain

$$\tilde{V}(\tilde{\alpha}_i; z) - \tilde{E}(\tilde{\alpha}_i) = f'^2\{V(\alpha_i; f(z)) - E(\alpha_i)\} + \frac{\hbar^2}{4m}\left\{\frac{3}{2}\left(\frac{f''}{f'}\right)^2 - \frac{f'''}{f'}\right\}. \tag{20}$$

By using the known values of $\tilde{E}(\tilde{\alpha}_i)$ and known functions $\tilde{\psi}_n(\tilde{\alpha}_i; z)$, from (16), (19) and (20) we can easily find the bound-state spectra $E_n(\alpha_i)$ and eigenfunctions $\psi_n(\alpha_i; x)$ of the given potential $V(\alpha_i; x)$.

Now we turn to map the ESP into PT-1. Using PCT

$$r = f(z) = -2a\ln[\cos(\alpha z)] \tag{21}$$

equation (1) becomes

$$V(r) = -(\lambda - \mu)\cot^2(\alpha z) + \mu\cot^4(\alpha z) = -\lambda\cot^2(\alpha z) + \mu\csc^2(\alpha z)\cot^2(\alpha z). \tag{22}$$

Substituting (21) and (22) into (20) we obtain

$$\tilde{V} - \tilde{E} = -4a^2\alpha^2(\lambda - E) - \left(4a^2\alpha^2 E + \frac{\alpha^2\hbar^2}{8m}\right)\sec^2(\alpha z) + \left(4a^2\alpha^2\mu + \frac{3\alpha^2\hbar^2}{8m}\right)\csc^2(\alpha z). \tag{23}$$

For PT-1 we have

$$\tilde{V} = -(A + B)^2 + A\left(A - \frac{\alpha\hbar}{\sqrt{2m}}\right)\sec^2(\alpha z) + B\left(B - \frac{\alpha\hbar}{\sqrt{2m}}\right)\csc^2(\alpha z) \tag{24}$$

$$\tilde{E} = \left(A + B + \frac{2na\alpha\hbar}{\sqrt{2m}}\right)^2 - (A+B)^2 \tag{25}$$

$$\tilde{\psi} = (1-y)^{\lambda/2}(1+y)^{s/2} P_n^{(\lambda-1/2,\,s-1/2)}(y),$$

$$\left(y = 1 - 2\sin^2(\alpha z), \qquad s = \frac{\sqrt{2m}\,A}{\alpha\hbar}, \qquad \lambda = \frac{\sqrt{2m}\,B}{\alpha\hbar}\right) \tag{26}$$

where $P_n^{(\alpha,\beta)}(x)$ is the Jacobi polynomial. Substituting (24) and (25) into the LHS of (23) and comparing the coefficients of the corresponding terms of the two sides, we obtain

$$A\left(A - \frac{\alpha\hbar}{\sqrt{2m}}\right) = -\left(4a^2\alpha^2 E + \frac{\alpha^2\hbar^2}{8m}\right) \tag{27}$$

$$B\left(B - \frac{\alpha\hbar}{\sqrt{2m}}\right) = 4a^2\alpha^2\mu + \frac{3\alpha^2\hbar^2}{8m} \tag{28}$$

$$\left(A + B + \frac{2na\alpha\hbar}{\sqrt{2m}}\right)^2 = 4a^2\alpha^2(\lambda - E). \tag{29}$$

Solving for (27) and (28) we obtain

$$A = 2a\alpha\sqrt{-E} + \frac{\alpha\hbar}{2\sqrt{2m}}, \qquad B = \frac{3\alpha\hbar}{2\sqrt{2m}} + \frac{2l\alpha\hbar}{\sqrt{2m}}. \tag{30}$$

Substituting (30) into (29) we obtain once again equation (14). Now we start to find the eigenfunctions of the ESP from (16) and (26).

$$y = 1 - 2\sin^2(\alpha z) = 2e^{-r/a} - 1, \qquad s = \frac{\sqrt{2m}\,A}{\alpha\hbar} = \frac{1}{2} + 2p_{n+1}, \qquad \lambda = \frac{\sqrt{2m}\,B}{\alpha\hbar} = \frac{3}{2} + 2l \tag{31}$$

hence from (26) we have

$$\tilde{\psi} = [2(1 - e^{-r/a})]^{(3/4+l)}[2e^{-r/a}]^{(p_{n+1}1/4)} P_n^{(1+2l,\,2p_{n+1})}(2e^{-r/a} - 1). \tag{32}$$

From (21) and (31) we obtain

$$f' = 2a\alpha\tan(\alpha z) = 2a\alpha\left(\frac{1-y}{1+y}\right)^{1/2} = 2a\alpha\left(\frac{1-e^{-r/a}}{e^{-r/a}}\right)^{1/2}. \tag{33}$$

From (16), (19), (32) and (33) we obtain

$$\psi = c\sqrt{f'}\,\tilde{\psi} = \text{const}(1 - e^{-r/a})^{(l+1)}e^{-p_{n+1}r/a} P_n^{(1+2l,\,2p_{n+1})}(2e^{-r/a} - 1). \tag{34}$$

Using the formula

$$P_n^{(\alpha,\beta)}(x) = \frac{\Gamma(n+\alpha+1)}{\Gamma(\alpha+1)\Gamma(n+1)} F\left(-n, \alpha+1+\beta+n, \alpha+1; \frac{1-x}{2}\right) \tag{35}$$

$$F(\alpha, \beta, \gamma; z) = F(\beta, \alpha, \gamma; z) \tag{36}$$

from (34) we obtain

$$\psi = C_n(e^{-r/a})^{p_{n+1}}(1 - e^{-r/a})^{(l+1)} F(-n, 2l + 2 + 2p_{n+1} + n, 2l + 2; 1 - e^{-r/a}) \tag{37}$$

which coincides completely with (27) of [1].

For the famous Hulthén potential, which is just the special case of the ESP when $l = 0$, i.e. $\mu = 0$,

$$V(r) = -V_0 \frac{e^{-r/a}}{1 - e^{-r/a}}.$$ (38)

Hence we obtain

$$E_n = -\frac{\hbar^2}{2ma^2}\left[\frac{b^2 - (n+1)^2}{2(n+1)}\right]^2 = -\frac{\hbar^2 p_{n+1}^2}{2ma^2}, \qquad n = 0, 1, 2,$$ (39)

$$\psi = C_n(e^{-r/a})^{p_{n+1}}(1 - e^{-r/a})F(-n, 2 + 2p_{n+1} + n, 2; 1 - e^{-r/a})$$ (40)

which coincides completely with the results of [12].

Hence the ESP belongs to the same subclass of SIP1 as PT-I; this class contains some very famous SIPs, such as the Rosen–Morse potential, the Eckart potential, the Scarf potential and the Pöschl–Teller potential. Their eigenfunctions all correspond to hypergeometric functions and can be mapped to each other by PCT.

4. Conclusion

The ESP is of considerable importance in nuclear physics and elementary particle physics. In this paper we have discussed the SUSY and SI of the ESP; we have found the energy levels and energy eigenfunctions of this potential. First we showed that the ESP has SI and belongs to SIP1, thence the energy levels of this potential are obtained from SUSYQM; of course we can also find the energy eigenfunctions by this method. Furthermore, by using the method of PCT, we reveal the relationship between the ESP and other SIPs in class SIP1: we find that the ESP belongs to the same subclass of SIP1 as PT-I; meanwhile we find the energy eigenfunctions of this potential. The results obtained can readily be applied to the special case of the ESP, the famous Hulthén potential.

References

[1] Boundjedaa T, Chétouani L and Guéchi L 1991 *J. Math. Phys.* **32** 441
[2] Dutt R, Khare A and Sukhatme U P 1988 *Am. J. Phys.* **56** 163
[3] Cooper F, Khare A and Sukhatme U P 1995 *Phys. Rep.* **251** 267
    Junker G 1996 *Supersymmetric Methods in Quantum and Statistical Physics* (Berlin: Springer)
[4] Witten E 1981 *Nucl. Phys.* B **188** 513
[5] Cooper F and Freeman B 1983 *Ann. Phys., NY* **146** 262
[6] Gendenstein L 1983 *JETP Lett.* **38** 356
[7] Jafarizadel M A and Fakhri H 1997 *Phys. Lett.* A **230** 164
[8] Khare A and Sukhatme U 1993 *J. Phys. A: Math. Gen.* **26** L901
[9] Barclay D *et al* 1993 *Phys. Rev.* A **48** 2786
[10] Qian S W and Gu Z Y 2001 *Commun. Theor. Phys.* **36** 391–4
[11] De R, Dutt R and Sukhatme U 1992 *J. Phys. A: Math. Gen.* **25** L843
[12] Cai J M, Cai P Y and Inomata A 1986 *Phys. Rev.* A **34** 4621

Appendix E

Knotted Pictures of Entangled States

Knotted pictures of entangled states

Shang-Wu Qian[1,2] **and Zhi-Yu Gu**[2,3]

[1] Physics Department, Peking University, Beijing 100871, People's Republic of China
[2] CCAST, World Laboratory, PO Box 8730, Beijing 100080, People's Republic of China
[3] Physics Department, Capital Normal University, Beijing 100037, People's Republic of China

Received 2 December 2001, in final form 14 February 2002
Published 12 April 2002
Online at stacks.iop.org/JPhysA/35/3733

Abstract

From the comparison of the correlation tensor in the theory of quantum networks and the Alexander relation matrix in the theory of knot crystals we find that there is a one-to-one correspondence between four Bell bases and four oriented links of the linkage 4_1 in knot theory. Meanwhile, we show that the inversion relations under the action of Pauli matrices are the same for the four Bell bases and their corresponding links. Similarly we show that there is also a one-to-one correspondence between 2^m GHZ states of m nodes and 2^m oriented links of m components.

PACS numbers: 03.65.Bz, 02.40.+m, 03.67.−a

1. Introduction

Since the pioneering work of Einstein *et al* [1], the concept of quantum entanglement has become more and more important in quantum mechanics [2–11]; it plays an essential role in quantum teleportation, quantum computation and the modern understanding of quantum phenomena. In this paper we shall study this problem from the point of view of knot crystals [12–14]. In the theory of quantum networks [15], the criterion of entanglement is the existence of nonzero elements of the correlation tensor [15]. Hence the correlation tensor plays a central role for studying the entangled state in quantum mechanics. On the other hand we find that the Alexander relation matrix [12] plays a central role in studying the entanglement of an oriented link with m components. From the comparison of the correlation tensor in the theory of quantum networks and the Alexander relation matrix in the theory of knot crystals we find that there is a one-to-one correspondence between four Bell bases and four oriented links of the linkage 4_1 in knot theory. Meanwhile, we show that the inversion relations under the action of Pauli matrices [16] are the same for the four Bell bases and their corresponding links. From the above arguments we have sufficient reasons to say that such correspondence is convincing and not accidental. Similarly we find that there is also a one-to-one correspondence between 2^m GHZ states of m nodes and 2^m oriented links of m components, and point out that the inversion relations under the action of Pauli matrices are the same for the 2^m GHZ states of

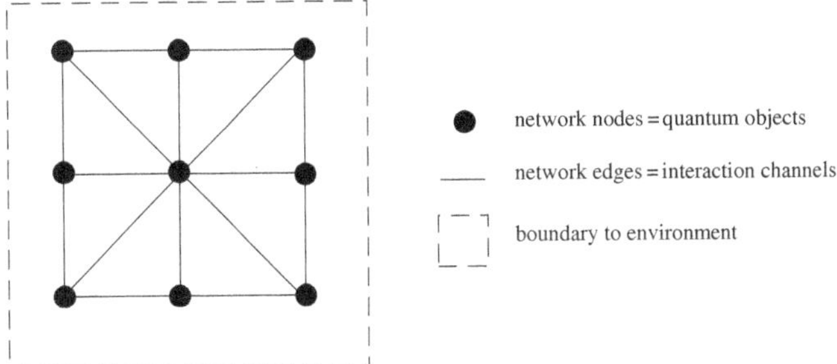

Figure 1. Quantum network.

m nodes and their corresponding 2^m oriented links. In section 2 we shall briefly introduce the correlation tensor and the covariant correlation tensor for a quantum network of two nodes. In section 3 we shall briefly introduce fundamental notions of knot theory and then derive the Alexander relation matrix for four oriented links of two components with four crossings and show that these links correspond to four Bell bases for the quantum network of two nodes; furthermore, we show that the inversion relations under the action of Pauli matrices are the same for the four Bell bases and their corresponding links. In section 4 we discuss GHZ states in general. In section 5 we discuss a quantum network of three nodes and generalize our results to a quantum network of m ($m \geqslant 4$) nodes.

2. Correlation tensor and covariance correlation tensor

A quantum network is a system consisting of N subsystems ('nodes') interacting with each other in n-dimensional Hilbert space. The subsystems of the network are special quantum objects denoted as 'network nodes'. Schematically this is represented by the diagram shown in figure 1; the interaction channels are represented by network edges.

The generating operators of $SU(n)$ are denoted by $\hat{\lambda}_j$. Let v be the ordinal numbers of the nodes and s be the dimensionality of the generating operator $\hat{\lambda}_j$, where $s = n^2 - 1$. In the case of a single node, i.e. $v = 1$ only, in terms of the generating operators, the density operator $\hat{\rho}$ is [13]

$$\hat{\rho} = \frac{1}{n}\hat{1} + \frac{1}{2}\sum_{j=1}^{s}\lambda_j\hat{\lambda}_j \tag{1}$$

where λ_j is the expectation value of $\hat{\lambda}_j$, i.e.

$$\lambda_j = \langle\hat{\lambda}_j\rangle = \mathrm{Tr}\{\hat{\rho}\hat{\lambda}_j\}. \tag{2}$$

In the case $n = 2$, $s = n^2 - 1 = 3$, $\hat{\lambda}_j$ is just the Pauli operator $\hat{\sigma}_j$, whereas

$$\lambda_j = \langle\hat{\lambda}_j\rangle = \langle\hat{\sigma}_j\rangle = P_j \tag{3}$$

where P_j is a component of the polarization vector, $j = x, y, z$. Substituting (3) into (1) we obtain the density matrix:

$$\rho = \frac{1}{2}\begin{pmatrix} 1 + P_z & P_x - iP_y \\ P_x + iP_y & 1 - P_z \end{pmatrix}. \tag{4}$$

In the case of two nodes, $\nu = 1, 2$, the density operator $\hat{\rho}$ is [13]

$$\hat{\rho} = \frac{1}{n_1 n_2}\hat{I} + \frac{1}{2n_2}\sum_{j=1}^{s_1}\lambda_j(1)[\hat{\lambda}_j(1) \otimes \hat{I}(2)] + \frac{1}{2n_1}\sum_{k=1}^{s_2}\lambda_k(2)[\hat{I}(1) \otimes \hat{\lambda}_k(2)]$$

$$+ \tfrac{1}{4}\sum_{j,k=1}^{s_1,s_2} K_{jk}(1,2)[\hat{\lambda}_j(1) \otimes \hat{\lambda}_k(2)] \tag{5}$$

where $s_1 = n_1^2 - 1$ and $s_2 = n_2^2 - 1$. The correlation tensor $K_{jk}(1,2)$ is defined as

$$K_{jk}(1,2) = \langle \hat{\lambda}_j(1)\hat{\lambda}_k(2)\rangle = \text{Tr}\{\hat{\rho}[\hat{\lambda}_j(1) \otimes \hat{\lambda}_k(2)]\}. \tag{6}$$

The covariance correlation tensor $M_{jk}(1,2)$ is introduced as

$$M_{jk}(1,2) = K_{jk}(1,2) - \lambda_j(1)\lambda_k(2). \tag{7}$$

Equations (6) and (7) are the definitions of the correlation tensor and covariant correlation tensor respectively. In terms of the covariant correlation tensor (5) can be written as

$$\hat{\rho} = \hat{\rho}(1) \otimes \hat{\rho}(2) + \tfrac{1}{4}\sum_{j,k=1}^{s_1,s_2} M_{jk}(1,2)[\hat{\lambda}_j(1) \otimes \hat{\lambda}_k(2)]. \tag{8}$$

When $M_{jk}(1,2) = 0$ (for all possible j and k), (8) becomes the product state $\hat{\rho} = \hat{\rho}(1) \otimes \hat{\rho}(2)$, that is, there is no entanglement, i.e. no inter-node coherence. Hence the tensor $M_{jk}(1,2)$ allows the term entanglement to be defined in a precise way. The criterion of an entangled state for a composite system consisting of two nodes is that $M_{jk}(1,2) = 0$ is not true for all possible j and k, i.e. there exist nonzero elements of the covariance correlation tensor $M_{jk}(1,2)$.

Now let us discuss the system consisting of two spin-$\frac{1}{2}$ particles. For this system, there are four well known Bell bases:

$$|\Psi^{\pm}\rangle = \frac{1}{\sqrt{2}}(|\uparrow\downarrow\rangle \pm |\downarrow\uparrow\rangle)$$

$$|\Phi^{\pm}\rangle = \frac{1}{\sqrt{2}}(|\uparrow\uparrow\rangle \pm |\downarrow\downarrow\rangle). \tag{9}$$

Calculating the density matrix of the Bell bases and using (8) we obtain the corresponding covariance correlation tensors as follows:

$$M(\Psi^+) = \begin{pmatrix} 1 & 0 & 0 \\ 0 & 1 & 0 \\ 0 & 0 & -1 \end{pmatrix} \qquad M(\Psi^-) = \begin{pmatrix} -1 & 0 & 0 \\ 0 & -1 & 0 \\ 0 & 0 & -1 \end{pmatrix}$$

$$M(\Phi^+) = \begin{pmatrix} 1 & 0 & 0 \\ 0 & -1 & 0 \\ 0 & 0 & 1 \end{pmatrix} \qquad M(\Phi^-) = \begin{pmatrix} -1 & 0 & 0 \\ 0 & 1 & 0 \\ 0 & 0 & 1 \end{pmatrix}. \tag{10}$$

Since there are nonzero elements in (10), Bell bases are entangled states.

Now let us discuss the inversion relations of Bell bases under the action of the Pauli operator [14]:

$$\sigma_1 = \begin{pmatrix} 0 & 1 \\ 1 & 0 \end{pmatrix} = \sigma_x \qquad \sigma_2 = \begin{pmatrix} 0 & 1 \\ -1 & 0 \end{pmatrix} = i\sigma_y \qquad \sigma_3 = \begin{pmatrix} 1 & 0 \\ 0 & -1 \end{pmatrix} = \sigma_z. \tag{11}$$

There are inversions of phase and spin, where σ_1 only causes inversion of spin and σ_3 only causes inversion of phase, but σ_2 can simultaneously cause inversion of phase and spin. The inversion relationship under the action of the Pauli operator is shown in figure 2.

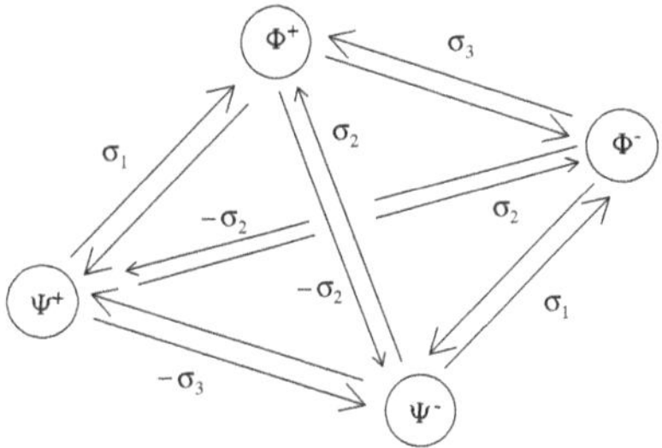

Figure 2. Inversion relationship of Bell bases.

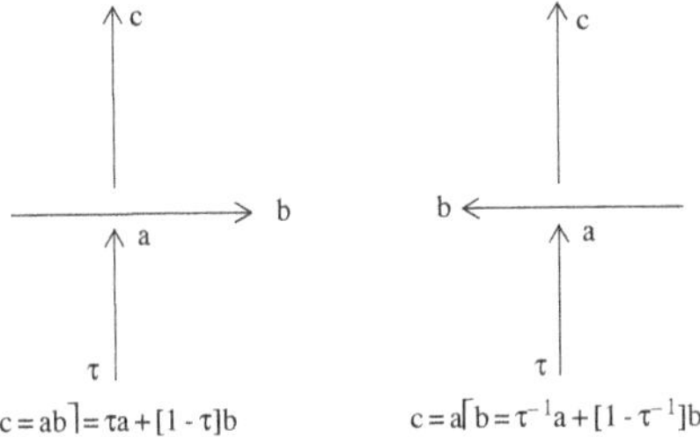

$$c = ab\rceil = \tau a + [1 - \tau]b \qquad\qquad c = a\lceil b = \tau^{-1}a + [1 - \tau^{-1}]b$$

Figure 3. Two modes of crossing.

3. Alexander relation matrix and inversion relations

Firstly we briefly introduce fundamental notions of knot theory. A knot is a closed piece of string without loose ends; the length, thickness and precise shape are of no interest. A link is a collection of two or more knots, or components, which may or may not be physically intertwined. Knots and links can be projected onto a plane and thus represented by planar diagram. A knot, or a link, is oriented if its loops are directed; otherwise, the knot is unoriented. There are two kinds of oriented crossing. For every crossing, one oriented line segment situates over the crossing and the other oriented line segment situates under the crossing: when the direction from the overcrossing oriented line segment to the undercrossing oriented line segment is anticlockwise, the crossing is called a positive crossing; otherwise, i.e. clockwise, the crossing is called a negative crossing. For example, the left crossing shown in figure 3 is a positive crossing, whereas the right crossing shown in the same figure is a negative crossing.

Now we introduce Alexander relation matrices (ARMs) and derive them for a two-component oriented link with four crossings, where the two-component unoriented link with four crossings in knot theory is denoted by 4_1, where 4 denotes the number of crossings and the subscript denotes the order of such links [12]. For example, there are three types for a two-component unoriented link with six crossings: they are $6_1, 6_2, 6_3$; thus the order of a link tells us which link of the class with known crossings it is. For a two-component unoriented link with four crossings, there is only one type; this is 4_1. Then we shall show that their nonzero submatrices have a one-to-one correspondence with the covariance correlation tensor of a system consisting of two nodes. Hence we can give the knotted picture of four Bell bases.

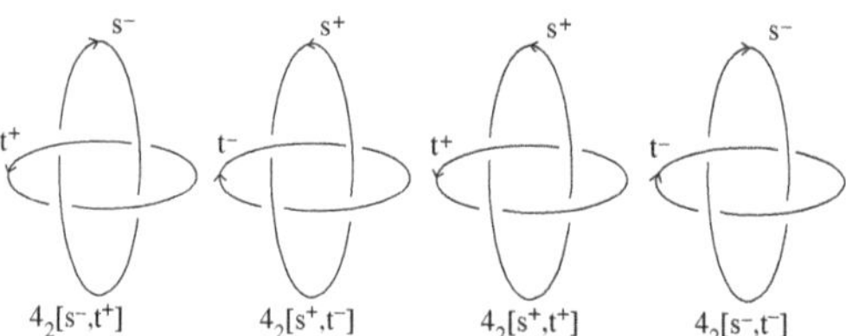

Figure 4. Four kinds of oriented link for a two-component link.

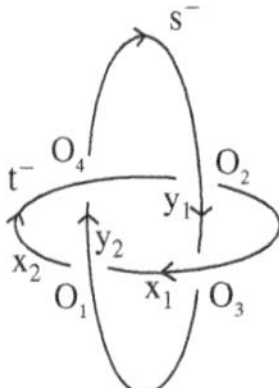

Figure 5. Diagram showing $4_2(s^-, t^-)$.

There are two modes of crossing, the right cross $b\rceil$ and the left cross $\lceil b$; they will serve to encode the two oriented types of diagrammatic crossing as shown in figure 3.

As illustrated in figure 3, the arc c emanating from an undercrossing is regarded as the result of the overcrossing arc b, acting ($b\rceil$ or $\lceil b$) on the incoming undercrossing arc a. Thus $c = a\, b\rceil$ corresponds to a right cross whereas $c = a\lceil b$ corresponds to a left cross. $a\, b\rceil$ and $a\lceil b$ are defined by the following equations:

$$ab\rceil = \tau a + (1 - \tau)b \qquad a\lceil b = \tau^{-1}a + (1 - \tau^{-1})b \tag{12}$$

where $a, b \in M$, M is any module over the ring $Z(\tau, \tau^{-1})$ and τ is the ring parameter.

For the convenience of discussion of a system consisting of more than two nodes, we use the notations $4_2(s^\pm, t^\pm)$ to represent the four oriented links, where the number 4 denotes that the number of oriented links is equal to four, the subscript 2 denotes that the number of components is equal to two and $t^\pm$ and $s^\pm$ denote the ring parameters of the four oriented components respectively: the positive sign denotes the anticlockwise direction of the corresponding component, whereas the negative sign denotes the clockwise direction of the corresponding component. Thus there are four kinds of oriented link for a two-component link: they are $4_2(s^+, t^+)$, $4_2(s^-, t^-)$, $4_2(s^-, t^+)$, $4_2(s^+, t^-)$, which are shown in figure 4.

Now we shall use (12) to derive the ARMs of the four oriented links $4_2(s^\pm, t^\pm)$. From the requirement that the Alexander polynomial of 4_1 is $P_A(s, t) = s + t$, irrespective of the directions of two components [12], i.e. the Alexander polynomial of the four oriented links $4_2(s^\pm, t^\pm)$ must have the same value as $P_A(s, t) = s + t$, we find that

$$s^+ = \frac{1}{s^-} = s \qquad t^- = \frac{1}{t^+} = t. \tag{13}$$

3.1. ARM of $4_2(s^-, t^-)$

$4_2(s^-, t^-)$ is shown in figure 5.

Using (12) for the four crossings we obtain

$$\text{For crossing } O_1 : x_2 = x_1 y_2] = t^- x_1 + (1 - t^-) y_2$$
$$\text{For crossing } O_2 : x_1 = x_2 y_1] = t^- x_2 + (1 - t^-) y_1$$
$$\text{For crossing } O_3 : y_2 = y_1 x_1] = s^- y_1 + (1 - s^-) x_1$$
$$\text{For crossing } O_4 : y_1 = y_2 x_2] = s^- y_2 + (1 - s^-) x_2.$$

(14)

Using (13) to arrange (14) we obtain

$$-t x_1 + x_2 + (t - 1) y_2 = 0 \qquad x_1 - t x_2 + (t - 1) y_1 = 0$$
$$(s^{-1} - 1) x_1 - s^{-1} y_1 + y_2 = 0 \qquad (s^{-1} - 1) x_2 + y_1 - s^{-1} y_2 = 0.$$

(15)

The coefficient matrix of (15) after eliminating negative powers is

$$A^{--}(s, t) = \begin{pmatrix} A_{11}^{--} & A_{12}^{--} \\ A_{21}^{--} & A_{22}^{--} \end{pmatrix}$$

(16)

where the matrices A_{ij}^{--} are all 2×2 matrices:

$$A_{11}^{--} = \begin{pmatrix} -t & 1 \\ 1 & -t \end{pmatrix} \qquad A_{12}^{--} = \begin{pmatrix} 0 & t - 1 \\ t - 1 & 0 \end{pmatrix}$$
$$A_{21}^{--} = \begin{pmatrix} 1 - s & 0 \\ 0 & 1 - s \end{pmatrix} \qquad A_{22}^{--} = \begin{pmatrix} s & -1 \\ -1 & s \end{pmatrix}.$$

(17)

The 4×4 matrix (16) is the ARM of $4_2(s^-, t^-)$; from (16) we can easily obtain the nonzero 3×3 submatrix:

$$A^{--} = \begin{pmatrix} -1 & 0 & 0 \\ 0 & 1 & 0 \\ 0 & 0 & 1 \end{pmatrix}.$$

(18)

3.2. ARM of $4_2(s^+, t^-)$

Using the above method we readily obtain the ARM of $4_2(s^+, t^-)$:

$$A^{+-}(s, t) = \begin{pmatrix} A_{11}^{+-} & A_{12}^{+-} \\ A_{21}^{+-} & A_{22}^{+-} \end{pmatrix}$$

(19)

where

$$A_{ij}^{+-} = \sigma_2 A_{ij}^{--}.$$

(20)

The nonzero submatrix of (19) is

$$A^{+-} = \begin{pmatrix} 1 & 0 & 0 \\ 0 & 1 & 0 \\ 0 & 0 & -1 \end{pmatrix}.$$

(21)

3.3. ARM of $4_2(s^+, t^+)$

Similarly we obtain the ARM of $4_2(s^+, t^+)$:

$$A^{++}(s, t) = \begin{pmatrix} A_{11}^{++} & A_{12}^{++} \\ A_{21}^{++} & A_{22}^{++} \end{pmatrix}$$

(22)

where

$$A_{ij}^{++} = \sigma_3 A_{ij}^{--}.$$

(23)

The nonzero submatrix of (22) is

$$A^{++} = \begin{pmatrix} 1 & 0 & 0 \\ 0 & -1 & 0 \\ 0 & 0 & 1 \end{pmatrix}.$$

(24)

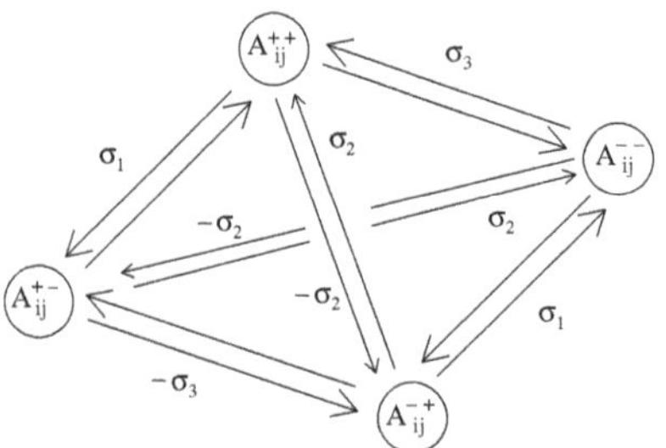

Figure 6. Inversion relationships of four oriented links.

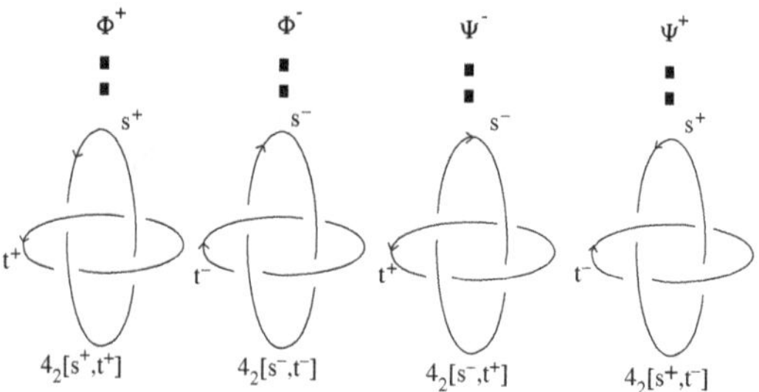

Figure 7. Correspondence between Bell bases and oriented links.

3.4. ARM of $4_2(s^-, t^+)$

Finally we obtain the ARM of $4_2(s^-, t^+)$:

$$A^{-+}(s, t) = \begin{pmatrix} A_{11}^{-+} & A_{12}^{-+} \\ A_{21}^{-+} & A_{22}^{-+} \end{pmatrix} \tag{25}$$

where

$$A_{ij}^{-+} = \sigma_1 A_{ij}^{--}. \tag{26}$$

The nonzero submatrix of (25) is

$$A^{-+} = \begin{pmatrix} -1 & 0 & 0 \\ 0 & -1 & 0 \\ 0 & 0 & -1 \end{pmatrix}. \tag{27}$$

3.5. Correspondence between four Bell bases and four oriented links of 4_1 in knot theory

From (20), (23) and (26) we know that there are inversion relationships between the ARMs of four oriented links under the action of the Pauli operator; these relations are shown in figure 6.

Comparing figure 6 with 2 and meanwhile comparing (18), (21), (24) and (27) with (10), obviously we have one-to-one correspondence between four Bell bases and four oriented links $4_2(s^+, t^+)$, $4_2(s^-, t^-)$, $4_2(s^-, t^+)$, $4_2(s^+, t^-)$. This correspondence is shown in figure 7.

4. GHZ states

Now we shall generalize our previous results for a quantum system of two nodes to a quantum system of m ($m \geqslant 2$) nodes. In the case of a system of two nodes, we know that there are four product states, which are $|\downarrow\downarrow\rangle = |1\rangle, |\downarrow\uparrow\rangle = |2\rangle, |\uparrow\downarrow\rangle = |3\rangle, |\uparrow\uparrow\rangle = |4\rangle$. From these four product states we construct four Bell bases, which are $|\Phi^{\pm}\rangle = (1/\sqrt{2})(|4\rangle \pm |1\rangle)$ and $|\Psi^{\pm}\rangle = (1/\sqrt{2})(|3\rangle \pm |2\rangle)$. Actually, Bell bases are GHZ states for $m = 2$. In the case of three nodes, there are eight product states, which are $|\downarrow\downarrow\downarrow\rangle = |1\rangle, |\downarrow\downarrow\uparrow\rangle = |2\rangle, |\downarrow\uparrow\downarrow\rangle = |3\rangle, |\downarrow\uparrow\uparrow\rangle = |4\rangle, |\uparrow\downarrow\downarrow\rangle = |5\rangle, |\uparrow\downarrow\uparrow\rangle = |6\rangle, |\uparrow\uparrow\downarrow\rangle = |7\rangle, |\uparrow\uparrow\uparrow\rangle = |8\rangle$. From these eight product states we can construct eight GHZ states for $m = 3$:

$$|\Phi_1^{\pm}\rangle = \frac{1}{\sqrt{2}}(|8\rangle \pm |1\rangle) \qquad |\Phi_2^{\pm}\rangle = \frac{1}{\sqrt{2}}(|7\rangle \pm |2\rangle)$$

$$|\Phi_3^{\pm}\rangle = \frac{1}{\sqrt{2}}(|6\rangle \pm |3\rangle) \qquad |\Phi_4^{\pm}\rangle = \frac{1}{\sqrt{2}}(|5\rangle \pm |4\rangle). \tag{28}$$

The general GHZ states for $m \geqslant 2$ states are defined as

$$|\Phi_j^{\pm}\rangle = \frac{1}{\sqrt{2}}(|2^m - j + 1\rangle \pm |j\rangle) \qquad j = 1, 2, \ldots, 2^{m-1}. \tag{29}$$

The eight GHZ states for $m = 3$ have a similar inversion relationship as the case for $m = 2$; they are

$$\begin{aligned}
\sigma_3 \Phi_j^+ &= \Phi_j^- & \sigma_3 \Phi_j^- &= \Phi_j^+ & j &= 1, 2, 3, 4 \\
\sigma_1 \Phi_1^{\pm} &= \Phi_2^{\pm} & \sigma_1 \Phi_2^{\pm} &= \Phi_1^{\pm} & \sigma_2 \Phi_1^{\pm} &= \Phi_2^{\mp} \\
\sigma_2 \Phi_1^{\mp} &= \Phi_2^{\pm} & \sigma_2 \Phi_3^{\pm} &= \Phi_4^{\mp} & \sigma_2 \Phi_4^{\mp} &= \Phi_3^{\pm}.
\end{aligned} \tag{30}$$

Hence, as in the case for $m = 2$, σ_3 only causes inversion of phase and σ_1 only causes inversion of spin, but σ_2 can simultaneously cause inversion of phase and spin. The above statement can be generalized to GHZ states for any value of $m \geqslant 4$.

5. One-to-one correspondence between GHZ states and oriented links

In the case of a quantum network of three nodes, there are eight GHZ states: we shall show that they correspond to eight oriented links $8_3(s^{\pm}, t^{\pm}, u^{\pm})$, where the number 8 denotes that the number of oriented links is equal to eight, the subscript 3 denotes that the number of components is equal to three and s, t and u denote the ring parameters of the three components respectively; the positive sign denotes the anticlockwise direction of the corresponding component, whereas the negative sign denotes the clockwise direction of the corresponding component. Since any two components of the link have four crossings, there are 12 crossings in this case. The eight kinds of oriented link for a three-component link with 12 crossings are $8_3(s^+, t^-, u^-), 8_3(s^-, t^+, u^+), 8_3(s^+, t^+, u^+), 8_3(s^-, t^-, u^-), 8_3(s^+, t^+, u^-), 8_3(s^-, t^-, u^+), 8_3(s^+, t^-, u^+), 8_3(s^-, t^+, u^-)$. For example, the oriented link $8_3(s^+, t^-, u^-)$ is one of the eight oriented links with three components; the direction of the first component knot is anticlockwise, whereas the directions of the second and third component knots are clockwise. Using the same method as in section 3, of course the calculation is very tedious; we found that there is a one-to-one correspondence between eight GHZ states and eight oriented 8_3 links:

$$
\begin{aligned}
\Phi_1^+ &\longleftrightarrow 8_3(s^+, t^+, u^+) & \quad \Phi_1^- &\longleftrightarrow 8_3(s^-, t^-, u^-) \\
\Phi_2^+ &\longleftrightarrow 8_3(s^+, t^+, u^-) & \quad \Phi_2^- &\longleftrightarrow 8_3(s^-, t^-, u^+) \\
\Phi_3^+ &\longleftrightarrow 8_3(s^+, t^-, u^+) & \quad \Phi_3^- &\longleftrightarrow 8_3(s^-, t^+, u^-) \\
\Phi_4^+ &\longleftrightarrow 8_3(s^+, t^-, u^-) & \quad \Phi_4^- &\longleftrightarrow 8_3(s^-, t^+, u^+).
\end{aligned}
\tag{31}
$$

For $m \geqslant 4$, since any two components of the link have four crossings, there are $4C_2^m = 4m(m-1)/2! = 2m(m-1)$ crossings. Though the calculation is exceedingly tedious, the method is essentially the same; we can still find the one-to-one correspondence between 2^m GHZ states and the 2^m oriented links. Details of this work will be shown in another paper.

Therefore, from the comparison of the correlation tensor of the 2^m GHZ states for an m-node quantum network on the one hand, and the ARM of the 2^m oriented m-component links with $2m(m-1)$ crossings on the other hand, we have found that there is a one-to-one correspondence between the 2^m GHZ states and the 2^m oriented links; thus we give the entangled state a knotted picture, which is very useful in handling and analysing entangled states. We have found that the knotted picture is very useful in discussion of the degree of entanglement, the problem of quantum teleportation and other related phenomema; all these problems we shall systematically discuss in other papers.

References

[1] Einstein A, Podolsky B and Rosen N 1935 *Phys. Rev.* **47** 777

[2] Bennett C H 1993 *Phys. Rev. Lett.* **70** 1895

[3] Ekert A and Josza R 1996 *Rev. Mod. Phys.* **68** 733

[4] DiVincenzo D P 1995 *Science* **270** 255

[5] Huches C H *et al* 1997 *Phys. Rev.* A **56** 1163

[6] Horodecki M and Horodecki P 1999 *Phys. Rev.* A **59** 4206

[7] Cerf N J, Adami C and Gingrich R M 1999 *Phys. Rev.* A **60** 898

[8] Hardy L and Song D D 2000 *Phys. Rev.* A **62** 052315
      Fan H 2001 *Phys. Lett.* A **286** 81

[9] Thapliyal A V 1999 *Phys. Rev.* A **59** 3336

[10] Vedral V *et al* 1997 *Phys. Rev. Lett.* **78** 2275

[11] Horodecki P and Lewenstein M 2000 *Phys. Rev. Lett.* **85** 2657

[12] Kauffman L H 1993 *Knots and Physics* 2nd edn (Singapore: World Scientific)
      Kleinert H 1990 *Path Integrals in Quantum Mechanics and Polymer Physics* (Singapore: World Scientific)

[13] Yang C N and Ge M L (ed) 1989 *Braid Group, Knot Theory and Statistical Mechanics* (New Jersey: World Scientific)

[14] Qian S W and Gu Z Y 2001 *Commun. Theor. Phys. (Beijing)* **35** 431

[15] Mahler G and Weberrufz V A 1998 *Quantum Networks* 2nd edn (revised and enlarged) (Berlin: Springer)

[16] Bennett C H *et al* 1996 *Phys. Rev.* A **54** 3824

Knotted Picture of the Whole Complete Quantum Measurement Process of Quantum Teleportation

Knotted picture of the whole complete quantum measurement process of quantum teleportation

Gu Zhi-Yu(顾之雨)[a)]   and   Qian Shang-Wu(钱尚武)[b)]

[a)] *Physics Department, Capital Normal University, Beijing 100048, China*
[b)] *Physics Department, Peking University, Beijing 100871, China*

(Received 20 October 2009; revised manuscript received 12 March 2010)

Based on the previous work about the knotted pictures of quantum states, quantum logic gates and unitary transformations, this paper further gives the whole complete quantum measurement process of quantum teleportation from the viewpoint of knot theory.

Keywords: Bell bases, single particle quantum state, knotted picture, flip operation, inversion operation, composition operation

PACC: 0367, 0240

1. Introduction

Quantum entanglement[1−12] plays an essential role in quantum teleportation, quantum computation and the understanding of quantum phenomena. Because of its importance we try to find its mathematical structure and physical background. Knot theory is a powerful mathematical tool for researching physical phenomena.[13−16] As to the mathematical structure of quantum entanglement, we have found that there is an intimate relationship between quantum entanglement and knot theory,[17−19] and we have successfully proved that there is a one-to-one correspondence between four Bell bases and four oriented links of the linkage 4_1 in knot theory,[17,18] in general, a one-to-one correspondence between GHZ states and the oriented links.[19] As to the physical background we try to use quantum network theory[20] to determine the state of entanglement for a given quantum state.[21,22] We have carried out a series of studies of quantum information from the point of view of knot theory,[23−28] such as the knotted pictures of some quantum states, some quantum logic gates and some unitary transformations. It is well known that Bennett *et al.*[1] suggested a scheme for teleporting a quantum state from one system to another, i.e. the process of quantum teleportation. To the best of our knowledge, up to now no researcher has successfully given the knotted picture of the process of quantum teleportation. In fact, our previous studies mentioned above have prepared all the prerequisites for the description of quantum

teleportation in knot theory. Based on these studies, we find that this problem can be easily solved at this time. Hence in this paper we will discuss the knotted picture of the whole complete quantum measurement process of quantum teleportation. To perform quantum teleportation of the initial quantum state carried by particle A, the receiver Bob and the sender Alice share an entangled pair of particles 1 and 2, which they have obtained from some source of entangled particles in, say, state $|\Phi^+\rangle_{12}$, one of the well-known Bell bases. Before, Alice cannot say anything about the state of particle 1 on its own. Nor does she know the state of particle 1. She only knows that these particles are orthogonal to each other. First, particle A, which carries the state to be sent to Bob, is given to Alice. Alice now measures particles A and 2 together, by projecting them onto the Bell-state bases. After projecting the two particles into an entangled state, Alice only knows the correlations between the two particles, she does not know anything about the individual states of them. In the process of quantum teleportation, Bell basis is chosen to be the quantum channel. It is well known that for the system consisting of two spin-$\frac{1}{2}$ particles 1 and 2, there are four well-known entangled Bell bases, which are[18]

$$|\Phi^+\rangle_{12} = \frac{1}{\sqrt{2}}(|\uparrow\uparrow\rangle_{12} + |\downarrow\downarrow\rangle_{12}),$$

$$|\Phi^-\rangle_{12} = \frac{1}{\sqrt{2}}(|\uparrow\uparrow\rangle_{12} - |\downarrow\downarrow\rangle_{12}),$$

$$|\Psi^+\rangle_{12} = \frac{1}{\sqrt{2}}(|\uparrow\downarrow\rangle_{12} + |\downarrow\uparrow\rangle_{12}),$$

$$|\Psi^-\rangle_{12} = \frac{1}{\sqrt{2}}(|\uparrow\downarrow\rangle_{12} - |\downarrow\uparrow\rangle_{12}). \qquad (1)$$

Previously we have introduced a unified expression $|\text{Bell}\rangle_{12}^i$ to express the above four Bell bases:[25]

$$i = 1 \longleftrightarrow |\Phi^+\rangle_{12},$$
$$i = 2 \longleftrightarrow |\Phi^-\rangle_{12},$$
$$i = 3 \longleftrightarrow |\Psi^+\rangle_{12},$$
$$i = 4 \longleftrightarrow |\Psi^-\rangle_{12}. \qquad (2)$$

In the process of quantum teleportation, supposing that the quantum channel state is chosen to be $|\Phi^+\rangle_{12}$, the unknown quantum state of particle A is $|\phi\rangle_A$, then before any measurement, the entire system, comprising Alice's unknown particle A and the EPR pair (particle 1 and particle 2), is in a pure product state $|\phi\rangle_A |\Phi^+\rangle_{12}$. The process of teleportation is composed of two parts: Alice's measurement on particle A and particle 2, abbreviated as M(A,2); Bob's measurement on particle 1, abbreviated as M(1). Actually this process is a process of re-entanglement, it is equivalent to the expansion of product state $|\phi\rangle_A |\Phi^+\rangle_{12}$ in terms of the four Bell bases $|\text{Bell}\rangle_{A2}^i$, for which $|\text{Bell}\rangle_{A2}^1 = |\Phi^+\rangle_{A2}$, $|\text{Bell}\rangle_{A2}^2 = |\Phi^-\rangle_{A2}$, $|\text{Bell}\rangle_{A2}^3 = |\Psi^+\rangle_{A2}$, $|\text{Bell}\rangle_{A2}^4 = |\Psi^-\rangle_{A2}$, i.e.

$$|\phi\rangle_A |\Phi^+\rangle_{12} = \sum_{i=1}^{4} |\text{Bell}\rangle_{A2}^i |\phi\rangle_1^i. \qquad (3)$$

The corresponding knotted equation is

$$\{|\phi\rangle_A\}\{|\Phi^+\rangle_{12}\} = \sum_{i=1}^{4}\{|\text{Bell}\rangle_{A2}^i\}\{|\phi\rangle_1^i\}, \qquad (4)$$

where $\{|\phi\rangle_A\}$, $\{|\Phi^+\rangle_{12}\}$, $\{|\text{Bell}\rangle_{A2}^i\}$, and $\{|\phi\rangle_1^i\}$ represent the knotted pictures of the corresponding quantum states respectively. The rest of the present paper is organized as follows. In Section 2 we shall discuss the knotted picture of M(A,2), i.e. $\{|\text{Bell}\rangle_{A2}^i\}$. In Section 3 we shall discuss knotted picture of M(1), i.e. $\{|\phi\rangle_1^i\}$. In Section 4 we shall point out the one-one correspondence between the knotted pictures of the unitary transformations involved in the two measurements M(A,2) and M(1). In Section 5 we shall discuss the completion of the process of quantum teleportation.

2. Knotted picture of measurement M(A,2)

Introducing four unitary operators based on the three Pauli operators,[25] namely U_i ($i = 1, 2, 3, 4$),

where

$$U_1 = I_2 = \begin{pmatrix} 1 & 0 \\ 0 & 1 \end{pmatrix}, \quad U_2 = \sigma_z = \begin{pmatrix} 1 & 0 \\ 0 & -1 \end{pmatrix},$$

$$U_3 = \sigma_x = \begin{pmatrix} 0 & 1 \\ 1 & 0 \end{pmatrix}, \quad U_4 = i\sigma_y = \begin{pmatrix} 0 & -1 \\ 1 & 0 \end{pmatrix}, (5)$$

and using Eqs. (1) and (2), then we can easily obtain the expression of $|\text{Bell}\rangle_{12}^i$ in terms of single particle states $\begin{pmatrix} |\uparrow\rangle_1 \\ |\downarrow\rangle_1 \end{pmatrix}$ and $\begin{pmatrix} |\uparrow\rangle_2 \\ |\downarrow\rangle_2 \end{pmatrix}$ as follows:[25]

$$|\text{Bell}\rangle_{12}^i = \frac{1}{\sqrt{2}}(|\uparrow\rangle_1, |\downarrow\rangle_1)U_i \begin{pmatrix} |\uparrow\rangle_2 \\ |\downarrow\rangle_2 \end{pmatrix}. \qquad (6)$$

From Eq. (6) we can obtain the corresponding knotted equation

$$\{|\text{Bell}\rangle_{12}^i\} = \{(|\uparrow\rangle_1, |\downarrow\rangle_1)\}\tilde{U}_i \left\{ \begin{pmatrix} |\uparrow\rangle_2 \\ |\downarrow\rangle_2 \end{pmatrix} \right\}, \qquad (7)$$

where $\tilde{U}_i$ is the corresponding knotted operator of unitary operator U_i.

The links corresponding to $\{|\text{Bell}\rangle_{12}^i\}$ are shown in Fig. 1.[18]

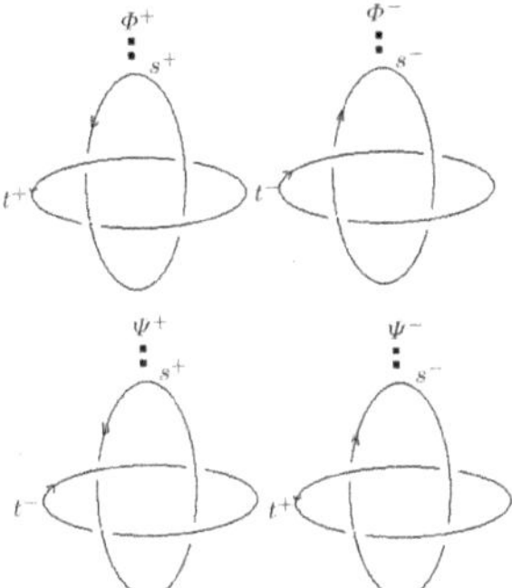

Fig. 1. Knotted pictures of the four Bell bases $\{|\text{Bell}\rangle_{12}^i\}$.

The knotted picture of quantum states $(|\uparrow\rangle_1, |\downarrow\rangle_1)$ and $\begin{pmatrix} |\uparrow\rangle_2 \\ |\downarrow\rangle_2 \end{pmatrix}$ can be given by the oriented knots shown in Fig. 2, which are represented by $\{(|\uparrow\rangle_1, |\downarrow\rangle_1)\}$ and $\left\{ \begin{pmatrix} |\uparrow\rangle_2 \\ |\downarrow\rangle_2 \end{pmatrix} \right\}$ respectively.[25]

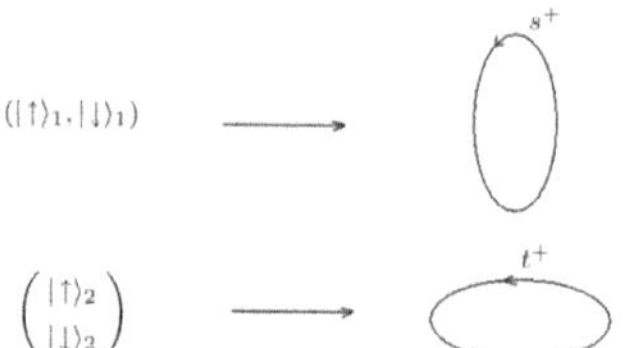

Fig. 2. Knotted pictures of quantum states $(|\uparrow\rangle_1, |\downarrow\rangle_1)$ and $\begin{pmatrix} |\uparrow\rangle_2 \\ |\downarrow\rangle_2 \end{pmatrix}$.

Similarly, we have

$$\{|\text{Bell}\rangle^i_{A2}\} = \{(|\uparrow\rangle_A , |\downarrow\rangle_A)\}\tilde{U}_i\left\{\begin{pmatrix} |\uparrow\rangle_2 \\ |\downarrow\rangle_2 \end{pmatrix}\right\}. \qquad (8)$$

In order to obtain the knotted operators $\tilde{U}_i$ of the four unitary operators U_i, we introduce the composition operator $\tilde{C}$, which is composed of the following three surgical operations.[25]

1) Cut

Cut knot $\left\{\begin{pmatrix} |\uparrow\rangle_2 \\ |\downarrow\rangle_2 \end{pmatrix}\right\}$, the point being cut c becomes two unconnected points c' and c'', then the resulted knot will become an open cycle which can pass through knot $\{(|\uparrow\rangle_1, |\downarrow\rangle_1)\}$.

2) Pass through

Let the being cut knot $\left\{\begin{pmatrix} |\uparrow\rangle_2 \\ |\downarrow\rangle_2 \end{pmatrix}\right\}$ pass through knot $\{(|\uparrow\rangle_1, |\downarrow\rangle_1)\}$, reach the required position.

3) Glue

Glue two unconnected points c' and c'' of knot $\left\{\begin{pmatrix} |\uparrow\rangle_2 \\ |\downarrow\rangle_2 \end{pmatrix}\right\}$ into one point c, thus obtain the required link.

These three steps are shown in Figs. (3a), (3b) and (3c) respectively.

Using composition operator $\tilde{C}$ we can easily obtain[25]

$$\{(|\uparrow\rangle_1, |\downarrow\rangle_1)\}\tilde{C}\left\{\begin{pmatrix} |\uparrow\rangle_2 \\ |\downarrow\rangle_2 \end{pmatrix}\right\} = \{|\Phi^+\rangle_{12}\}. \qquad (9)$$

Obviously if we change the direction of orientation of knot $\{(|\uparrow\rangle_1, |\downarrow\rangle_1)\}$, from counterclockwise to clockwise, i.e. from positive direction to negative direction, then we can use the resulting inversed knot $\{(|\uparrow\rangle_1, |\downarrow\rangle_1)\}_{\text{inv}}$ to obtain the following equation:[25]

$$\{(|\uparrow\rangle_1, |\downarrow\rangle_1)\}_{\text{inv}}\tilde{C}\left\{\begin{pmatrix} |\uparrow\rangle_2 \\ |\downarrow\rangle_2 \end{pmatrix}\right\} = \{|\Psi^-\rangle_{12}\}. \qquad (10)$$

Similarly, we have

$$\{(|\uparrow\rangle_1, |\downarrow\rangle_1)\}\tilde{C}\left\{\begin{pmatrix} |\uparrow\rangle_2 \\ |\downarrow\rangle_2 \end{pmatrix}\right\}_{\text{inv}} = \{|\Psi^+\rangle_{12}\}, \qquad (11)$$

and

$$\{(|\uparrow\rangle_1, |\downarrow\rangle_1)\}_{\text{inv}}\tilde{C}\left\{\begin{pmatrix} |\uparrow\rangle_2 \\ |\downarrow\rangle_2 \end{pmatrix}\right\}_{\text{inv}} = \{|\Phi^-\rangle_{12}\}. \qquad (12)$$

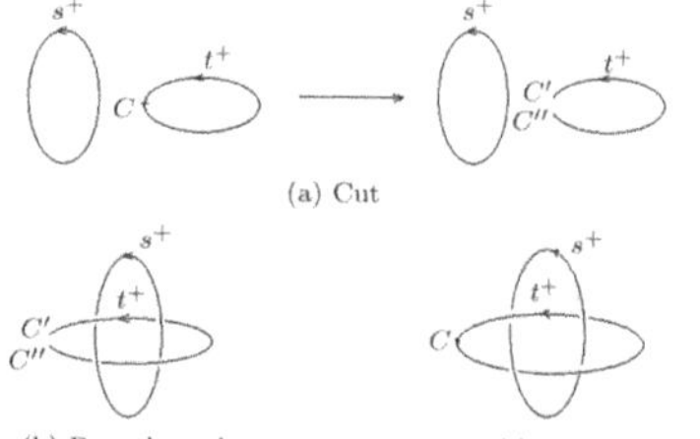

(a) Cut

(b) Pass through   (c) Glue

Fig. 3. Three steps of composition operation.

In order to give a unified form of Eqs. (9)–(12) we can write $\tilde{U}_i$ as

$$\tilde{U}_i = \tilde{l}_i\tilde{C}\tilde{r}_i, \qquad (13)$$

where $\tilde{l}_i$ is the left operator operating on the left knot, and $\tilde{r}_i$ is the right operator operating on the right knot. There are only two kinds of left (right) operators which are identity operator I and inversion operator τ, I keeps the orientation of the left (right) knot invariant, whereas τ changes the orientation of the left (right) knot from positive direction to negative direction and *vice versa*. Obviously $\tau^2 = \tau\tau = I$. Left and right operations are shown in Figs. 4 and 5 respectively.[25]

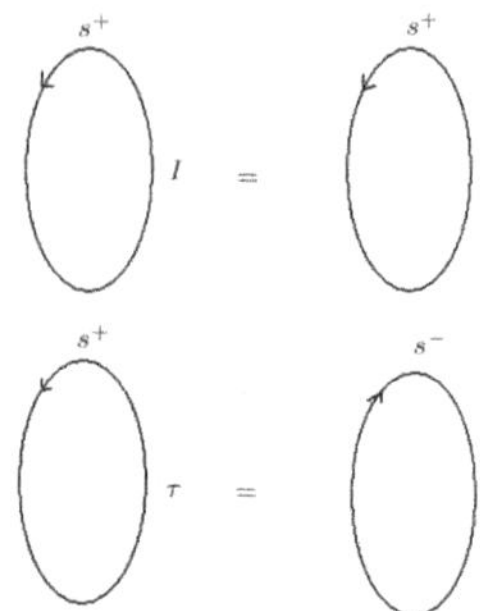

Fig. 4. Left operations of the operators I and τ.

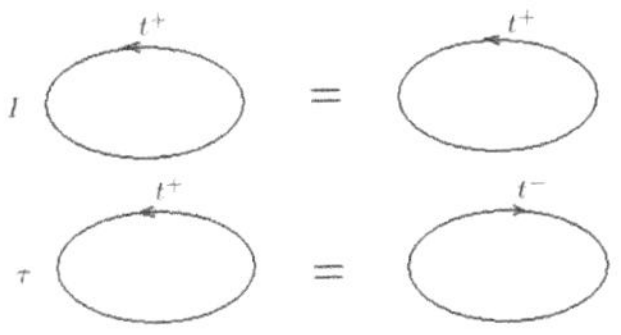

Fig. 5. Right operations of the operators I and τ.

From Eq. (13) we readily obtain

$$\tilde{U}_1 = I\tilde{C}I, \quad \tilde{U}_2 = \tau\tilde{C}\tau, \quad \tilde{U}_3 = I\tilde{C}\tau,$$

and

$$\tilde{U}_4 = \tau\tilde{C}I. \tag{14}$$

3. Knotted picture of M(1) on the quantum channel

Now we discuss the knotted picture of M(1), i.e. $\{|\phi\rangle_1^i\}$, where $|\phi\rangle_1^i$ are the following quantum states:

$$|\phi\rangle_1^1 = \begin{pmatrix} a \\ b \end{pmatrix}_1 = a|\uparrow\rangle_1 + b|\downarrow\rangle_1,$$

$$|\phi\rangle_1^2 = \begin{pmatrix} a \\ -b \end{pmatrix}_1 = a|\uparrow\rangle_1 - b|\downarrow\rangle_1,$$

$$|\phi\rangle_1^3 = \begin{pmatrix} b \\ a \end{pmatrix}_1 = a|\downarrow\rangle_1 + b|\uparrow\rangle_1,$$

$$|\phi\rangle_1^4 = \begin{pmatrix} -b \\ a \end{pmatrix}_1 = a|\downarrow\rangle_1 - b|\uparrow\rangle_1. \tag{15}$$

Quantum states $|\phi\rangle_1^i$ shown in Eq. (15) are superposed states of two eigenstates $|\uparrow\rangle_1$ and $|\downarrow\rangle_1$, there are three characteristics which must be shown in their corresponding knotted pictures. The first characteristic is the two eigenstates $|\uparrow\rangle_1$ and $|\downarrow\rangle_1$ involved in the quantum states $|\phi\rangle_1^i$, which can be represented by two oriented circles: $|\uparrow\rangle_1$ corresponding to the circle with counterclockwise orientation, and $|\downarrow\rangle_1$ corresponding to the circle with clockwise orientation. Since $|\uparrow\rangle_1$ and $|\downarrow\rangle_1$ are the eigenstates of the same particle 1, they must connect each other through the surgical operations cut and glue described above. If we locate the oriented circle representing $|\uparrow\rangle_1$ on the left side, and arrange, the oriented circle representing $|\downarrow\rangle_1$ on the right side, then by the surgical operations, cut and glue, we can connect the two circles into one twisted knot as shown in Fig. 6. There are two possible ways of connection, one corresponds to positive cross connecting two oriented circles (left part of Fig. 6), the other corresponds to a negative cross connecting two oriented circles (right part of Fig. 6).[26]

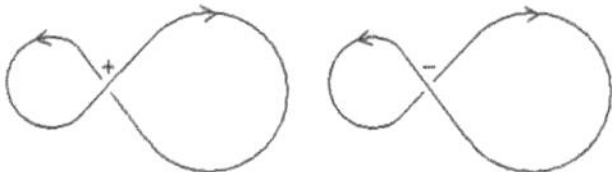

Fig. 6. Two possible ways of connecting two oppositely oriented circles into one twisted knot.

The second characteristic is the plus or minus sign between the two composed eigenstates. Naturally we may use the positive cross to represent the positive sign, and the negative cross to denote the negative sign. The third characteristic is the value of two coefficients before the two eigenstates $|\uparrow\rangle_1$ and $|\downarrow\rangle_1$. For the purpose of representation of these coefficients we must use non-topological method, let the area of the orientated circle equal the square of the coefficient before each of corresponding eigenstates separately, i.e., for the quantum state $|\phi\rangle_1^1 = \begin{pmatrix} a \\ b \end{pmatrix}_1 = a|\uparrow\rangle_1 + b|\downarrow\rangle_1$, we let $\pi R_a^2 = a^2$, and $\pi R_b^2 = b^2$, where R_a and R_b are the radii of the corresponding circles. Using this method we can successfully obtain the four quasi-knots for quantum states $|\phi\rangle_1^i$ as shown in Fig. 7.[26]

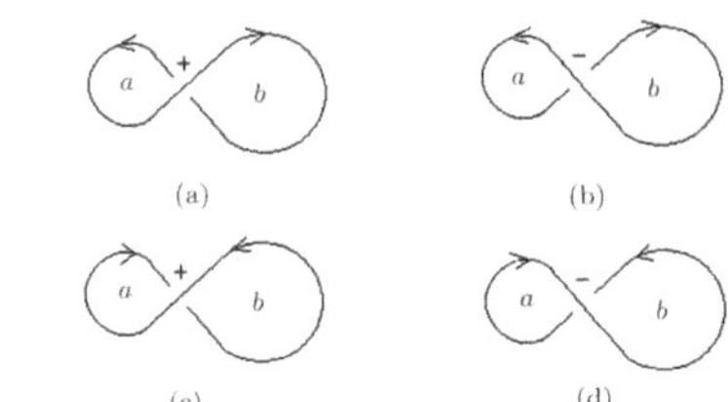

Fig. 7. Four kinds of quasi-knots for the knotted picture of single particle quantum state.

where figure 7(a) shows the knotted picture of $|\phi\rangle_1^1$, i.e.,

$$\{|\phi\rangle_1^1\} = \{a|\uparrow\rangle_1 + b|\downarrow\rangle_1\} = \left\{ \begin{pmatrix} a \\ b \end{pmatrix}_1 \right\},$$

figure 7(b) shows the knotted picture of $|\phi\rangle_1^2$, i.e.,

$$\{|\phi\rangle_1^2\} = \{a|\uparrow\rangle_1 - b|\downarrow\rangle_1\} = \left\{ \begin{pmatrix} a \\ -b \end{pmatrix}_1 \right\},$$

figure 7(c) shows the knotted picture of $|\phi\rangle_1^3$, i.e.,

$$\{|\phi\rangle_1^3\} = \{a|\downarrow\rangle_1 + b|\uparrow\rangle_1\} = \left\{ \begin{pmatrix} b \\ a \end{pmatrix}_1 \right\},$$

and figure 7(d) shows the knotted picture of $|\phi\rangle_1^4$, i.e.,

$$\{|\phi\rangle_1^4\} = \{a|\downarrow\rangle_1 - b|\uparrow\rangle_1\} = \left\{\begin{pmatrix} -b \\ a \end{pmatrix}_1\right\}.$$

It can be easily verified that the four quantum states $|\phi\rangle_1^i$ are interrelated to each other through the four unitary operators shown in Eq. (5),[26] i.e.

$$\begin{pmatrix} a \\ b \end{pmatrix}_1 = U_1 \begin{pmatrix} a \\ b \end{pmatrix}_1 = \begin{pmatrix} 1 & 0 \\ 0 & 1 \end{pmatrix}\begin{pmatrix} a \\ b \end{pmatrix}_1,$$

$$\begin{pmatrix} a \\ -b \end{pmatrix}_1 = U_2 \begin{pmatrix} a \\ b \end{pmatrix}_1 = \begin{pmatrix} 1 & 0 \\ 0 & -1 \end{pmatrix}\begin{pmatrix} a \\ b \end{pmatrix}_1,$$

$$\begin{pmatrix} b \\ a \end{pmatrix}_1 = U_3 \begin{pmatrix} a \\ b \end{pmatrix}_1 = \begin{pmatrix} 0 & 1 \\ 1 & 0 \end{pmatrix}\begin{pmatrix} a \\ b \end{pmatrix}_1,$$

$$\begin{pmatrix} -b \\ a \end{pmatrix}_1 = U_4 \begin{pmatrix} a \\ b \end{pmatrix}_1 = \begin{pmatrix} 0 & -1 \\ 1 & 0 \end{pmatrix}\begin{pmatrix} a \\ b \end{pmatrix}_1. \quad (16)$$

Equation (16) can be abbreviated to

$$|\phi\rangle_1^i = U_i|\phi\rangle_1^1. \quad (17)$$

The corresponding knotted equation of Eq. (17) is

$$\{|\phi\rangle_1^i\} = \tilde{U}_i\{|\phi\rangle_1^1\}, \quad (18)$$

where $\{|\phi\rangle_1^i\}$ denotes the knotted picture of $|\phi\rangle_1^i$, and $\tilde{U}_i$ represents the knotted operation of U_i. We can express $\tilde{U}_i$ by introducing the following three surgical operations: I, f and τ, which are shown in Fig. 8.

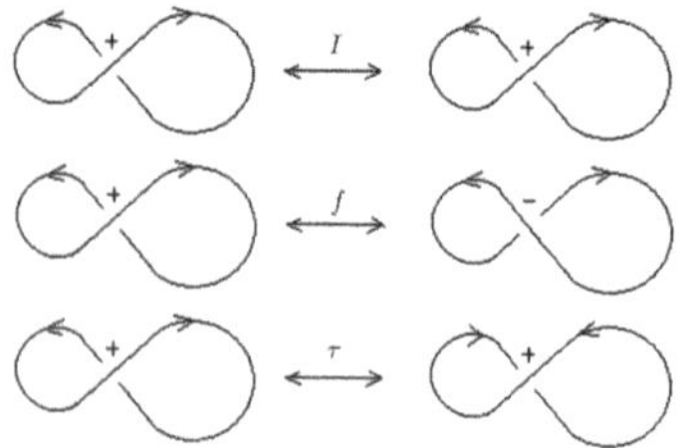

Fig. 8. Three surgical operations: identity, flip and inversion.

Identity operation I does not induce any change of the knot, flip operation f changes the sign of the cross, inversion operation τ changes the orientation of the knot. In terms of these three surgical operations, we have

$$\tilde{U}_1 = I, \quad \tilde{U}_2 = f, \quad \tilde{U}_3 = \tau, \quad \tilde{U}_4 = f\tau. \quad (19)$$

Thus we have found the surgical knotted operations $\tilde{U}_i$ of unitary operators U_i. The knotted pictures $\{|\phi\rangle_1^i\}$ of replicas $|\phi\rangle_1^i$ appearing in measurement M(1) in the process of quantum teleportation are already shown in Fig. 7.

In order to differentiate the two-side two-qubit operator $\tilde{U}_i$ in Eq. (14) and the one-side one-qubit operator $\tilde{U}_i$ in Eq. (19) for the same unitary operator U_i, we use symbol $\tilde{U}_2^i$ to replace $\tilde{U}_i$ in Eq. (14), and use symbol $\tilde{U}_1^i$ to replace $\tilde{U}_i$ in Eq. (19), hence we have

$$\tilde{U}_2^1 = I\check{C}I, \quad \tilde{U}_2^2 = \tau\check{C}\tau, \quad \tilde{U}_2^3 = I\check{C}\tau, \quad \tilde{U}_2^4 = \tau\check{C}I, (20)$$

and

$$\tilde{U}_1^1 = I, \quad \tilde{U}_1^2 = f, \quad \tilde{U}_1^3 = \tau, \quad \tilde{U}_1^4 = f\tau. \quad (21)$$

4. One-one correspondence between knotted pictures of unitary transformations involved in two measurements M(A,2) and M(1)

During measurements M(A,2) and M(1), pure product state $|\phi\rangle_A |\Phi^+\rangle_{12}$ collapses stochastically into one of the four product states $|\text{Bell}\rangle_{A2}^i|\phi\rangle_1^i$ ($i = 1, 2, 3, 4$). For any one product state, entangled state $|\text{Bell}\rangle_{A2}^i$ and single particle state $|\phi\rangle_1^i$ must correspond to the same number i. From Eq. (8) and the replacement of $\tilde{U}_i$ by $\tilde{U}_2^i$, we have

$$\{|\text{Bell}\rangle_{A2}^i\} = \{(|\uparrow\rangle_A, |\downarrow\rangle_A)\}\tilde{U}_2^i\left\{\begin{pmatrix} |\uparrow\rangle_2 \\ |\downarrow\rangle_2 \end{pmatrix}\right\}.$$

Similarly from Eq. (18) and the replacement of $\tilde{U}_i$ by $\tilde{U}_1^i$, we have $\{|\phi\rangle_1^i\} = \tilde{U}_1^i\{|\phi\rangle_1^1\}$. Since the results of measurements M(A,2) and M(1) correspond to the same value of number i, we obtain one-one correspondence $\tilde{U}_2^i \longleftrightarrow \tilde{U}_1^i$. For the case $i = 1$, $|\text{Bell}\rangle_{A2}^1 = |\Phi^+\rangle_{A2}$ and $\tilde{U}_2^1 = I\check{C}I$, the corresponding one-side one-qubit operator is $\tilde{U}_1^1 = I$; for the case $i = 2$, $|\text{Bell}\rangle_{A2}^2 = |\Phi^-\rangle_{A2}$ and $\tilde{U}_2^2 = \tau\check{C}\tau$, the corresponding one-side one-qubit operator is $\tilde{U}_1^2 = f$; for the case $i = 3$, $|\text{Bell}\rangle_{A2}^3 = |\Psi^+\rangle_{A2}$ and $\tilde{U}_2^3 = I\check{C}\tau$, the corresponding one-side one-qubit operator is $\tilde{U}_1^3 = \tau$; for the case $i = 4$, $|\text{Bell}\rangle_{A2}^4 = |\Psi^-\rangle_{A2}$ and $\tilde{U}_2^4 = \tau\check{C}I$, the corresponding one-side one-qubit operator is $\tilde{U}_1^4 = f\tau$.

5. Completion of the process of quantum teleportation

After Alice completed her measurement M(A,2) she knew the entangled state $|\text{Bell}\rangle^i_{A2}$ and thus the value of number i, then she told this value to Bob through the classical channel. After Bob completed his measurement M(1) and received the message about the value of i through the classical channel, he readily knew that his measured result on particle 1 was the quantum state $|\phi\rangle^i_1$. Then from knotted equation (18), i.e., $\{|\phi\rangle^i_1\} = \tilde{U}^i_1\{|\phi\rangle^1_1\}$, he readily knew the knotted picture of the unknown quantum state $\{|\phi\rangle^1_1\} = (\tilde{U}^i_1)^{-1}\{|\phi\rangle^i_1\}$, and hence the unknown quantum state $|\phi\rangle^1_1 = \binom{a}{b}_1 = a|\uparrow\rangle_1 + b|\downarrow\rangle_1$ brought by particle 1. This state is just the teleported state sent by Alice, originally brought by particle A, i.e. $|\phi\rangle_A = |\phi\rangle^1_A = a|\uparrow\rangle_A + b|\downarrow\rangle_A$. Hence, from the measurements M(A,2), M(1) and the requirement of the same value of number i, Bob can easily know the teleported unknown state sent by Alice, thus the quantum teleportation process is completed. For example, if the result of measurement M(A,2) is $|\text{Bell}\rangle^2_{A2} = |\Phi^-\rangle_{A2}$, i.e. $i = 2$, after Alice told this result to Bob, Bob readily knew that his measured result was $\{|\phi\rangle^2_1\}$, from the relation $\{|\phi\rangle^2_1\} = \tilde{U}^2_1\{|\phi\rangle^1_1\} = f\{|\phi\rangle^1_1\}$ he easily knew $\{|\phi\rangle^1_1\} = f^{-1}\{|\phi\rangle^2_1\} = f\{|\phi\rangle^2_1\}$, since $f^{-1} = f$. This process is schematically shown in Fig. 9.

$$\{|\phi\rangle^2_1\} = \quad\xrightarrow{f^{-1}=f}\quad = \{|\phi\rangle^1_1\}$$

Fig. 9. Finding the unknown quantum state in the process of quantum teleportation for case $i = 2$.

Though from quantum channel, Bob receives only the replica of the knotted picture of the unknown quantum state to be sent by Alice, i.e. $\{|\phi\rangle^i_1\}$, from the classical channel he knows the knotted picture of the unitary transformation which is needed for him, i.e. $\tilde{U}^i_2$, then by one-one correspondence $\tilde{U}^i_2 \longleftrightarrow \tilde{U}^i_1$ Bob can readily know the original quantum state which Alice wants to send to him. Therefore we have completed the whole process of the quantum teleportation from the viewpoint of knot theory.

References

[1] Bennett C H 1993 *Phys. Rev. Lett.* **70** 1895

[2] Ekert A and Josza R 1996 *Rev. Mod. Phys.* **68** 733

[3] DiVincenzo D P 1995 *Science* **270** 255

[4] Fuchs C A, Gisin M, Griffiths R B, Niu C S and Peres A 1997 *Phys. Rev. A* **56** 1163

[5] Horodecki M and Horodecki P 1999 *Phys. Rev. A* **59** 4206

[6] Cerf N J, Adami C and Gingrich R M 1999 *Phys. Rev. A* **60** 898

[7] Hardy L and Song D D 2001 *Phys. Rev. A* **62** 052315

[8] Fan H 2000 *Phys. Lett. A* **286** 81

[9] Thapliyal A V 1999 *Phys. Rev. A* **59** 3336

[10] Vedral V, Plenio M B, Rippin M A and Knight P L 1997 *Phys. Rev. Lett.* **78** 2275

[11] Horodecki P and Lewenstein M 2000 *Phys. Rev. Lett.* **85** 2657

[12] Bennett C H, DiVincen Z O, Smolin J A and Wootters W K 1996 *Phys. Rev. A* **54** 3824

[13] Kauffman L H 1993 *Knots and Physics* 2nd ed. (Singapore: World Scientific)

[14] Kleinert H 1990 *Path Integrals in Quantum Mechanics and Polymer Physics* (Singapore: World Scientific)

[15] Kauffman L H and Lomonaco S J 2002 *New Journal of Physics* 73.1–73.8

[16] Yang C N and Ge M L 1989 *Braid Group, Knot Theory and Statistical Mechanics* (New Jersey: World Scientific)

[17] Qian S W and Gu Z Y 2002 *Commun. Theor. Phys.* (Beijing, China) **37** 659

[18] Qian S W and Gu Z Y 2002 *J. Phys. A: Math. Gen.* **35** 3733

[19] Qian S W and Gu Z Y 2002 *Commun. Theor. Phys.* (Beijing, China) **38** 421

[20] Mahler G and Weberrufz V A 1998 *Quantum Networks* second revised and enlarged edition (Berline: Springer)

[21] Qian S W and Gu Z Y 2003 *Commun. Theor. Phys.* (Beijing, China) **39** 15

[22] Gu Z Y and Qian S W 2003 *Commun. Theor. Phys.* (Beijing, China) **39** 421

[23] Qian S W and Gu Z Y 2004 *Commun. Theor. Phys.* (Beijing, China) **41** 201

[24] Gu Z Y and Qian S W 2004 *Commun. Theor. Phys.* (Beijing, China) **41** 531

[25] Gu Z Y and Qian S W 2008 *Commun. Theor. Phys.* (Beijing, China) **49** 65

[26] Gu Z Y and Qian S W 2008 *Commun. Theor. Phys.* (Beijing, China) **49** 1163

[27] Gu Z Y and Qian S W 2009 *Commun. Theor. Phys.* (Beijing, China) **51** 769

[28] Gu Z Y and Qian S W 2009 *Commun. Theor. Phys.* (Beijing, China) **51** 967

Appendices of Essentials of Relativity

Appendix G

Einstein's Field Equations

By the principle of equivalence, the gravitational field is equivalent to metric field, i.e. gravitational field should influence spacetime metric. By principle of general covariance, all physical laws must be written in the form of tensor equations, and by principle of minimal coupling, the method of obtaining tensor equations in the presence of gravitation is just replacing $\eta_{\mu\nu}$ by $g_{\mu\nu}$ and ordinary derivative by covariant derivative from the corresponding equations in the absence of gravitation (i.e. which hold in the local inertial system), thus we can find all influences from the gravitational field. However, thus far we have not yet touched upon the equations of motion of the gravitational field, i.e. the relations between the dynamic quantities $g_{\mu\nu}$, and the distribution of matter along with its motion, i.e. the energy-momentum tensor of matter. When we know $g_{\mu\nu}$, we know all influences of the gravitational field, this is the aim and reason why we want to know the field equations discovered by Einstein in the year 1916.

G.1   Riemann curvature tensor

G.1.1   *Non-commutativity of covariant derivative*

In general $A_{\beta;\mu;\nu}$ is not equal to $A_{\beta;\nu;\mu}$, i.e. covariant derivatives are not commutative. This property of covariant derivative comes from the path dependence of parallel transport. Now let us calculate the difference between $A_{\beta;\mu;\nu}$ and $A_{\beta;\nu;\mu}$.

231

We know that

$$
\begin{aligned}
A_{\beta;\mu;\nu} &= A_{\beta;\mu,\nu} - \Gamma^{\sigma}_{\beta\nu}A_{\sigma;\mu} - \Gamma^{\sigma}_{\mu\nu}A_{\beta;\sigma} \\
&= (A_{\beta,\mu} - \Gamma^{\alpha}_{\beta\mu}A_{\alpha})_{,\nu} - \Gamma^{\sigma}_{\beta\nu}(A_{\sigma,\mu} - \Gamma^{\alpha}_{\sigma\mu}A_{\alpha}) \\
&\quad - \Gamma^{\sigma}_{\mu\nu}(A_{\beta,\sigma} - \Gamma^{\alpha}_{\beta\sigma}A_{\alpha}) \\
&= [A_{\beta,\mu,\nu} - \Gamma^{\alpha}_{\beta\mu}A_{\alpha,\nu} - \Gamma^{\sigma}_{\beta\nu}A_{\sigma,\mu} - \Gamma^{\sigma}_{\mu\nu}(A_{\beta,\sigma} - \Gamma^{\alpha}_{\beta\sigma}A_{\alpha})] \\
&\quad - \Gamma^{\alpha}_{\beta\mu,\nu}A_{\alpha} + \Gamma^{\sigma}_{\beta\nu}\Gamma^{\alpha}_{\sigma\mu}A_{\alpha}.
\end{aligned}
\tag{G.1}
$$

Exchanging μ and ν in Eq. (G.1), we readily obtain the expression of $A_{\beta;\nu;\mu}$. Since the quantities in the square bracket of Eq. (G.1) are symmetrical about μ and ν, they do not contribute to the difference. Hence we obtain

$$
A_{\beta;\mu;\nu} - A_{\beta;\nu;\mu} = (-\Gamma^{\alpha}_{\beta\mu,\nu} + \Gamma^{\alpha}_{\beta\nu,\mu} + \Gamma^{\sigma}_{\beta\nu}\Gamma^{\alpha}_{\sigma\mu} - \Gamma^{\sigma}_{\beta\mu}\Gamma^{\alpha}_{\sigma\nu})A_{\alpha}, \tag{G.2}
$$

which can be rewritten as

$$
A_{\beta;\mu;\nu} - A_{\beta;\nu;\mu} = R^{\alpha}_{\beta\mu\nu}A_{\alpha}, \tag{G.3}
$$

where

$$
R^{\alpha}_{\beta\mu\nu} = -\Gamma^{\alpha}_{\beta\mu,\nu} + \Gamma^{\alpha}_{\beta\nu,\mu} + \Gamma^{\sigma}_{\beta\nu}\Gamma^{\alpha}_{\sigma\mu} - \Gamma^{\sigma}_{\beta\mu}\Gamma^{\alpha}_{\sigma\nu}. \tag{G.4}
$$

G.1.2   *Curvature tensor and its properties*

The left-hand side of Eq. (G.3) is a tensor, the quantity on the right-hand side is an arbitrary covariant vector, hence the geometric quantity $R^{\alpha}_{\beta\mu\nu}$, which has $4 \times 4 \times 4 \times 4 = 256$ components, is a fourth-rank general tensor. This tensor is called the Riemann curvature tensor. Lowering the index α, we can obtain a totally covariant tensor:

$$
R_{\alpha\beta\mu\nu} = g_{\alpha\sigma}R^{\sigma}_{\beta\mu\nu}. \tag{G.5}
$$

The tensor $R_{\alpha\beta\mu\nu}$ possesses the following algebraic properties:

(1) anti-symmetric in the first two indices, i.e.

$$
R_{\alpha\beta\mu\nu} = -R_{\beta\alpha\mu\nu}; \tag{G.6}
$$

(2) anti-symmetric in the last two indices, i.e.

$$
R_{\alpha\beta\mu\nu} = -R_{\alpha\beta\nu\mu}; \tag{G.7}
$$

(3) symmetric in the exchange of the first two indices with the last two indices, i.e.

$$R_{\alpha\beta\mu\nu} = R_{\mu\nu\alpha\beta};$$ (G.8)

(4) cyclic in the last three indices, i.e.

$$R_{\alpha\beta\mu\nu} + R_{\alpha\mu\nu\beta} + R_{\alpha\nu\beta\mu} = 0.$$ (G.9)

With the above four properties, even though $R_{\alpha\beta\mu\nu}$ has 256 components, only 20 of them are independent. This is because from Eq. (G.6), we know that $\alpha\beta$ only have 6 independent components; similarly from Eq. (G.7) $\mu\nu$ only have 6 independent components. Furthermore, from Eq. (G.8), we know that among $6 \times 6 = 36$ components of $R_{\alpha\beta\mu\nu}$, only 21 components are independent. Finally with Eq. (G.9), we are only left with 20 independent components.

It can be proved that the curvature tensor is the only tensor which is composed of only $g_{\mu\nu}$, its first-rank derivative, and the linear term of its second-rank derivative.

Besides those algebraic properties, curvature tensor $R_{\alpha\beta\mu\nu}$ also satisfies an important differential identity:

$$R^{\alpha}{}_{\beta\mu\nu;\sigma} + R^{\alpha}{}_{\beta\nu\sigma;\mu} + R^{\alpha}{}_{\beta\sigma\mu;\nu} = 0,$$ (G.10)

which is the so-called Bianchi identity.

Contracting the first and the last indices of $R^{\alpha}{}_{\beta\mu\nu}$, we get a second-rank tensor $R_{\beta\mu}$:

$$R_{\beta\mu} \equiv R^{\alpha}{}_{\beta\mu\alpha}.$$ (G.11)

This is called the Ricci tensor, which is symmetric in indices β and μ, i.e.

$$R_{\beta\mu} = R_{\mu\beta}.$$ (G.12)

Contracting the above tensor again, we obtain a scalar:

$$R \equiv R^{\beta}{}_{\beta} = R^{\alpha\beta}{}_{\beta\alpha}.$$ (G.13)

The obtained scalar R is called the curvature scalar. From Bianchi identity (G.10) we can obtain the following identity:

$$\left(R^{\mu}{}_{\nu} - \frac{1}{2}\delta^{\mu}{}_{\nu} \right)_{;\mu} = 0.$$ (G.14)

G.1.3 *Curvature tensor and parallel transport*

To illustrate the inseparable relationship between the curvature tensor $R^\alpha{}_{\beta\mu\nu}$ and the curved property of spacetime, we use the method of parallel transport to move the vector a^α from P to P' through two different paths: one is from P to P_1 then from P_1 to P', another is from P to P_2 then from P_2 to P', as shown in Fig. G.1. In the following we shall prove that the two vectors obtained from two different paths are different and the difference between them is determined by the curvature tensor, thereby show that $R^\alpha{}_{\beta\mu\nu}$ is the measure of the curvature of the spacetime.

Moving the vector a^α from P to P_1 by means of parallel transport, we obtain the vector at P_1:

$$a^\alpha - \Gamma^\alpha{}_{\beta\mu}(P)a^\beta \mathrm{d}\xi^\mu.$$

Moving again this vector from P_1 to P' we obtain the vector at P':

$$[a^\alpha - \Gamma^\alpha{}_{\beta\mu}(P)a^\beta \mathrm{d}\xi^\mu] - \Gamma^\alpha{}_{\sigma\nu}(P_1)[a^\sigma - \Gamma^\sigma{}_{\beta\mu}(P)a^\beta \mathrm{d}\xi^\mu]\mathrm{d}s^\nu. \qquad \text{(G.15)}$$

Since

$$\Gamma^\alpha{}_{\sigma\nu}(P_1) \approx \Gamma^\alpha{}_{\sigma\nu}(P) + \Gamma^\alpha{}_{\sigma\nu,\mu}(P)\mathrm{d}\xi^\mu,$$

neglecting third power of the infinitesimal displacement we can write Eq. (G.15) as

$$a^\alpha - \Gamma^\alpha{}_{\beta\mu}a^\beta \mathrm{d}\xi^\mu - \Gamma^\alpha{}_{\sigma\nu}a^\sigma \mathrm{d}s^\nu + \Gamma^\alpha{}_{\sigma\nu}\Gamma^\sigma{}_{\beta\mu}a^\beta \mathrm{d}\xi^\mu \mathrm{d}s^\nu - \Gamma^\alpha{}_{\sigma\nu,\mu}a^\sigma \mathrm{d}\xi^\mu \mathrm{d}s^\nu.$$

$$\text{(G.16)}$$

Equation (G.16) is the obtained vector when we move the vector a^α along the path PP_1P' to P' by means of parallel transport. Interchanging $\mathrm{d}\xi$

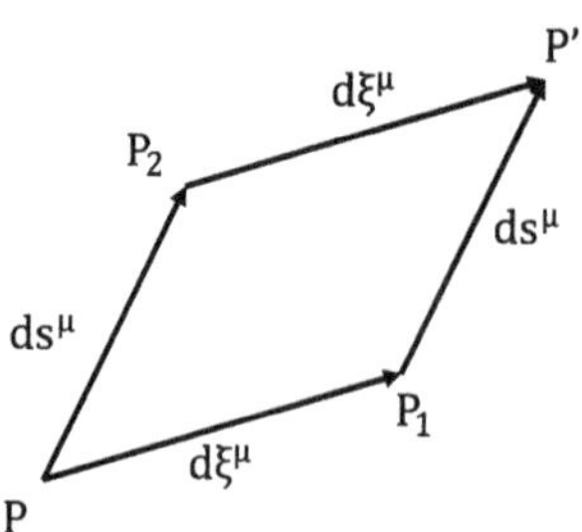

Fig. G.1 Parallel transport by two different paths.

and ds, we can readily obtain the vector when we move a^α along the path PP_2P' to P' by means of parallel transport:

$$a^\alpha - \Gamma^\alpha_{\beta\mu}a^\beta ds^\mu - \Gamma^\alpha_{\sigma\nu}a^\sigma d\xi^\nu + \Gamma^\alpha_{\sigma\nu}\Gamma^\sigma_{\beta\mu}a^\beta ds^\mu d\xi^\nu - \Gamma^\alpha_{\sigma\nu,\mu}a^\sigma ds^\mu d\xi^\nu.$$
(G.17)

Interchanging the two final dummy indices μ and ν in Eq. (G.17), then computing Eq. (G.17) $-$ Eq. (G.16) we obtain the difference between the two contravariant vectors obtained from the contravariant vector a^α along two different paths by means of parallel transport:

$$\Delta a^\alpha = \left(\Gamma^\alpha_{\beta\nu,\mu} - \Gamma^\alpha_{\beta\mu,\nu} + \Gamma^\alpha_{\sigma\mu}\Gamma^\sigma_{\beta\nu} - \Gamma^\alpha_{\sigma\nu}\Gamma^\sigma_{\beta\mu}\right)a^\beta d\xi^\mu ds^\nu$$

$$= R^\alpha_{\beta\mu\nu}a^\beta d\xi^\mu ds^\nu.$$
(G.18)

Using a similar method, we can prove that the difference between the two covariant vectors obtained from a covariant vector b_α along two different paths by means of parallel transport is given by

$$\Delta b_\alpha = R^\beta_{\alpha\mu\nu}b_\beta d\xi^\mu ds^\nu.$$
(G.19)

For flat space-time, $R^\alpha_{\beta\mu\nu} = 0$. On the contrary, we can also say if in the whole spacetime $R^\alpha_{\beta\mu\nu} = 0$, then the spacetime is flat. In a flat spacetime, even though in curvilinear coordinates $g_{\mu\nu} \neq \eta_{\mu\nu}$, but because $R^\alpha_{\beta\mu\nu} = 0$, we can always change the metric tensor $g_{\mu\nu}$ such that it equals to $\eta_{\mu\nu}$ everywhere through coordinate transformations. Therefore, whether the curvature tensor equals zero is the indicator of the existence or absence of gravitational field.

G.2 Einstein's equation of gravitational field

G.2.1 *Process of obtaining Einstein's equation of gravitational field*

By principle of equivalence, gravitational field is equivalent to metric field, hence the metric property should be determined by the distribution and motion of matter. By principle of general covariance, the equation which determines the dynamical quantities $g_{\mu\nu}$ must be a tensor equation. The tensor which determines the distribution and motion of matter is the energy-momentum tensor $T_{\mu\nu}$. Therefore, on the basis of the above considerations, the equation of gravitational field we want to find should have the following form:

$$G_{\mu\nu} = kT_{\mu\nu},$$

where $G_{\mu\nu}$ is the tensor determined by the geometrical properties of the spacetime, k is a proportional coefficient.

Then, how to find $G_{\mu\nu}$ and k? There are two fundamental principles which we based on, one is the existence of Newtonian limit, i.e. when the field is both weak and static, the above field equation must reduce to Poison's equation

$$\nabla^2 \phi = 4\pi G\rho,$$

where ϕ is the gravitational potential, G is the gravitational constant. Another principle is the conservation of energy and momentum. In the flat spacetime we already know that the energy-momentum tensor of the whole system obeys the following conservation law:

$$\partial_\nu T^{\mu\nu} = 0 \quad \text{or} \quad T^{\mu\nu}{}_{,\nu} = 0.$$

On the basis of principle of minimum coupling, in the curved spacetime the above equation should be replaced by the following equation:

$$T^{\mu\nu}{}_{;\nu} = 0 \tag{G.20}$$

or

$$T_\mu{}^\nu{}_{;\nu} = 0. \tag{G.21}$$

Equation (G.20) or (G.21) is the expression of conservation of energy and momentum in curved spacetime.

Since the left-hand side of Poisson's equation contains second-rank derivatives of the gravitational potential ϕ, and under Newton's approximation, it has been proved that $g_{00} = -(1 + 2\phi)$, hence we can simply recognize that $G_{\mu\nu}$ only contains second-rank derivatives of the metric tensor $g_{\mu\nu}$.

In order to assure the uniqueness of the obtained solution, we also require that $G_{\mu\nu}$ only contains linear term of the second-rank derivatives. We have already mentioned before that the curvature tensor $R_{\alpha\beta\mu\nu}$ is the only tensor which is composed of only $g_{\mu\nu}$, its first-rank derivative and the linear term of its second-rank derivative. Since $T_{\mu\nu}$ is a second-rank covariant tensor, $G_{\mu\nu}$ also must be a second-rank covariant tensor. Therefore the most ordinary second-rank covariant tensor satisfying the former requirements should be the linear superposition of the following three quantities: $R_{\mu\nu}$, $g_{\mu\nu}R$, and $g_{\mu\nu}$, i.e.

$$G_{\mu\nu} = aR_{\mu\nu} + bg_{\mu\nu}R + \Lambda g_{\mu\nu}. \tag{G.22}$$

Since $T_{\mu\nu}$ is symmetric, $G_{\mu\nu}$ also should be symmetric. Equation (G.22) satisfies this requirement.

From Eq. (G.21) we know

$$G_{\mu}^{\ \nu}{}_{;\nu} = 0.$$

Hence from Eq. (G.22) we obtain

$$(aR_{\mu}^{\ \nu} + b\delta_{\mu}^{\ \nu}R)_{;\nu} = 0.$$

Moreover, from identity (G.14) (which is called the reduced rank Bianchi identity), we know

$$\left(R_{\mu}^{\ \nu} - \frac{1}{2}\delta_{\mu}^{\ \nu}R\right)_{;\nu} = 0.$$

Hence we obtain

$$b = -\frac{1}{2}a.$$

Therefore Eq. (G.22) becomes

$$G_{\mu\nu} = a\left(R_{\mu\nu} - \frac{1}{2}g_{\mu\nu}R\right) + \Lambda g_{\mu\nu}.$$

Thence the gravitational field equation should be

$$a\left(R_{\mu\nu} - \frac{1}{2}g_{\mu\nu}R\right) + \Lambda g_{\mu\nu} = kT_{\mu\nu}.$$

Finally, based on the requirement that in the case of weak and static field, the above equation should become the Poisson's equation (to be discussed in Sec. G.2.3), we can obtain the following results:

$$\Lambda = 0, \quad \frac{k}{a} = -8\pi G.$$

Thus we obtain Einstein's field equations

$$R_{\mu\nu} - \frac{1}{2}g_{\mu\nu}R = -8\pi GT_{\mu\nu}, \tag{G.23}$$

or

$$G_{\mu\nu} = -8\pi GT_{\mu\nu}, \tag{G.24}$$

where

$$G_{\mu\nu} = R_{\mu\nu} - \frac{1}{2}g_{\mu\nu}R \qquad (G.25)$$

is the Einstein tensor.

G.2.2   *Discussions*

(1) The field equation is nonlinear for $g_{\mu\nu}$. Field equation (G.24) is very complex. Though the second-rank derivatives of metric field tensor linearly emerge in the equation, i.e., Eq. (G.24) is a system of second-rank quasi-linear partial differential equations, but the first-rank derivatives of $g_{\mu\nu}$ emerge in a quadratic manner. Moreover the dependence of Eq. (G.24) upon $g_{\mu\nu}$ is somehow even more complex since the left-hand side contains the inverse of $g_{\mu\nu}$, i.e. $g^{\mu\nu}$. The functional relationship between $g^{\mu\nu}$ and $g_{\mu\nu}$ is very complex. Since the gravitational field equation is nonlinear for $g_{\mu\nu}$, the principle of linear superposition is not applicable for gravitational field. In particular, it has self interaction. This is because gravitational field itself also has energy and momentum, thence it is also the source of gravitational field, just this peculiarity makes gravitational field equation to be a nonlinear equation.

(2) The energy-momentum tensor $T_{\mu\nu}$ includes only the energy and momentum of all the masses but not that of the gravitational field; the latter part is included in the left-hand side of the gravitational field equation.

(3) From the equation $T_{\mu}{}^{\nu}{}_{;\nu} = 0$ we can readily obtain the geodesic equation of the motion of the particle. This is the important work done in 1936 by Einstein, Infeld and Hoffman.

(4) Coordinate conditions. Since $G_{\mu\nu}$ is symmetric, Eq. (G.24) gives 10 equations. But since $G_{\mu}{}^{\nu}{}_{;\nu} = 0$ gives four identities, only six of these equations are independent. These are not sufficient to determine the 10 unknown components of the metric tensor $g_{\mu\nu}(x)$. Therefore, we must add four coordinate conditions to determine $g_{\mu\nu}(x)$. This is very natural since field equation can only determine the geometric properties of spacetime, it cannot point out the coordinate system we must choose. From the viewpoint of general relativity, all reference systems are equivalent to each other, the field equation cannot determine a special reference system chosen for describing the so-called geometric properties of the spacetime. Among the frequently used coordinate

conditions, there is the harmonic condition:

$$G^{\mu\nu}\Gamma^{\alpha}_{\mu\nu} = 0. \tag{G.26}$$

Equation (G.26) includes four differential equations. Also we can apply purely algebraic coordinate condition, e.g. time-orthogonal coordinates (Gaussian coordinates), for which

$$g_{00} = -1, \quad g_{0k} = 0. \tag{G.27}$$

The role of coordinate conditions in general relativity is similar to the role of gauge conditions in electrodynamics.

G.2.3 *Weak field linear approximation*

The condition for weak field is that the difference between $g_{\mu\nu}$ and $\eta_{\mu\nu}$ is very small, i.e.

$$g_{\mu\nu} = \eta_{\mu\nu} + h_{\mu\nu}, \quad |h_{\mu\nu}| \ll 1.$$

In the following we shall linearize Einstein's field equation (G.24) under the condition of weak field, thus finding the linearized field equation and obtaining the Poison's equation under the condition of static field, thereby finding the value of the coupling constant k, i.e. $k = -8\pi G$.

Einstein's field equation is

$$G_{\mu\nu} = kT_{\mu\nu}.$$

Hence

$$G^{\mu}_{\ \nu} = R^{\mu}_{\ \nu} - \frac{1}{2}\delta^{\mu}_{\ \nu}R = kT^{\mu}_{\ \nu}.$$

After contraction we obtain

$$R - \frac{1}{2} \times 4R = kT,$$

i.e.

$$R = -kT.$$

Therefore, we can rewrite Einstein's field equation $G_{\mu\nu} = kT_{\mu\nu}$ as follows:

$$R_{\mu\nu} = kT_{\mu\nu} + \frac{1}{2}g_{\mu\nu}R = k\left(T_{\mu\nu} - \frac{1}{2}g_{\mu\nu}T\right). \tag{G.28}$$

G.2.3.1 *Linearized Ricci tensor*

Firstly, under the weak field condition, we neglect second-order small quantity (we recognize that $h_{\mu\nu}$ and its various kinds of derivatives are all first-order small quantities), thence find the Ricci tensor.

Under such approximations, the affine connection is

$$\Gamma^{\alpha}_{\;\beta\mu} = \frac{1}{2}g^{\alpha\sigma}\left(\frac{\partial g_{\sigma\mu}}{\partial x^{\beta}} + \frac{\partial g_{\sigma\beta}}{\partial x^{\mu}} - \frac{\partial g_{\beta\mu}}{\partial x^{\sigma}}\right)$$

$$\approx \frac{1}{2}\eta^{\alpha\sigma}\left(\frac{\partial h_{\sigma\mu}}{\partial x^{\beta}} + \frac{\partial h_{\sigma\beta}}{\partial x^{\mu}} - \frac{\partial h_{\beta\mu}}{\partial x^{\sigma}}\right).$$

The curvature tensor is

$$R^{\alpha}_{\;\beta\mu\nu} \approx \frac{\partial}{\partial x^{\mu}}\Gamma^{\alpha}_{\;\beta\nu} - \frac{\partial}{\partial x^{\nu}}\Gamma^{\alpha}_{\;\beta\mu}$$

$$= \frac{1}{2}\left[\frac{\partial^{2}h^{\alpha}_{\;\nu}}{\partial x^{\mu}\partial x^{\beta}} - \frac{\partial^{2}h^{\alpha}_{\;\mu}}{\partial x^{\nu}\partial x^{\beta}} + \eta^{\alpha\sigma}\left(\frac{\partial^{2}h_{\beta\mu}}{\partial x^{\nu}\partial x^{\sigma}} - \frac{\partial^{2}h_{\beta\nu}}{\partial x^{\mu}\partial x^{\sigma}}\right)\right].$$

Thence we obtain the Ricci tensor as follows:

$$R_{\beta\mu} \approx \frac{1}{2}\left(\frac{\partial^{2}h}{\partial x^{\mu}\partial x^{\beta}} - \frac{\partial^{2}h^{\alpha}_{\;\mu}}{\partial x^{\alpha}\partial x^{\beta}} + \eta^{\alpha\sigma}\frac{\partial^{2}h_{\beta\mu}}{\partial x^{\alpha}\partial x^{\sigma}} - \frac{\partial^{2}h^{\sigma}_{\;\beta}}{\partial x^{\mu}\partial x^{\sigma}}\right). \tag{G.29}$$

Since we neglect second-order small quantities, we use the Minkowski metric tensor to raise or lower the indices,

$$\eta^{\alpha\sigma}h_{\alpha\beta} = h^{\sigma}_{\;\beta}, \quad \eta^{\alpha\beta}h_{\alpha\beta} = h^{\beta}_{\;\beta} = h.$$

G.2.3.2 *Coordinate condition*

In this case the harmonic condition (G.26) reduces to

$$0 = G^{\mu\nu}\Gamma^{\alpha}_{\;\mu\nu} \approx \frac{1}{2}\eta^{\mu\nu}\eta^{\alpha\sigma}\left(\frac{\partial h_{\sigma\mu}}{\partial x^{\nu}} + \frac{\partial h_{\sigma\nu}}{\partial x^{\mu}} - \frac{\partial h_{\mu\nu}}{\partial x^{\sigma}}\right)$$

$$= \frac{1}{2}\eta^{\mu\nu}\left(\frac{\partial h^{\alpha}_{\;\mu}}{\partial x^{\nu}} + \frac{\partial h^{\alpha}_{\;\nu}}{\partial x^{\mu}} - \eta^{\alpha\sigma}\frac{\partial h_{\mu\nu}}{\partial x^{\sigma}}\right)$$

$$= \frac{1}{2}\left(\frac{\partial h^{\alpha\nu}}{\partial x^{\nu}} + \frac{\partial h^{\alpha\mu}}{\partial x^{\mu}} - \eta^{\alpha\sigma}\frac{\partial h}{\partial x^{\sigma}}\right).$$

Hence

$$\eta^{\alpha\sigma}\frac{\partial h}{\partial x^{\sigma}} = 2\frac{\partial h^{\alpha\sigma}}{\partial x^{\sigma}}$$

$$\eta_{\tau\alpha}\eta^{\alpha\sigma}\frac{\partial h}{\partial x^{\sigma}} = 2\eta_{\tau\alpha}\frac{\partial h^{\alpha\sigma}}{\partial x^{\sigma}}$$

$$\frac{\partial h}{\partial x^{\tau}} = 2\frac{\partial h^{\sigma}_{\ \tau}}{\partial x^{\sigma}},$$

i.e.

$$\frac{\partial h^{\mu}_{\ \nu}}{\partial x^{\mu}} = \frac{1}{2}\frac{\partial h}{\partial x^{\nu}}. \tag{G.30}$$

Using coordinate condition (G.30) we can further simplify Eq. (G.29). From Eq. (G.30) we obtain

$$\frac{\partial^2 h^{\alpha}_{\ \mu}}{\partial x^{\beta}\partial x^{\alpha}} = \frac{\partial}{\partial x^{\beta}}\left(\frac{1}{2}\frac{\partial h}{\partial x^{\mu}}\right) = \frac{1}{2}\frac{\partial^2 h}{\partial x^{\mu}\partial x^{\beta}}, \tag{G.31}$$

$$\frac{\partial^2 h^{\sigma}_{\ \beta}}{\partial x^{\mu}\partial x^{\sigma}} = \frac{\partial}{\partial x^{\mu}}\left(\frac{1}{2}\frac{\partial h}{\partial x^{\beta}}\right) = \frac{1}{2}\frac{\partial^2 h}{\partial x^{\mu}\partial x^{\beta}}. \tag{G.32}$$

Substituting Eqs. (G.31) and (G.32) into Eq. (G.29), we obtain

$$R_{\beta\mu} = \frac{1}{2}\eta^{\alpha\sigma}\frac{\partial^2 h_{\beta\mu}}{\partial x^{\alpha}\partial x^{\sigma}} = -\frac{1}{2}\Box^2 h_{\beta\mu}. \tag{G.33}$$

This is the linearized Ricci tensor satisfying the coordinate condition (G.30).

G.2.3.3 *Energy-momentum tensor*

In the text we have already mentioned that the energy-momentum tensor of a system of non-interacting particles is

$$T^{\mu\nu} = \rho_0 u^{\mu}u^{\nu}.$$

For porous material, the above formula approximately holds, thus we have

$$T = g_{\mu\nu}T^{\mu\nu} = \rho_0(g_{\mu\nu}u^{\mu}u^{\nu}) = \rho_0,$$

and

$$T_{\mu\nu} = g_{\mu\alpha}g_{\nu\beta}T^{\alpha\beta}$$

$$= \rho_0 g_{\mu\alpha}g_{\nu\beta}u^\alpha u^\beta$$

$$= \rho_0 g_{\mu\alpha}g_{\nu\beta}\frac{\mathrm{d}x^\alpha}{\mathrm{d}s}\frac{\mathrm{d}x^\beta}{\mathrm{d}s}.$$

Under the condition of slow motion, we have

$$\frac{\mathrm{d}x^1}{\mathrm{d}s} \ll 1, \quad \frac{\mathrm{d}x^2}{\mathrm{d}s} \ll 1, \quad \frac{\mathrm{d}x^3}{\mathrm{d}s} \ll 1.$$

Therefore

$$T_{\mu\nu} = \rho_0 g_{\mu 0}g_{\nu 0}\frac{\mathrm{d}x^0}{\mathrm{d}s}\frac{\mathrm{d}x^0}{\mathrm{d}s} = \rho_0 \frac{g_{\mu 0}g_{\nu 0}}{g_{00}}.$$

Under the condition of weak field, we have $T_{00} = \rho_0$, all other components of $T_{\mu\nu}$ equal zero.

G.2.3.4　*Field equation*

Coordinate condition (G.30) can be written as

$$\psi^\mu{}_{\nu;\mu} = 0, \tag{G.34}$$

where

$$\psi^\mu{}_\nu = h^\mu{}_\nu - \frac{1}{2}\delta^\mu{}_\nu h, \tag{G.35}$$

or equivalently

$$\psi^{\mu\nu}{}_{,\mu} = 0, \tag{G.36}$$

where

$$\psi^{\mu\nu} = h^{\mu\nu} - \frac{1}{2}\eta^{\mu\nu}h. \tag{G.37}$$

When we apply coordinate condition (G.36), under the condition of weak field and linear approximation, the Einstein's field equation

$$G_{\mu\nu} \equiv R_{\mu\nu} - \frac{1}{2}g_{\mu\nu}R = kT_{\mu\nu}$$

becomes

$$\partial_\lambda\partial^\lambda\left(h^{\mu\nu} - \frac{1}{2}\eta^{\mu\nu}h\right) = 2kT^{\mu\nu},$$

i.e.

$$\Box^2 \psi^{\mu\nu} = 2kT^{\mu\nu}. \tag{G.38}$$

Under the conditions of weak field and linear approximation, the coordinate condition (G.36) and field equation (G.38) are the components of the fundamental system of differential equations for the gravitational field.

G.2.3.5 *Newton's approximation*

We have mentioned before that for slow-motion particle moving in weak and static gravitational field, we have

$$h_{00} = -2\phi. \tag{G.39}$$

In Eq. (G.28), considering the component $\mu = \nu = 0$, for the static case we have

$$R_{00} = -\frac{1}{2}\Box^2 h_{00} = \frac{1}{2}\nabla^2 h_{00}$$

$$= k\left(T_{00} - \frac{1}{2}\eta_{00}T\right) = \frac{1}{2}k\rho_0.$$

Substituting Eq. (G.39) into the above formula we readily obtain

$$\nabla^2\phi = -\frac{1}{2}k\rho_0.$$

Comparing the above formula with Poison's equation

$$\nabla^2\phi = 4\pi G\rho,$$

we readily obtain

$$k = -8\pi G.$$

This ensures that in weak and static field Einstein's field equation is exactly Poison's equation.

G.2.3.6 *External spherically symmetric static mass*

In the external part of the spherically symmetric mass, we have $T^{\mu\nu} = 0$. Hence field equation (G.38) reduces to

$$\nabla^2\psi^{\mu\nu} = 0. \tag{G.40}$$

The spherically symmetric solution of the above equation is

$$\psi^{\mu\nu} = \frac{C_{\mu\nu}}{r},$$
(G.41)

where $C_{\mu\nu}$ is a constant matrix.

Substituting Eq. (G.41) into coordinate condition (G.36), we obtain

$$C^{i\nu}\partial_i\left(\frac{1}{r}\right) = C^{i\nu}\frac{-x^i}{r^3} = 0,$$

where Latin index $i = 1, 2, 3$. If we want the above formula to hold for all x^i, then we must have $C^{i\nu} = 0$, hence the only nonzero component of $C^{\mu\nu}$ is C^{00}. Since from Eq. (G.37) , we have $\psi = h - \frac{1}{2} \times 4h = -h$, we obtain

$$\psi^{\mu\nu} = h^{\mu\nu} + \frac{1}{2}\eta^{\mu\nu}\psi,$$

i.e.

$$h^{\mu\nu} = \psi^{\mu\nu} - \frac{1}{2}\eta^{\mu\nu}\psi,$$

or

$$h_{\mu\nu} = \psi_{\mu\nu} - \frac{1}{2}\eta_{\mu\nu}\psi.$$
(G.42)

From Eq. (G.42) we obtain

$$h_{00} = \psi_{00} - \frac{1}{2}\eta_{00}\psi = \frac{C_{00}}{\gamma} - \frac{1}{2}\frac{C_{00}}{\gamma} = \frac{C_{00}}{2\gamma}.$$

Since

$$h_{00} = -2\phi = \frac{2GM}{\gamma} = \frac{C_{00}}{2\gamma},$$

we have

$$C_{00} = 4GM.$$

Again from Eq. (G.42) we obtain

$$h_{11} = h_{22} = h_{33} = \frac{1}{2}\psi = \frac{1}{2}\psi_{00} = \frac{C_{00}}{2\gamma} = \frac{2GM}{\gamma}.$$

Therefore, under weak field approximation, the matrix $h_{\mu\nu}$ around a spherical symmetric static mass is

$$h_{\mu\nu} = \begin{pmatrix} \frac{2GM}{\gamma} & 0 & 0 & 0 \\ 0 & \frac{2GM}{\gamma} & 0 & 0 \\ 0 & 0 & \frac{2GM}{\gamma} & 0 \\ 0 & 0 & 0 & \frac{2GM}{\gamma} \end{pmatrix}.$$

The corresponding four-dimensional line element expression is

$$\mathrm{d}s^2 = -\left(1 - \frac{2GM}{\gamma}\right)\mathrm{d}t^2 + \left(1 + \frac{2GM}{\gamma}\right)(\mathrm{d}x^2 + \mathrm{d}y^2 + \mathrm{d}z^2). \qquad \text{(G.43)}$$

G.3 Schwarzschild solution and its consequences

Since the birth of general relativity in 1916, for the first half-century of its life, we can say that it is the paradise of theorists, but on the other side we can also say it is the hell of experimentalists. No theory is thought to be more beautiful than general relativity, but also no theory is thought to be more difficult to test than it. Such situations have changed a lot since the second half-century of the life of GR. We had the following four famous experimental verification:

(1) gravitational redshift of spectral lines;
(2) deflection of light ray propagating nearby the sun;
(3) precession of perihelion of planets;
(4) retardation of radio return waves.

Very recently, in the year 2016, the successful detection of gravitational waves was a very exciting event for whole world.

All these observations and experiments are conducted in the space without matter, i.e. they only test the Einstein vacuum equation. More precisely, they only concern spherically symmetric static field. The precise solution of spherically symmetric vacuum field of static mass was given by Schwarzschild in 1916, usually known as the Schwarzschild solution. For the cases (1), (2) and (4), actually we only need to use weak-field linear approximation. Only for case (3) we need to use second-order approximation, i.e. we need to use the Schwarzschild solution. Hence, for a strict statement, we can at most say that the so-called four experimental tests of general relativity confirm the Schwarzschild solution, not general relativity itself.

G.3.1 *Schwarzschild solution*

G.3.1.1 *General expression of the metric of spherically symmetric static mass*

The characteristics of static field are the independence upon time and the existence of temporal symmetry, i.e. the symmetry about time inversion. Since there is time-inversion symmetry for static field, in the expression for the square of four-dimensional line element there are no such cross terms as $\mathrm{d}x\mathrm{d}t$, $\mathrm{d}y\mathrm{d}t$, $\mathrm{d}z\mathrm{d}t$, there exists only $\mathrm{d}t^2$ and cross terms consist of any two of $\mathrm{d}x$, $\mathrm{d}y$ or $\mathrm{d}z$.

The characteristics of spherical symmetry is the form invariance of the square of four-dimensional line element, i.e. in the expression of $\mathrm{d}s^2$, x, y, z, $\mathrm{d}x$, $\mathrm{d}y$, $\mathrm{d}z$ can only appear in the form of $\sqrt{x^2 + y^2 + z^2}$, $\mathrm{d}x^2 + \mathrm{d}y^2 + \mathrm{d}z^2$ and $x\mathrm{d}x + y\mathrm{d}y + z\mathrm{d}z$. If we use spherical coordinates, then they can only appear in the form of r, $\mathrm{d}r^2 + r^2\mathrm{d}\theta^2 + r^2\sin^2\theta\mathrm{d}\phi^2$ and $r\mathrm{d}r$, i.e.

$$\mathrm{d}s^2 = -A(r)\mathrm{d}t^2 + B(r)(\mathrm{d}r^2 + r^2\mathrm{d}\theta^2 + r^2\sin^2\theta\mathrm{d}\phi^2) + C(r)\mathrm{d}r^2. \quad (\text{G.44})$$

We can also introduce a new radial coordinate r' where $r'^2 = r\sqrt{B(r)}$, then Eq. (G.44) can be simplified as

$$\mathrm{d}s^2 = -A'(r')\mathrm{d}t^2 + B'(r')\mathrm{d}r'^2 + r'^2\mathrm{d}\theta^2 + r'^2\sin^2\theta\mathrm{d}\phi^2. \quad (\text{G.45})$$

For convenience, we can write r' as r and write $A'(r')$ and $B'(r')$ in the forms of exponential function, finally we obtain

$$\mathrm{d}s^2 = -e^{N(r)}\mathrm{d}t^2 + e^{L(r)}\mathrm{d}r^2 + r^2\mathrm{d}\theta^2 + r^2\sin^2\theta\mathrm{d}\phi^2. \quad (\text{G.46})$$

Equation (G.46) is the general expression of the square of four-dimensional line element for a spherically symmetric static mass.

If we let $x^0 \equiv t$, $x^1 \equiv r$, $x^2 \equiv \theta$, $x^3 \equiv \phi$, then the corresponding metric tensor is

$$g_{\mu\nu} = \begin{pmatrix} -e^N & 0 & 0 & 0 \\ 0 & e^L & 0 & 0 \\ 0 & 0 & r^2 & 0 \\ 0 & 0 & 0 & r^2\sin^2\theta \end{pmatrix}. \quad (\text{G.47})$$

G.3.1.2   *Christoffel symbol*

In this case all nonzero Christoffel symbols are

$$\Gamma^0{}_{01} = \Gamma^0{}_{10} = \frac{1}{2}N', \quad \Gamma^1{}_{00} = \frac{1}{2}N'e^{N-L},$$

$$\Gamma^1{}_{11} = \frac{1}{2}L', \quad \Gamma^1{}_{22} = -re^{-L}, \quad \Gamma^1{}_{33} = -r\sin^2\theta\, e^{-L},$$

$$\Gamma^1{}_{12} = \Gamma^1{}_{21} = \frac{1}{r},$$

$$\Gamma^2{}_{33} = -\sin\theta\cos\theta,$$

$$\Gamma^3{}_{13} = \Gamma^3{}_{31} = \frac{1}{r}, \quad \Gamma^3{}_{23} = \Gamma^3{}_{32} = \cot\theta, \tag{G.48}$$

where $N' = \frac{\mathrm{d}N}{\mathrm{d}r}$, $L' = \frac{\mathrm{d}L}{\mathrm{d}r}$.

G.3.1.3   *Einstein tensor*

Einstein's vacuum field equations are

$$G_\mu{}^\nu = 0.$$

From Eq. (G.48) we find that only diagonal components of $G_\mu{}^\nu$ are nonzero. They are

$$G_0{}^0 = -e^{-L}\left(\frac{L'}{r} - \frac{1}{r^2}\right)\frac{1}{r^2},$$

$$G_1{}^1 = -e^{-L}\left(\frac{N'}{r} + \frac{1}{r^2}\right)\frac{1}{r^2},$$

$$G_2{}^2 = G_3{}^3 = -e^{-L}\left(\frac{N''}{2} - \frac{L'N'}{4} + \frac{N'^2}{4} + \frac{N'-L'}{2}\right). \tag{G.49}$$

G.3.1.4   *Solution of the vacuum field equations*

From vacuum field equation $G_0{}^0 = 0$, we obtain

$$e^{-L}(-rL' + 1) = -1.$$

Let $e^L = x$,

$$\frac{\mathrm{d}x}{x - x^2} = \frac{\mathrm{d}r}{r}$$

$$\frac{\mathrm{d}x}{x} + \frac{\mathrm{d}x}{1 - x} = \frac{\mathrm{d}r}{r}.$$

Thus the solution is

$$e^L = -\frac{1}{1 - c/r}.$$ (G.50)

From vacuum field equation $G_1{}^1 = 0$ we minus $G_0{}^0 = 0$, we obtain $G_1{}^1 - G_0{}^0 = 0$, then from this equation we get

$$L' = -N',$$ (G.51)

hence,

$$L = -N + \text{const.}$$

Considering that when $r \to \infty$, $g_{\mu\nu} \to \eta_{\mu\nu}$, i.e. $L \to 0$, $N \to 0$. Thence we find that const. $= 0$, i.e.

$$L = -N.$$ (G.52)

Hence we get

$$e^{N(r)} = -\left(1 - \frac{c}{r}\right).$$ (G.53)

From Eqs. (G.50) and (G.53) we get

$$ds^2 = -\left(1 - \frac{c}{r}\right) dt^2 + \frac{dr^2}{1 - c/r} + r^2 d\theta^2 + r^2 \sin^2\theta d\phi^2.$$ (G.54)

Since under weak field approximation

$$g_{00} = -\left(1 - \frac{c}{r}\right) = -(1 + 2\phi) = -\left(1 - \frac{2GM}{r}\right),$$

hence

$$c = 2GM.$$

Finally we get

$$ds^2 = -\left(1 - \frac{2GM}{r}\right) dt^2 + \frac{dr^2}{1 - 2GM/r} + r^2 d\theta^2 + r^2 \sin^2\theta d\phi^2.$$ (G.55)

Equation (G.55) is the external solution of a spherically symmetric static mass, i.e. the so-called Schwarzschild solution.

G.3.1.5 *Birkhoff theorem*

In the previous section, under the static field and spherical symmetry these two conditions, we obtain the vacuum field solution, i.e. the so-called

Schwarzschild solution. Birkhoff at 1923 proved that the static field condition is superfluous. Any vacuum field solution of spherically symmetric mass is always the Schwarzschild solution. This conclusion is known as the Birkhoff theorem.

Theorem G.1 (Birkhoff theorem). *Any spherically symmetric vacuum solution of Einstein's equation must be static and must agree with the Schwarzschild solution.*

We only need to take N and L as a function of r and t, then we can prove this conclusion. As to the term $\mathrm{d}r\mathrm{d}t$, we can eliminate this term by coordinate transformation. When we eliminate the term $\mathrm{d}r\mathrm{d}t$, the differences of the general expressions of the metric tensor from Eq. (G.47) is because in this case N and L are functions of r and t, not functions of only r. From the Birkhoff theorem we draw the following conclusion: for a pulsing mass distribution along radial direction, since its metric tensor is static, it cannot produce gravitational wave.

G.3.2 *Motion of planets: Perihelion precession of Mercury*

Now we shall study the motion of particle in the Schwarzschild field. This problem directly corresponds to the motion of planets in the gravitational field of the Sun. From 19th century astronomers already knew that there is discrepancy in the precession of perihelion of Mercury between observational values and theoretically calculated values from classical mechanics. The observational value is $5599''.74$ per century, but among that $43''$ cannot be explained by theory. Though this discrepancy is very small, but it is already hundreds of times the observational errors. Therefore it is truly a discrepancy which is unable to be explained by theory. The first attempt to explain this discrepancy is by assuming a new planet, Vulcan, inside the orbit of Mercury. There were numerous theorists who tried to find the position of this new planet, but all their efforts failed. The hypothetical planet was abandoned in 1915 when Einstein successfully explained the observed effect with general relativity.

G.3.2.1 *Differential equation of the orbit of planet*

Equation of motion of a mass particle in the gravitational field is

$$\frac{\mathrm{d}^2 x^\mu}{\mathrm{d}s^2} + \Gamma^\mu{}_{\nu\lambda} \frac{\mathrm{d}x^\nu}{\mathrm{d}s} \frac{\mathrm{d}x^\lambda}{\mathrm{d}s} = 0.$$

In the particular case of Shwarzschild field, equation of motion is

$$\ddot{r}^2 = \frac{1}{2}N'e^{N-L}\dot{t}^2 + \frac{1}{2}L'\dot{r}^2 - re^{-L}\dot{\theta} - r^2\sin^2\theta e^{-L}\dot{\phi}^2 = 0, \qquad \text{(G.56)}$$

$$\ddot{\theta} + \frac{2}{r}\dot{r}\dot{\theta} - \cos\theta\sin\theta\dot{\phi}^2 = 0, \qquad \text{(G.57)}$$

$$\ddot{\phi} + \frac{2}{r}\dot{r}\dot{\theta} + 2\cot\theta\dot{\phi}\dot{\theta} = 0, \qquad \text{(G.58)}$$

where dot represents differentiation with respect to s.

Assume that the orbit of planet is in the plane $\theta = \frac{\pi}{2}$. Equation (G.57) illustrates that, if at the initial moment $\theta = \frac{\pi}{2}$, $\dot{\theta} = 0$, then $\ddot{\theta} = 0$, i.e. the orbit will stay on the plane $\theta = \frac{\pi}{2}$. Directly integrating Eqs. (G.57) and (G.58), we get

$$\dot{\phi} = \frac{A}{r^2}, \qquad \text{(G.59)}$$

$$\dot{t} = Be^{-N}, \qquad \text{(G.60)}$$

where A and B are constants. Since $g_{\mu\nu}\frac{\mathrm{d}x^\nu}{\mathrm{d}s}\frac{\mathrm{d}x^\nu}{\mathrm{d}s} = 1$, we get

$$-e^N\dot{t}^2 + e^L\dot{r}^2 + r^2\dot{\theta}^2 + r^2\sin^2\theta\dot{\phi}^2 = 1. \qquad \text{(G.61)}$$

Substituting Eqs. (G.59) and (G.60) into Eq. (G.61), considering that $\theta = \frac{\pi}{2}$, $e^{-L} = e^N = 1 - \frac{2GM}{r}$, we obtain

$$-\frac{B^2}{1 - 2GM/r} + \frac{\dot{r}^2}{1 - 2GM/r} + \frac{A^2}{r^2} = 1. \qquad \text{(G.62)}$$

If we introduce $u = \frac{1}{r}$, then

$$\dot{r} = \frac{\mathrm{d}r}{\mathrm{d}\phi}\dot{\phi} = -\frac{1}{u^2}\frac{\mathrm{d}u}{\mathrm{d}\phi}\frac{A}{r^2} = -\frac{\mathrm{d}u}{\mathrm{d}\phi}A,$$

and Eq. (G.62) becomes

$$-B^2 + A^2\left(\frac{\mathrm{d}u}{\mathrm{d}\phi}\right)^2 + A^2u^2(1 - 2GMu) = 1 - 2GMu. \qquad \text{(G.63)}$$

Differentiating with respect to ϕ, we readily obtain the second-order differential equation of the orbit:

$$\frac{\mathrm{d}^2u}{\mathrm{d}\phi^2} + u - \frac{GM}{A^2} - 3GMu^2 = 0. \qquad \text{(G.64)}$$

G.3.2.2 *Precession of perihelion of Mercury*

The difference between Eq. (G.64) and the Binet formula in classical mechanics is only the additional term $3GMu^2$. Just the existence of this term caused the precession of perihelion of Mercury.

The term $3GMu^2$ in Eq. (G.64) is much less than GMA^2,

$$\frac{3GMu^2}{\frac{GM}{A^2}} = 3A^2u^2 = 3(r^2\dot\phi^2)u^2 \approx 3\left(r\frac{\mathrm{d}\phi}{\mathrm{d}t}\right)^2.$$

Since $r\frac{\mathrm{d}\phi}{\mathrm{d}t}$ is the transverse velocity of the planet, it is very much less than the light speed, i.e. very much less than 1, $r\frac{\mathrm{d}\phi}{\mathrm{d}t} \ll 1$. In the case of Mercury,

$$3\left(r\frac{\mathrm{d}\phi}{\mathrm{d}t}\right)^2 = 7.7 \times 10^{-8}.$$

Therefore, the last term of Eq. (G.64) is only a perturbation, hence we can use the method of perturbation theory to solve it. Neglecting the last term, we obtain the solution

$$u = \frac{GM}{A^2}[1 + \epsilon\cos(\phi - \phi_0)]. \tag{G.65}$$

This is the equation of ellipse with eccentricity ϵ, and the position of its perihelion is at $\phi = \phi_0$. Substituting Eq. (G.65) into the last term of Eq. (G.64), we obtain

$$\frac{\mathrm{d}^2u}{\mathrm{d}\phi^2} + u - \frac{GM}{A^2} - \frac{3(GM)^3}{A^4} - \frac{6\epsilon(GM)^3}{A^4}\cos(\phi - \phi_0)$$

$$-3\epsilon^2\frac{(GM)^3}{A^4}\cos^2(\phi - \phi_0) = 0. \tag{G.66}$$

For orbit nearing circle, ϵ is very small, hence we can neglect the term containing ϵ^2. The constant term $(GM)^3/A^4$ can also be neglected in comparison with GM/A^2. Hence Eq. (G.66) becomes

$$\frac{\mathrm{d}^2u}{\mathrm{d}\phi^2} + u - \frac{GM}{A^2} - \frac{6\epsilon(GM)^3}{A^4}\cos(\phi - \phi_0) = 0. \tag{G.67}$$

The solution is

$$u = \frac{GM}{A^2}[1 + \epsilon\cos(\phi - \phi_0)] + \frac{3\epsilon(GM)^3}{A^4}\phi\sin(\phi - \phi_0). \tag{G.68}$$

$(GM)^2\phi/A^2$ is small, Eq. (G.68) becomes

$$u \approx \frac{GM}{A^2}\left[1 + \epsilon \cos\left(\phi - \phi_0 - \frac{3(GM)^2}{A^2}\phi\right)\right]. \tag{G.69}$$

Equation (G.69) represents a precessing elliptic orbit. The argument of the cosine function changes by 2π when ϕ changes by

$$\Delta\phi = 2\pi\left[1 - \frac{3(GM)^2}{A^2}\right]^{-1} \approx 2\pi\left[1 + \frac{3(GM)^2}{A^2}\right]. \tag{G.70}$$

From Eq. (G.70) we can see that the angular distance between one perihelion and the next is larger than 2π by $6\pi(GM)^2/A^2$. For orbit nearing circle, from Eq. (G.65) we obtain

$$\frac{GM}{A^2} \approx \frac{1}{r}. \tag{G.71}$$

Hence the angular distance between one perihelion and the next is larger than 2π by (when we use CGS unit system)

$$6\pi\frac{GM}{rc^2}. \tag{G.72}$$

If we calculate in detail, then in Eq. (G.72) we should replace $\frac{1}{r}$ by $\frac{1}{r_{max}} + \frac{1}{r_{min}}$. For Mercury the calculated value of is $43.03''$ per century, which coincides very well with the observed value.

G.3.3 *Deflection of light ray in the Schwarzschild field*

The zero geodesics of light ray is $ds = 0$. Previously we have mentioned that the differential equation of the orbit of a massive particle in the Schwarzschild field is

$$\frac{d^2u}{d\phi^2} + u - \frac{GM}{A^2} - 3GMu^2 = 0,$$

i.e.

$$\frac{d^2u}{d\phi^2} + u - \frac{GM}{r^4}\left(\frac{ds}{d\phi}\right)^2 - 3GMu^2 = 0.$$

Because the deflection of light ray $d\phi$ in the Schwarzschild field does not equal to zero, but for light ray $ds = 0$, the differential equation of the orbit of light ray is

$$\frac{\mathrm{d}^2 u}{\mathrm{d}\phi^2} + u - 3GMu^2 = 0. \tag{G.73}$$

Since $3GMu^2 \ll u$, we can use the method of perturbation theory to calculate the deflection of light ray. Firstly we find the solution of the following differential equation:

$$\frac{\mathrm{d}^2 u}{\mathrm{d}\phi^2} + u = 0.$$

The solution is

$$u = \frac{1}{r} = \frac{\cos\phi}{R}. \tag{G.74}$$

Equation (G.74) is the equation of a straight line, where R is the perpendicular distance of this straight line from the Sun. Substituting Eq. (G.74) into (G.73) we get

$$\frac{\mathrm{d}^2 u}{\mathrm{d}\phi^2} + u = 3GM\frac{\cos^2\phi}{R^2}. \tag{G.75}$$

This equation has a particular solution

$$u_1 = \frac{GM}{R^2}(\cos^2\phi + 2\sin^2\phi), \tag{G.76}$$

hence the solution of Eq. (G.73) is

$$\frac{1}{r} = u = \frac{\cos\phi}{R} + \frac{GM}{R^2}(\cos^2\phi + 2\sin^2\phi). \tag{G.77}$$

This can be rewritten as

$$R = r\cos\phi + \frac{GM}{Rr}(r^2\cos^2\phi + 2r^2\sin^2\phi)$$

$$= x + \frac{GM}{R}\frac{x^2 + 2y^2}{\sqrt{x^2 + y^2}}. \tag{G.78}$$

In Eq. (G.78), when we let $y \gg x$, we readily get the equations of two approaching lines as follows:

$$x = R - \frac{2GM}{R}(\pm y). \tag{G.79}$$

Hence we obtain the deflection angle of the light ray Δ as

$$\Delta = \frac{4GM}{R}.$$

When we use CGS unit system, the deflection angle of light ray is

$$\Delta = \frac{4GM}{Rc^2}.$$
(G.80)

For star light passing through periphery of the Sun, we use the following data:

$$M = M_\odot = 1.97 \times 10^{33} \text{ g},$$

$$R = R_\odot = 6.95 \times 10^{10} \text{ cm},$$

$$G = 6.67 \times 10^{-8} \text{ dyne cm}^2\text{g}^2,$$

$$c = 3.00 \times 10^{10} \text{ cm}.$$

The result is

$$\Delta = 1.75'',$$

which coincides very well with the observed value obtained by Eddington, Dyson *et al.* in 1919.

G.3.4 *Time delay of the return wave of radar signal*

Since the gravitational field reduces the speed of propagation of light signals, there is a time delay suffered by a radar signal sent from Earth to a planet and reflected back to Earth. In the flat spacetime, as shown in Fig. G.2, the time needed for the radar signal sent from Earth, E, to a planet, P, should be $t = 2(a + b)/c$.

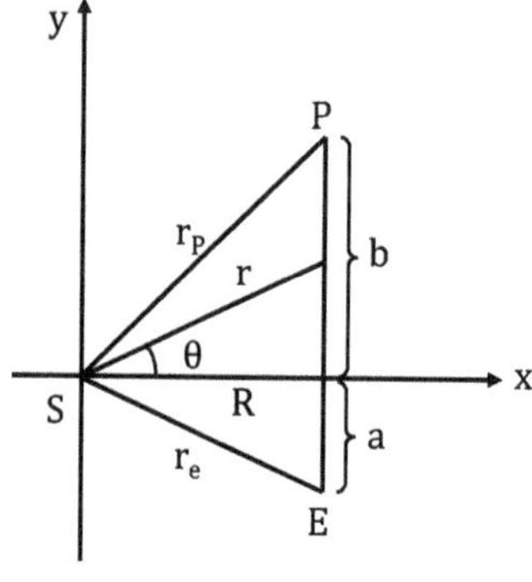

Fig. G.2 Radar signal sent from Earth to a planet and reflected back to Earth.

Since there is gravitational field of the Sun, the spacetime is curved, hence there should be time delay for the return wave of radar signal. In the Schwarzschild solution if we let $ds = 0$, we readily get the following equation of light (since light is in the (x, y) plane, $d\phi = 0$):

$$ds^2 = -\left(1 - \frac{2GM}{r}\right)dt^2 + \frac{dr^2}{1 - 2GM/r} + r^2 d\theta^2 = 0. \qquad \text{(G.81)}$$

From Fig. G.2 we get

$$r = \frac{R}{\cos\theta}, \quad dr = \frac{R\sin\theta d\theta}{\cos^2\theta}. \qquad \text{(G.82)}$$

Substituting into Eq. (G.81) and eliminating r and dr, we obtain

$$dt = \frac{Rd\theta}{c}\left(\frac{1}{\cos^2\theta} + \frac{GM}{R}\frac{1 + \sin^2\theta}{\cos\theta}\right). \qquad \text{(G.83)}$$

Integrating Eq. (G.83) and multiplying it by 2, we readily obtain the time needed for one round trip of the radar signal as

$$t = \frac{2(a+b)}{c} + GM\left[2\ln\frac{(r_p + b)(r_e + a)}{R^2} - \left(\frac{b}{r_p} + \frac{a}{r_e}\right)\right]. \qquad \text{(G.84)}$$

The second term in Eq. (G.84) is just the time delay of the return wave of radar signal. In 1969–1970 Shapiro *et al.* successfully completed this experiment. The results agreed very well with the above theoretical results, hence this experimental devise is often referred to as the fourth experimental verification of general relativity.

G.3.5 *Gravitational redshift*

We have mentioned before that the proper time interval is an invariant, but it is not an exact differential. For general relativity, on the world line of the motion of massive particle, the proper time is just the four-dimensional line element. It is an invariant, i.e. it is independent upon general coordinate transformations, but it is not an exact differential, i.e. its value is dependent upon the shape of the world line between two events. Hence when we consider the emission process of an atom, the proper time of the signal sent by an atom depends upon the position of the observer.

At position 1 the proper time interval is

$$ds_1 = \sqrt{-(g_{00})_1}dt. \qquad \text{(G.85)}$$

However, at position 2 the proper time interval is

$$\mathrm{d}s_2 = \sqrt{-(g_{00})_2}\,\mathrm{d}t. \tag{G.86}$$

Since the coordinate time interval $\mathrm{d}t$ is an exact differential, which is independent upon the shape of the world line, its value is the same for both positions. Hence in Eqs. (G.85) and (G.86), $\mathrm{d}t$ is equal to each other, but $\mathrm{d}s$ is different from each other.

Let atom emit signal at position 1 and receive the signal at position 2, then the corresponding frequency of the signal emitted in proper time $\mathrm{d}s_1$ is

$$\nu_1 = \frac{n}{\mathrm{d}s_1},$$

where n is the number of wave trains in the signal. This signal is received at position 2 in proper time $\mathrm{d}s_2$, the corresponding frequency is

$$\nu_2 = \frac{n}{\mathrm{d}s_2}.$$

From the above two equations we readily obtain

$$\frac{\nu_2}{\nu_1} = \sqrt{\frac{(g_{00})_1}{(g_{00})_2}}. \tag{G.87}$$

Equation (G.87) is the redshift formula in general relativity. Under the approximation of weak and static field, we have

$$\frac{\nu_2}{\nu_1} = \sqrt{\frac{1 + 2\phi_1}{1 + 2\phi_2}}.$$

Thence we get

$$\frac{\nu_2 - \nu_1}{\nu_1} \approx \phi_1 - \phi_2,$$

i.e.

$$\frac{\Delta\nu}{\nu_1} \approx \Delta\phi. \tag{G.88}$$

If we consider the light emitted from the atom at the surface of the Sun, and is received at the surface of Earth, then since the gravitational potential ϕ_1 is more negative than the gravitational potential ϕ_2, $\phi_1 - \phi_2 < 0$, therefore $\Delta\nu < 0$. There happens gravitational redshift, but the value is very small, $\Delta\nu/\nu_1 \approx 2 \times 10^{-6}$.

G.3.6 *Schwarzschild geometry*

G.3.6.1 *Singularity and pseudo-singularity*

In the Schwarzschild solution (G.55) when $r \to 2GM$, there is a singularity. At the point $r = r_s \equiv 2GM$,

$$g_{00} = -\left(1 - \frac{2GM}{r}\right) \to 0, \tag{G.89}$$

$$g_{11} = \frac{1}{1 - 2GM/R} \to \infty. \tag{G.90}$$

$r_s = 2GM$ is called the Schwarzschild radius. Using CGS system, we have

$$r_s = \frac{2GM}{c^2}. \tag{G.91}$$

For the Sun, $M = M_\odot = 2.0 \times 10^{33}$ g, we get

$$r_s = 3.0 \text{ km.}$$

If the real radius is larger than r_s then there is no need to worry about Eqs. (G.89) and (G.90), since Eq. (G.55) is only applied to the exterior region of the star. If the mass of a star $M \sim M_\odot$ and its radius is less than r_s, then its mass density must be $\rho \sim 10^{16}$ g/cm^3. This mass density is greater than the mass density of a neutron star (the radius of neutron star is around 10 km, the mass density is around 10^{14} g/cm^3). When a star collapses until its radius is less than r_s, it cannot stay in the equilibrium state and will continue to collapse. It should be pointed out that the critical density of the star with radius less than Schwarzschild radius r_s decreases as the increase of the mass, i.e. the larger the mass, then the smaller the critical density. For the whole galaxy, the critical density is only 10^{-6} g/cm^3.

From the redshift formula

$$\frac{\nu_2}{\nu_1} = \sqrt{\frac{(g_{00})_1}{(g_{00})_2}} = \sqrt{\frac{(1 - \frac{2GM}{r})_1}{(1 - \frac{2GM}{r})_2}}.$$

Hence, at the surface $r = r_s$ the frequency of the light emitted by the atom is ν_1, at $r > r_s$ the frequency ν_2 becomes zero, the wavelength approaches infinity, i.e. an infinite redshift. Therefore the surface $r = r_s$ is the infinite redshift surface.

The proper time interval expressed by the clock at the infinite redshift surface is

$$ds = \sqrt{1 - \frac{2GM}{r}}\, dt = 0.$$

That is, in comparison with the clock at infinity (which records coordinate time), this clock progresses infinitely slow. Since the characteristics of the world line of the light signal is $ds = 0$, only light signal (along the radius and point out externally) can stay at rest on the $r = r_s$ surface.

In the region $0 < r < r_s$, there is no singularity for the Schwarzschild solution, but in this region,

$$g_{00} = -\left(1 - \frac{r_s}{r}\right) > 0,$$

$$g_{11} = \frac{1}{1 - r_s/R} < 0. \tag{G.92}$$

That is, the signs of g_{00} and g_{11} are the opposite of the exterior case. Hence, in the region $r < r_s$, t is a space-like coordinate, whereas r is a time-like coordinate. Since the metric is a function of r, when $r < r_s$ the measure of the progress of time is r, in the region the metric is related with time, i.e. the so-called time-dependent metric.

The Schwarzschild singularity at $r = r_s$ is not a real singularity, actually it arises because of the improper choice of the coordinates. When a spaceship crosses this surface, nothing abnormal happens. The Riemann curvature tensor at $r = r_s$ is still finite. Kruskal in 1960 introduced a set of coordinates (u, v, θ, ϕ) to remove this pseudo-singularity. In the interior of Schwarzschild radius,

$$u = \sqrt{1 - \frac{r}{r_s}} \, \frac{e^r}{2r_s} \sinh\left(\frac{t}{2r_s}\right) \quad (r < r_s),$$

$$v = \sqrt{1 - \frac{r}{r_s}} \, \frac{e^r}{2r_s} \cosh\left(\frac{t}{2r_s}\right), \tag{G.93}$$

whereas in the exterior region,

$$u = \sqrt{\frac{r}{r_s} - 1} \, \frac{e^r}{2r_s} \cosh\left(\frac{t}{2r_s}\right),$$

$$v = \sqrt{\frac{r}{r_s} - 1} \, \frac{e^r}{2r_s} \sinh\left(\frac{t}{2r_s}\right). \tag{G.94}$$

Using Kruskal coordinates, we have

$$ds^2 = \frac{4r_s^3}{r} \frac{e^{-r}}{r_s}(-dv^2 + du^2) + r^2 d\theta^2 + r^2 \sin^2\theta d\phi^2, \tag{G.95}$$

where u is a space-like coordinate, v is a time-like coordinate.

When we use this set of coordinates no singularity appears at $r = r_s$, hence the singularity at $r = r_s$ is a pseudo-singularity. However, the singularity at $r = 0$ is a real singularity, which cannot be eliminated by the transformation of coordinates. In 1966–1967 S.W. Hawking and R. Penrose proved that any metrics in general relativity has at least one singularity. The fact that the singularity cannot be avoided is the essential difficulty of general relativity, which cannot be eliminated in the original frame.

G.3.6.2 *Schwarzschild event horizon*

Though from local point of view the surface $r = r_s$ does not have any abnormal property, but from the viewpoint as a whole there is still some very abnormal properties. In the following we shall point out that the surface $r = r_s$ is an event horizon, that from the interior of the surface it is impossible to send out any signal, i.e. the surface $r = r_s$ is a one-way membrane, signal can only go in but cannot be sent out. The spacetime region inside the event horizon is called a black hole.

There is need to point out, that the existence or non-existence of infinite redshift surface, i.e. whether g_{00} is zero or not, is related to the choice of coordinates, but the existence of event horizon is independent of the choice of coordinates. All observers, no matter what coordinates they choose, all unanimously obtain the conclusion of existence of event horizon.

In the exterior of a black hole, the future light cone is parallel to the t-axis; but in the interior, since r is the time-like coordinate, the future light cone is parallel to the r-axis. Since the world line of any signal stay in the interior of the future light cone, in the interior of black hole it can only propagate along the direction where r decreases, thus it can never leave the black hole.

To illustrate the fact that from the local point of view the event horizon does not have any abnormal property, we are going to calculate the proper time which is needed for a spaceship to enter the black hole. Consider the radius motion directing towards the black hole, thus $\dot{\phi} = 0$. From Eq. (G.59), we know $\dot{\phi} = A/r^2$, hence we get $A = 0$. Thence from Eq. (G.62) we obtain

$$\frac{B^2}{1 - r_s/r} - \frac{\dot{r}^2}{1 - r_s/r} = 1,$$

i.e.

$$\left(\frac{\mathrm{d}r}{\mathrm{d}s}\right)^2 = \frac{r_s}{r} - (1 - B^2).$$

Integrating the equation above we get

$$s - s_0 = \frac{r[\frac{r_s}{r} - (1 - B^2)]^{1/2}}{1 - B^2} + \frac{r_s}{(1 - B^2)^{3/2}} \tan^{-1} \left[\frac{\frac{r_s}{r} - (1 - B^2)}{1 - B^2} \right]^{1/2},$$

$$(G.96)$$

where s_0 is an integrating constant.

Obviously, when there is finite change of r, the corresponding change of s is also finite. For example, suppose at initial time the spaceship is at rest at $r_0 = r_s/(1 - B^2)$, then the total proper time interval needed for arriving at $r = 0$ is

$$\Delta s = \frac{r_s}{(1 - B^2)^{3/2}} \frac{\pi}{2} = r_s \frac{\pi}{2} \left(\frac{r_0}{r_s} \right)^{3/2}. \qquad (G.97)$$

This is a finite quantity. Hence, there is no abnormal behavior when the spaceship crosses event horizon.

A Possible Modification of Einstein's Theory of General Relativity

A Possible Modification of Einstein's Theory of General Relativity

QIAN Shang-Wu

Physics Department, Peking University, Beijing 100871, China

(Received May 28, 2003)

Abstract *This article suggests a new metric theory of gravitation, in which metric field is determined not only by matter and nongravitational field but also by vector graviton field, and in principle there is no need to introduce the Einstein's tensor. In order to satisfy automatically the geodesic postulate, an additional coordinate condition is needed. For the spherically symmetric static field, it leads us to quite different conclusions from those of Einstein's general relativity in the interior region of the surface of infinite redshift. Accurate to the first order of GM/r, it obtains the same results about the four experimental tests of general relativity.*

PACS numbers: 04.20.-q, 04.20.Cv, 04.50.+h

Key words: general relativity, metric theory of gravitation, vector graviton field

1 Introduction

In general relativity (GR), the Einstein's field equation is known as

$$G_{\mu\nu} = R_{\mu\nu} - \frac{1}{2}g_{\mu\nu}R = -8\pi GT_{\mu\nu}\,, \tag{1}$$

where $R_{\mu\nu}$ is the Ricci tensor, R is the curvature scalar, $g_{\mu\nu}$ is the metric tensor, $T_{\mu\nu}$ is the energy-momentum tensor of matter (not including the gravitational field), the energy-momentum of the gravitational field is included implicitly on the left-hand side of this equation, and the Einstein's tensor $G_{\mu\nu}$ is introduced in order to satisfy automatically the geodesic postulate

$$T^{\mu\nu}_{;\nu} = 0\,. \tag{2}$$

The right-hand side (RHS) of Eq. (1) does not include explicitly the energy momentum of gravitational field, whereas the left-hand side (LHS) of Eq. (1) is not just $R_{\mu\nu}$ but an artificially introduced quantity $G_{\mu\nu}$, which is somehow unnatural and man-made.

For the spherically symmetric static field, in GR we have the famous Schwarzschild solution

$$\mathrm{d}s^2 = \left(1 - \frac{2GM}{r}\right)\mathrm{d}t^2 - \frac{\mathrm{d}r^2}{(1 - 2GM/r)}$$
$$- r^2\mathrm{d}\theta^2 - r^2\sin^2\theta\mathrm{d}\phi^2\,. \tag{3}$$

The Schwarzschild solution develops a "singularity" as $r \to 2GM$, and $r_s = 2GM$ is called the Schwarzschild radius of the mass M. The surface $r = r_s$ is a surface of infinite redshift (SIR) (relative to $r = \infty$). It is easily seen that when $r < r_s$, t from a timelike coordinate becomes a spacelike coordinate, whereas r from a spacelike coordinate becomes a timelike coordinate. It is quite an unsatisfactory and unnatural result.

Because of the above considerations, this paper suggests a possible modification of Einstein's field equation (1) and its consequences (3) to overcome these unnatural

and man-made results. We want to modify Eq. (1) such that its RHS includes explicitly the energy-momentum of gravitational field, i.e. the vector graviton field, whereas the LHS only contains the Ricci tensor $R_{\mu\nu}$. In effect, the suggested modification is also a metric theory of gravitation, the metric field $g_{\mu\nu}$ plays the same role for gravitational field as that in the Einstein's theory of general relativity, and the vector graviton field here only plays the same role as that of matter and other fields, i.e. generating the metric field $g_{\mu\nu}$. After the suggested modification, the LHS of the new field equation is just the Ricci tensor $R_{\mu\nu}$, whereas the RHS includes explicitly the energy-momentum of the vector graviton field. In order to ensure Eq. (2) we only need to add a coordinate condition. The modified metric theory of gravitation has been verified that it satisfies all the criteria for the viability of a theory: self-consistency, completeness, and agreement with past experiments. Actually, it is a new kind of metric theory, which suggests that there exists a vector graviton field $H^{\mu\nu}$. This field $H^{\mu\nu}$ should act together with the matter and other fields to generate the metric field $g_{\mu\nu}$, which plays the same role for gravitational field as that in the Einstein's theory of general relativity. It should be emphasized that the field $H^{\mu\nu}$ does not directly influence the motion of a particle in the vector graviton field, but only acts together with the matter and other fields to generate the metric field $g_{\mu\nu}$, and that just the metric field $g_{\mu\nu}$ directly influences the motion of a particle in the vector graviton field. This modification may be called as the vector graviton metric theory (VGM), which we have found predicts almost the same results as Einstein's theory about the four famous experimental tests of general relativity: light deflection, time-delay in radar propagation, gravitational redshift, and perihelion procession of planets. But it gives a quite different and reasonable result in the neighborhood of the surface of infinite redshift and some other interesting and reasonable results. In Sec. 2 this paper will give the explicit expression of the energy-momentum tensor of the vector graviton field. In Sec. 3 this paper will give the modified field equation in VGM and the additional coordinate condition to ensure the geodesic postulate Eq. (2). In Sec. 4 this paper derives the external solution for the spherical symmetric static mass, which is slightly different from the Schwarzschild solution. In Sec. 5 this paper discusses the four experimental tests of general relativity. In the last section this paper discusses the behavior of rod and clock in the neighborhood of the surface of infinite redshift, and shows that t is always a timelike coordinate whereas r is always a spacelike coordinate in this case.

2 Graviton Field Tensor and Graviton Energy-Momentum Tensor

Schwartz[1] has proved that Maxwell's equations could be derived from the Coulomb's law and the requirement of Lorentz invariance. Since Newton's gravitation law and Coulomb's law are completely similar in their form, hence we recognize that there exists a vector graviton field described by an antisymmetric tensor $H^{\mu\nu}$. For an observer in the flat space

$$H^{\mu\nu} = \begin{pmatrix} 0 & F_x & F_y & F_z \\ -F_x & 0 & \Omega_z & -\Omega_y \\ -F_y & -\Omega_z & 0 & \Omega_x \\ -F_z & \Omega_y & -\Omega_x & 0 \end{pmatrix}, \quad (4)$$

where (F_x, F_y, F_z) is the ordinary gravitational field strength, $(\Omega_x, \Omega_y, \Omega_z)$ is the gravitational vortex strength introduced by Carstoiu.[2] In flat space, $H^{\mu\nu}$ satisfies Maxwell-like equations

$$H^{\mu\nu}_{,\nu} = -4\pi G k^\mu, \qquad \{H_{\mu\nu,\lambda}\} = 0, \quad (5)$$

where k^μ is the four-dimensional mass current density vector, G is the gravitation constant, speed of light c is taken as 1, and $\{\}$ designates antisymmetrized sum over all permutations of the indices μ, ν, λ. It has to be emphasized that $H^{\mu\nu}$ is not the metric field $g^{\mu\nu}$ which is equivalent to the gravitational field. $H^{\mu\nu}$ plays the same role as other matters, which only influences metric field, not directly determines the motion of a particle, whereas under the action of metric field, particle moves along a geodesic line in the Riemannian space.

For photon field, energy-momentum tensor is[3]

$$T^{\mu\nu}_{(\mathrm{em})} = F^\mu_\alpha F^{\alpha\nu} + \frac{1}{4} g^{\mu\nu} F^{\alpha\beta} F_{\alpha\beta}, \quad (6)$$

where $F_{\mu\nu}$ is the electromagnetic field tensor. The vector graviton field of a body is similar to the electromagnetic field of a negative charge, and its energy-momentum tensor can be analogously written as

$$T^{\mu\nu}_{(\mathrm{gr})} = \frac{1}{4\pi G} \left(H^\mu_\alpha H^{\alpha\nu} + \frac{1}{4} g^{\mu\nu} H^{\alpha\beta} H_{\alpha\beta} \right). \quad (7)$$

Its covariant divergence obeys

$$T^{\mu\nu}_{(\mathrm{gr});\nu} = -H^\mu_\alpha k^\alpha. \quad (8)$$

When $k^\alpha = 0$, we have $T^{\mu\nu}_{(\mathrm{gr});\nu} = 0$. In curved space, $H^{\mu\nu}$ satisfies

$$H^{\mu\nu}_{;\nu} = -4\pi G k^\mu, \qquad \{H_{\mu\nu,\lambda}\} = 0. \quad (9)$$

3 Field Equations in VGM

In our new metric theory of gravitation VGM, $T^{\mu\nu}_{(\mathrm{gr})}$ plays the same role as $T^{\mu\nu}$, hence in RHS of Einstein's field equations we use $T_{\mu\nu} + T_{\mu\nu(\mathrm{gr})}$ to replace $T_{\mu\nu}$, furthermore in LHS of Eq. (1) we use $R_{\mu\nu}$ to replace the man-made quantity $G_{\mu\nu}$, finally we suggest the following field equations

$$R_{\mu\nu} = -8\pi G (T_{\mu\nu} + T_{\mu\nu(\mathrm{gr})}). \quad (10)$$

In order to satisfy Eq. (2), from Eqs. (9), (10), and the Bianchi identity

$$G^{\mu\nu}_{;\nu} = 0, \quad (11)$$

we obtain

$$\left(\frac{1}{2} g^{\mu\nu} R \right)_{;\nu} = -8\pi G T^{\mu\nu}_{(\mathrm{gr});\nu} = 8\pi G H^\mu_\alpha k^\alpha. \quad (12)$$

Equation (12) can be considered as an additional condition imposed on $g^{\mu\nu}$, which plays a similar role as the coordinate condition. In summary, the field equation of VGM is Eq. (10), where $T^{\mu\nu}_{(\mathrm{gr})}$ is given by Eq. (7), and there is an additional equation imposed on the metric tensor $g^{\mu\nu}$, which is Eq. (12). Equation (2) is automatically satisfied by using the additional coordinate condition Eq. (12).

For metric field in the vacuum, $T_{\mu\nu} = 0$, from Eq. (10) we have

$$R_{\mu\nu} = -8\pi G T_{\mu\nu(\mathrm{gr})}. \quad (13)$$

Considering that in the vacuum $k^\alpha = 0$, from Eqs. (8) and (13) we get $R^{\mu\nu}_{;\nu} = 0$, then from Eq. (11) we have $R_{;\nu} = R_{,\nu} = 0$. Hence $R = \mathrm{const}$. The simplest choice for R is $R = 0$, hence in the vacuum there exist only vector graviton field and the required metric field. The constant curvature scalar R is taken to be zero and there is no difference between $G_{\mu\nu}$ and $R_{\mu\nu}$ for the vacuum field, thence equation (13) can be written as

$$G_{\mu\nu} - R_{\mu\nu} = 8\pi G T_{\mu\nu(\mathrm{gr})}. \quad (14)$$

The coordinate condition Eq. (12) is automatically satisfied in this case. As to equation of motion of a particle in the gravitational field, from Eq. (2) we get the geodesic equation

$$\frac{d^2 x^\mu}{d\tau^2} + \Gamma^\mu_{\nu\lambda} \frac{dx^\nu}{d\tau} \frac{dx^\lambda}{d\tau} = 0, \quad (15)$$

where $\Gamma^\mu_{\nu\lambda}$ are Christoffel symbols and $d\tau$ is proper time interval.

4 External Solution for the Spherical Symmetric Static Mass

Considering the static character and the spherical symmetry of the external field, the space-time interval can be written as[4]

$$ds^2 = e^{N(r)}dt^2 - e^{L(r)}dr^2 - r^2 d\theta^2 - r^2\sin^2\theta d\phi^2 . \tag{16}$$

Similar to the argument in Sec. 13.1 of Ref. [3] for the field of a charged mass point, from Eqs. (4) and (5) we obtain the nonzero component of $H_{\mu\nu}$,

$$H_{10} = -H_{01} = \frac{GM}{r^2}\exp\left[\frac{1}{2}(N+L)\right], \tag{17}$$

thence from Eq. (7) we find

$$T_{\mu\nu(\mathrm{gr})} = \frac{1}{8\pi G}\left(\frac{GM}{r}\right)^2 \begin{pmatrix} e^{N(r)} & 0 & 0 & 0 \\ 0 & -e^{L(r)} & 0 & 0 \\ 0 & 0 & r^2 & 0 \\ 0 & 0 & 0 & r^2\sin^2\theta \end{pmatrix}. \tag{18}$$

Solving Eq. (14) we finally obtain

$$ds^2 = \left(1 - \frac{GM}{r}\right)^2 dt^2 - \frac{dr^2}{(1-GM/r)^2} - r^2 d\theta^2 - r^2\sin^2\theta d\phi^2 . \tag{19}$$

5 Four Experimental Tests of General Relativity

Comparing Eq. (19) with the Schwarzschild line element, we readily find that to the precision of the first order of GM/r, two expressions are completely coincident. Hence, for two experimental tests: light deflection and the time-delay in radar propagation, we get the same results as GR. As to gravitational redshift, from Eq. (19) we get

$$\frac{\nu_2}{\nu_1} = \frac{1-GM/r_1}{1-GM/r_2}, \tag{20}$$

whereas Schwarzschild's metric gives

$$\left(\frac{\nu_2}{\nu_1}\right)_{\mathrm{Sch}} = \sqrt{\frac{1-2GM/r_1}{1-2GM/r_2}} . \tag{21}$$

For weak field equations (20) and (21) give almost the same results, which in the strong field region they should give apparently different results, i.e. the surface of infinite redshift (SIR) is different for two theories. For VGM, it situates at $r = r_c = GM$, while for GR it situates at $r = r_s = 2GM = 2r_c$. In consideration of perihelion procession of Mercury, since it needs to consider the terms involved $(GM/r)^2$, there is apparent difference between VGM and GR: for GR the value is 43".03/century, while for VGM the value is 35".86/century. The recent observed value is 41".4/century (it needs further observation), which is not only due to relativistic gravity but also due to solar oblateness[5] and anomalous perihelion shifts[6] (it needs further consideration). The reasonable perihelion shifts due to relativistic gravity is perhaps around 39"/century, between the value predicted by VGM and the value predicted by GR. From the perihelion shifts itself, we cannot say at this time which theory is better, GR or VGM.

6 Neighborhood of SIR

Firstly we consider the proper time interval $d\tau$, for VGM a clock placed (at rest) near SIR ($r = r_c = GM$) shows a proper time

$$d\tau^2 = \left(1 - \frac{GM}{r}\right)^2 dt^2 . \tag{22}$$

Hence the clock runs slower and slower when it approaches SIR, $d\tau$ tends to zero as $r \to r_c$, i.e. the clock runs infinitely slow compared to a clock at infinity, when the clock crosses SIR it runs faster and faster when r becomes smaller and smaller. For GR

$$(d\tau^2)_{\mathrm{Sch}} = \left(1 - \frac{2GM}{r}\right)dt^2 . \tag{23}$$

When $r < r_s = 2GM$, the RHS of Eq. (23) is negative, and equation (23) is not usable, for t becomes spacelike coordinate, so it is very unnatural, whereas in VGM t keeps timelike coordinate when $r < r_c$.

Secondly we consider space interval dl. From Ref. [7] we know

$$dl^2 = \left(\frac{g_{0\alpha}g_{0\beta}}{g_{00}} - g_{\alpha\beta}\right)dx^\alpha dx^\beta , \tag{24}$$

where the repeated appeared Greek letters mean summation from 1 to 3. From Eq. (19) we get

$$dl^2 = \frac{dr^2}{(1-GM/r)^2} + r^2 d\theta^2 + r^2\sin^2\theta d\phi^2 . \tag{25}$$

For a rod placed along the radial direction,

$$dl^2 = \frac{dr^2}{(1-GM/r)^2} , \tag{26}$$

hence the rod contracts shorter and shorter when it approaches SIR, and dl tends to zero as $r \to r_c$, i.e. the rod contracts infinitely short compared to a rod at infinity. When the rod crosses SIR it elongates longer and longer when r becomes smaller and smaller. For GR

$$(\mathrm{d}l^2)_{\mathrm{Sch}} = \frac{\mathrm{d}r^2}{(1 - 2GM/r)} \,. \qquad (27)$$

When $r < r_s = 2GM$, the RHS of Eq. (27) is negative, and equation (27) is not usable, for r becomes timelike coordinate, and it is very unnatural, whereas in VGM r keeps spacelike coordinate when $r < r_c$.

From Eq. (19), we readily know that the radial speed of the light signal is

$$\left(\frac{\mathrm{d}r}{\mathrm{d}t}\right)_{\mathrm{light}} = \left(1 - \frac{r_c}{r}\right)^2 , \qquad (28)$$

when $r \to r_c$ it tends to zero, hence SIR is the surface at which radial speed of light equals zero. In the interior region of SIR, $(\mathrm{d}r/\mathrm{d}t)_{\mathrm{light}}$ increases as r decreases, when $r = r_c/2$, it increases to 1, when r further decreases, it will be greater than 1, i.e. in the region of strong gravitational field the speed of light can surpass c, the speed of light in the vacuum.

It is very important that in Eq. (19) when $r < r_c$, $g_{00} = (1 - r_c/r)^2$ is still positive, $g_{11} = -(1 - r_c/r)^{-2}$ is still negative, g_{00} and g_{11} do not change their sign, hence t is always a timelike coordinate, and r is always a spacelike coordinate. They do not change their roles when they cross SIR. Whereas in GR, they do change their roles when they cross SIR $r = r_s$. In GR, when $r > r_s$, future light cone directs towards the direction of increasing t, but when $r < r_s$ future light cone directs abruptly towards the direction of decreasing r (r becomes timelike coordinate). In the interior region of SIR $r = r_s$, anybody can only move in the direction of decreasing r, the reverse direction is impossible, and SIR acts as a one-way membrane. In VGM, future light cone always directs towards the direction of increasing t, and SIR never acts as a one-way membrane.

Thus it can be seen that when the radius of the spherical mass is less than $r_c = GM$, in the neighborhood of r_c, the results of VGM is quite different from those of GR. There is no one-way membrane, t is always a timelike coordinate whereas r is always a spacelike coordinate, i.e. the conclusions from VGM are quite different from those of Schwarzchild black hole.

References

[1] M. Schwartz, *Principles of Electrodynamics*, McGraw-Hill, New York (1972).

[2] J. Carstoiu, Compt. Rend. **268** (1969) 201.

[3] R. Adler, M. Bazin, and M. Schiffer, *Introduction to General Relativity*, McGraw-Hill, New York (1965).

[4] H.C. Ohanian, *Gravitation and Spacetime*, W.W. Norton & Company, New York (1976).

[5] C.W. Misner, K.S. Thorne, and J.A. Wheeler, *Gravitation*, W.H. Freeman & Company, San Francisco (1973).

[6] S.W. Hawking and W. Israel, *General Relativity*, Cambridge University Press, Cambridge, London, New York, Melbourne (1979) p. 60.

[7] L.D. Landau and E.M. Lifshitz, *The Classical Theory of Fields*, forth revised English edition, Pergamon Press, Oxford, New York (1975) p. 235.

A Possible Interpretation on Distance-Dependent Effect of Gravitational Constant in Newton's Theory of Gravitation

A Possible Interpretation on Distance-Dependent Effect of Gravitational Constant in Newton's Theory of Gravitation

QIAN Shang-Wu

Physics Department, Peking University, Beijing 100871, China

(Received October 8, 2004)

Abstract *Based on the new metric theory of gravitation suggested by the author of this article, it gives a possible theoretical interpretation on the famous experiment done by D.R. Long in 1976, i.e. the distance-dependent effect of the gravitational constant in Newton's theory of gravitation.*

PACS numbers: 04.20.-q, 04.20.Cv, 04.50.+h

Key words: general relativity, metric theory of gravitation, vector graviton field

In 1976 there was a famous experiment which was published in Nature[1] and propagated in Science News.[2] This experiment was done by D.R. Long and his group, who found that the gravitational constant G is not strictly a constant, but is slightly dependent on the distance between two mutually interacting bodies. Till now there is no satisfactory interpretation of this effect. Now we have found that this effect is quite easily explained by my recently suggested theory of gravitation,[3] i.e. the vector gravitation metric theory (VGM). In Ref. [3] we have derived the external solution for the spherical symmetrical static mass in VGM, which is

$$\mathrm{d}s^2 = \mathrm{e}^{N(r)}\mathrm{d}t^2 - \mathrm{e}^{L(r)}\mathrm{d}r^2 - r^2\mathrm{d}\theta^2 - r^2\sin^2\theta\mathrm{d}\phi^2$$

$$= \left(1 - \frac{GM}{r}\right)^2\mathrm{d}t^2 - \frac{\mathrm{d}r^2}{(1 - GM/r)^2}$$

$$- r^2\mathrm{d}\theta^2 - r^2\sin^2\theta\mathrm{d}\phi^2, \tag{1}$$

where G is the gravitational constant, which is equal to 6.670×10^{-11} m$^3\cdot$kg$^{-1}\cdot$s^{-1}. From the equation of motion of a particle in the gravitational field,

$$\frac{\mathrm{d}^2x^\mu}{\mathrm{d}s^2} + \Gamma^\mu_{\nu\lambda}\frac{\mathrm{d}x^\nu}{\mathrm{d}s}\frac{\mathrm{d}x^\lambda}{\mathrm{d}s} = 0, \tag{2}$$

we readily obtain the equation of motion of a static particle in the gravitational field

$$\frac{\mathrm{d}^2r}{\mathrm{d}s^2} + \Gamma^1_{00}\left(\frac{\mathrm{d}t}{\mathrm{d}s}\right)^2 = 0, \tag{3}$$

where the Christoffel symbol Γ^1_{00} is given by Ref. [4] [page 189, Eq. (6.20)],

$$\Gamma^1_{00} = \frac{N'}{2}\exp(N - L) = \frac{1}{2}g_{00}(g_{00})', \tag{4}$$

and $N' = \mathrm{d}N/\mathrm{d}r$, $(g_{00})' = \mathrm{d}g_{00}/\mathrm{d}r$. From Eqs. (3) and (4) we get

$$\frac{\mathrm{d}^2r}{\mathrm{d}s^2} = -\frac{1}{2}(g_{00})'. \tag{5}$$

Under Newton's approximation [Ref. [4], page 134, Eq. (4.142)] we have

$$g_{00} = 1 + 2\varphi, \tag{6}$$

where φ is the gravitational potential in the field of the spherical symmetrical static mass M. From Eqs. (1) and (6) we obtain

$$\varphi = -\frac{GM}{r} + \frac{1}{2}\left(\frac{GM}{r}\right)^2. \tag{7}$$

When we introduce a variable gravitational constant G^*, and formally let

$$\varphi = -\frac{G^*M}{r}, \tag{8}$$

then the gravitational force subjected by a static particle with unit mass is

$$f = -\frac{\mathrm{d}\varphi}{\mathrm{d}r} = -\frac{G^*M}{r^2}. \tag{9}$$

Then from Eqs. (7) and (8) we have

$$G^* = G\left(1 - \frac{GM}{2r}\right). \tag{10}$$

Equation (10) qualitatively explains the experimental result obtained by D.R. Long. The second term in Eq. (10) represents the effect of repulsive force, in general it is much smaller than the effect of attractive force, i.e. the first term in Eq. (10). But in the region where r is comparable with the Schwarzchild radius $r_s = 2GM$, the second term will play more and more important role than that of the first term.

Obviously, we cannot get the distance dependence of gravitational constant from the Einstein's theory of general relativity. The external solution for the spherical symmetrical static mass in general relativity is the famous Schwarzchild solution, i.e.

$$\mathrm{d}s^2 = \left(1 - \frac{2GM}{r}\right)\mathrm{d}t^2 - \frac{\mathrm{d}r^2}{(1 - 2GM/r)}$$

$$- r^2\mathrm{d}\theta^2 - r^2\sin^2\theta\mathrm{d}\phi^2. \tag{11}$$

In this case we have

$$g_{00} = \left(1 - \frac{2GM}{r}\right). \tag{12}$$

Hence, in general relativity, under Newton's approximation, from Eq. (6) we get $\varphi = -GM/r$ instead of Eq. (7), consequently we can only get a distance-independent gravitational constant G, not a distance-dependent gravitational constant G^*.

Furthermore, it is worth while to point out that equation (1) can be obtained directly from the Schwarzchild solution (11) by replacing G by G^*. Replacing G by G^*, from Eq. (11) we get

$$\mathrm{d}s^2 = \left(1 - \frac{2G^*M}{r}\right)\mathrm{d}t^2 - \frac{\mathrm{d}r^2}{(1 - 2G^*M/r)} - r^2\mathrm{d}\theta^2 - r^2\sin^2\theta\mathrm{d}\phi^2$$

$$= \left(1 - \frac{GM}{r}\right)^2\mathrm{d}t^2 - \frac{\mathrm{d}r^2}{(1 - GM/r)^2} - r^2\mathrm{d}\theta^2 - r^2\sin^2\theta\mathrm{d}\phi^2 \, . \tag{13}$$

Equation (13) is just the Eq. (1) in VGM .

It is interesting to note the fact that such a replacement (G by G^*) can be taken as a rule to get the required formula in the vacuum for VGM from the corresponding formula in general relativity. For example, from the famous Reissner–Nordstrom metric for the charged spherical mass,

$$\mathrm{d}s^2 = \left[\left(1 - \frac{2GM}{r}\right) + \frac{GQ^2}{r^2}\right]\mathrm{d}t^2 - \frac{\mathrm{d}r^2}{(1 - 2GM/r) + GQ^2/r^2} - r^2\mathrm{d}\theta^2 - r^2\sin^2\theta\mathrm{d}\phi^2 \, , \tag{14}$$

we can easily get the corresponding metric for the charged spherical mass in VGM,

$$\mathrm{d}s^2 = \left[\left(1 - \frac{GM}{r}\right)^2 + \frac{GQ^2}{r^2}\right]\mathrm{d}t^2 - \frac{\mathrm{d}r^2}{(1 - GM/r)^2 + GQ^2/r^2} - r^2\mathrm{d}\theta^2 - r^2\sin^2\theta\mathrm{d}\phi^2 \, , \tag{15}$$

simply by the replacement G by G^*, i.e. $(1 - 2GM/r)$ by $(1 - GM/r)^2$, where Q is the charge of the spherical mass. The main idea of the proof of this replacement in this case is given in the Appendix. Furthermore, we have checked that this replacement is also applicable for the rotating charged spherical mass, which we will show in another article.

Appendix

For the spherical symmetric charged static mass, the metric in the external region is

$$\mathrm{d}s^2 = \mathrm{e}^{N(r)}\mathrm{d}t^2 - \mathrm{e}^{L(r)}\mathrm{d}r^2 - r^2\mathrm{d}\theta^2 - r^2\sin^2\theta\mathrm{d}\phi^2 \, . \tag{A1}$$

The field equations in the external region are

$$R_{\mu\nu} = -8\pi G(T_{\mu\nu(\mathrm{em})} + T_{\mu\nu(\mathrm{gr})}) \, , \tag{A2}$$

where $T_{\mu\nu(\mathrm{em})}$ is the energy-momentum tensor of the electromagnetic field. From Ref. [3] we have

$$T_{\mu\nu(\mathrm{gr})} = \frac{1}{8\pi G}\left(\frac{GM}{r^2}\right)^2 \begin{pmatrix} \mathrm{e}^{N(r)} & 0 & 0 & 0 \\ 0 & -\mathrm{e}^{L(r)} & 0 & 0 \\ 0 & 0 & r^2 & 0 \\ 0 & 0 & 0 & r^2\sin^2\theta \end{pmatrix} \, . \tag{A3}$$

Similarly we have[4]

$$T_{\mu\nu(\mathrm{em})} = \frac{1}{8\pi}\left(\frac{Q}{r^2}\right)^2 \begin{pmatrix} \mathrm{e}^{N(r)} & 0 & 0 & 0 \\ 0 & -\mathrm{e}^{L(r)} & 0 & 0 \\ 0 & 0 & r^2 & 0 \\ 0 & 0 & 0 & r^2\sin^2\theta \end{pmatrix} \, . \tag{A4}$$

Solving Eq. (A2), after some tedious calculations, we can get the required formula (15).

References

[1] D.R. Long, Nature **260** (1976) 417.

[2] Science News, "Complicating the law of gravity", **109** (1976) 244.

[3] S.W. Qian, Commun. Theor. Phys. (Beijing, China) **41** (2004) 377.

[4] R. Adler, M. Bazin, and M. Schiffer, *Introduction to General Relativity*, McGraw-Hill, New York (1965) 1.

Metric of Rotating Charged Spherical Mass in Vacuum for Vector Graviton Metric Theory of Gravitation

Metric of Rotating Charged Spherical Mass in Vacuum for Vector Graviton Metric Theory of Gravitation

QIAN Shang-Wu,[1] ZHONG Zai-Zhe,[2] and GU Zhi-Yu[3]

[1] Physics Department, Peking University, Beijing 100871, China

[2] Mathematics Department, Liaoning Normal University, Dalian 116029, China

[3] Physics Department, Capital Normal University, Beijing 100037, China

(Received March 14, 2005)

Abstract *Based on the vector graviton metric theory of gravitation (VGM) suggested by one of the authors of this article, using the method of null tetrad and analytic continuation, this paper gives the metric of the rotating charged spherical mass in VGM. The result shows once again that a replacement of G by $G^* = G(1 - GM/2r)$ in general relativity will yield the corresponding result in VGM for the metric in vacuum.*

PACS numbers: 04.20.-q, 04.20.Cv, 04.50.+h

Key words: metric theory of gravitation, vector graviton field, tetrad, Kerr–Newman metric

1 Introduction

One of the authors of this article suggested a new metric theory of gravitation,[1] in which metric field is determined not only by matter and nongravitational field but also by vector graviton field, and in principle there is no need to introduce the Einstein's tensor. In order to satisfy automatically the geodesic postulate, an additional coordinate condition is needed. For the spherically symmetric static field, it leads us to quite different conclusions from that of Einstein's general relativity in the interior region of the surface of infinite redshift. Being accurate to the first order of GM/r, it obtains the same results about the four experimental tests of general relativity. Based on this new metric theory of gravitation [vector graviton metric theory (VGM)], we gave a possible theoretical interpretation of the famous experiment done by D.R. Long in 1976, i.e. the distance-dependent effect of the gravitational constant in Newton's theory of gravitation, in which the gravitational constant G becomes G^* in VGM,[2] where $G^* = G(1 - GM/2r)$. Furthermore, we have noted an interesting fact that such a replacement (G by G^*) can be taken as a rule to get the required formula in the vacuum for VGM from the corresponding formula in general relativity (GR). In Ref. [2] we gave an example about obtaining the metric for the static charged spherical mass, since in GR, this metric is given by the famous Reissner–Nordstrom metric for the charged spherical mass,

$$ds^2 = \left[\left(1 - \frac{2GM}{r}\right) + \frac{GQ^2}{r^2}\right]dt^2 - \frac{dr^2}{(1 - 2GM/r) + GQ^2/r^2} - r^2 d\theta^2 - r^2 \sin^2\theta d\phi^2\,, \tag{1}$$

hence by the above-mentioned rule we can readily get the corresponding metric for the charged spherical mass in VGM

$$ds^2 = \left[\left(1 - \frac{GM}{r}\right)^2 + \frac{GQ^2}{r^2}\right]dt^2 - \frac{dr^2}{(1 - GM/r)^2 + GQ^2/r^2} - r^2 d\theta^2 - r^2 \sin^2\theta d\phi^2\,, \tag{2}$$

simply by the replacement G by G^*, i.e. the replacement $(1 - 2GM/r)$ by $(1 - GM/r)^2$, where Q is the charge of the spherical mass. We have checked this result from the detailed calculation in VGM. Furthermore, in this article we shall give a detailed calculation about obtaining the metric of the rotating charged spherical mass in VGM. We shall verify that the above-mentioned replacement is also applicable for this rather complex case. In Sec. 2 we shall introduce a kind of coordinate transformation. In Sec. 3 we shall introduce two tetrads, which will be used to facilitate our calculation. In Sec. 4 we shall use the method of analytic continuation to obtain the desired results.

2 Coordinate Transformation

Let $m = GM$, $q = \sqrt{G}Q$ and perform the following coordinate transformation:

$$r' = r, \quad \theta' = \theta, \quad \phi' = \phi,$$

$$du = dt - \frac{dr}{(1 - m/r)^2 + q^2/r^2}\,, \tag{3}$$

then we can write Eq. (2) as

$$ds^2 = \left[\left(1 - \frac{m}{r}\right)^2 + \frac{q^2}{r^2}\right]du^2 + 2du\,dr - r^2(d\theta^2 + \sin^2\theta d\phi^2)\,. \tag{4}$$

From Eq. (4) we know

$$g_{\mu\nu} = \begin{pmatrix} (1-\frac{m}{r})^2 + \frac{q^2}{r^2} & 1 & 0 & 0 \\ 1 & 0 & 0 & 0 \\ 0 & 0 & -r^2 & 0 \\ 0 & 0 & 0 & -r^2\sin^2\theta \end{pmatrix}, \quad (5)$$

$$g = -r^4\sin^2\theta. \quad (6)$$

From Eqs. (5) and (6) we get

$$g^{\mu\nu} = \begin{pmatrix} 0 & 1 & 0 & 0 \\ 1 & -[(1-\frac{m}{r})^2 + \frac{q^2}{r^2}] & 0 & 0 \\ 0 & 0 & -\frac{1}{r^2} & 0 \\ 0 & 0 & 0 & -\frac{1}{r^2\sin^2\theta} \end{pmatrix}. \quad (7)$$

3 Tetrads

In order to calculate the metric of rotating charged spherical mass we introduce two tetrads[3] $\lambda_{(a)}^\mu$ and $\tau_{(a)}^\nu$ and let $g^{\mu\nu} = \lambda_{(a)}^\mu \tau_{(a)}^\nu$, where $a = 0,1,2,3$ and $\lambda_{(a)}^\mu = (l^\mu, n^\mu, -m^\mu, -\bar{m}^\mu)$, $\tau_{(a)}^\nu = (n^\nu, l^\nu, \bar{m}^\nu, m^\nu)$, hence

$$g^{\mu\nu} = l^\mu n^\nu + n^\mu l^\nu - m^\mu \bar{m}^\nu - \bar{m}^\mu m^\nu, \quad (8)$$

where $\bar{m}$ is the complex conjugate of m , and

$$l^\mu = \delta_1^\mu, \quad m^\mu = \frac{1}{\sqrt{2r}}\Big[\delta_2^\mu + \frac{i}{\sin\theta}\delta_3^\mu\Big],$$

$$n^\mu = \delta_0^\mu - \frac{1}{2}\Big[\Big(1-\frac{m}{r}\Big)^2 + \frac{q^2}{r^2}\Big]\delta_1^\mu, \quad (9)$$

where δ_a^μ are Kronecker delta, i.e. $\delta_a^\mu = 1$, when $\mu = a$, and $\delta_a^\mu = 0$ when $\mu \neq a$. From $l_\mu = g_{\mu\nu}l^\nu$ we get

$$l_0 = \Big[\Big(1-\frac{m}{r}\Big)^2 + \frac{q^2}{r^2}\Big]l^0 + l^1, \quad l_1 = l^0,$$

$$l_2 = -r^2 l^2, \quad l_3 = -r^2(\sin^2\theta)l^3, \quad (10)$$

hence we have

$$l^\mu l_\mu = 0. \quad (11)$$

Similarly we obtain

$$m^\mu m_\mu = n^\mu n_\mu = 0, \quad (12)$$

hence l^μ, n^μ, and m^μ all are null basis. Equation (8) is the metric of the charged sphere in the representation of vector coordinate frame.

4 Analytic Continuation of Coordinate r

When we make an analytic continuation of the coordinate r to the complex space, the null basis l^μ, n^μ, m^μ will be rewritten as

$$l^\mu = \delta_1^\mu, \quad m^\mu = \frac{1}{\sqrt{2\bar{r}}}\Big[\delta_2^\mu + \frac{i}{\sin\theta}\delta_3^\mu\Big],$$

$$n^\mu = \delta_0^\mu - \frac{1}{2}\Big[1 - \Big(\frac{1}{r}+\frac{1}{\bar{r}}\Big)m + \frac{m^2}{r\bar{r}} + \frac{q^2}{r\bar{r}}\Big]\delta_1^\mu. \quad (13)$$

Perform the following coordinate transformation:

$$r' = r + ia\cos\theta, \quad u' = u - ia\cos\theta, \quad (14)$$

or equivalently

$$dr' = dr - ia\sin\theta d\theta,$$

$$du' = du + ia\sin\theta d\theta, \quad (15)$$

where a is the angular momentum per unit mass , i.e. the transformation matrix is

$$(\alpha) = \begin{pmatrix} 1 & 0 & ia\sin\theta & 0 \\ 0 & 1 & -ia\sin\theta & 0 \\ 0 & 0 & 1 & 0 \\ 0 & 0 & 0 & 1 \end{pmatrix}. \quad (16)$$

Since l^μ, n^μ, m^μ are 4-vectors in coordinate space, hence from Eq. (16) we readily get their values after the above-mentioned coordinate transformation as follows:

$$l'^\mu = \delta_1^\mu, \quad m'^\mu = \frac{ia\sin\theta(\delta_0^\mu - \delta_1^\mu) + \delta_2^\mu + (i/\sin\theta)\delta_3^\mu}{\sqrt{2}(r' + ia\cos\theta)},$$

$$n'^\mu = \delta_0^\mu - \Big[\frac{1}{2} - \frac{mr' - m^2/2 - q^2/2}{r'^2 + a^2\cos^2\theta}\Big]\delta_1^\mu. \quad (17)$$

From Eq. (17) we readily get

$$g'^{\mu\nu} = \lambda_{(a)}'^\mu \tau_{(u)}'^\nu = l'^\mu n'^\nu + n'^\mu l'^\nu - m'^\mu \bar{m}'^\nu - \bar{m}'^\mu m'^\nu = \begin{pmatrix} -\gamma a^2\sin^2\theta & \gamma(r^2+a^2) & 0 & -\gamma a \\ \gamma(r^2+a^2) & -\gamma[(r-m)^2 + q^2 + a^2] & 0 & \gamma a \\ 0 & 0 & -\gamma & 0 \\ -\gamma a & \gamma a & 0 & -\gamma\sin^{-2}\theta \end{pmatrix}, \quad (18)$$

where $\gamma \equiv (r^2 + a^2\cos^2\theta)^{-1}$. From Eq. (18) we obtain

$$g'_{\mu\nu} = \begin{pmatrix} 1 - \gamma(2mr - m^2 - q^2) & 1 & 0 & \gamma(2mr - m^2 - q^2)a\sin^2\theta \\ 1 & 0 & 0 & -a\sin^2\theta \\ 0 & 0 & -\gamma^{-1} & 0 \\ \gamma(2mr - m^2 - q^2)a\sin^2\theta & -a\sin^2\theta & 0 & -\sin^2\theta[r^2 + a^2 + \frac{(2mr-m^2-q^2)a^2\sin^2\theta}{r^2+a^2\cos^2\theta}] \end{pmatrix}. \quad (19)$$

When we let

$$\Delta^* = (r-m)^2 + q^2 + a^2, \quad \rho^2 = r^2 + a^2\cos^2\theta, \quad (20)$$

equation (19) can be written as

$$g'_{\mu\nu} = \begin{pmatrix} \frac{\Delta^*}{\rho^2} - \frac{a^2\sin^2\theta}{\rho^2} & 1 & 0 & \frac{a\sin^2\theta}{\rho^2}(r^2+a^2-\Delta^*) \\ 1 & 0 & 0 & -a\sin^2\theta \\ 0 & 0 & -\rho^2 & 0 \\ \frac{a\sin^2\theta}{\rho^2}(r^2+a^2-\Delta^*) & -a\sin^2\theta & 0 & \frac{\sin^2\theta}{\rho^2}[\Delta^*a^2\sin^2\theta-(r^2+a^2)^2] \end{pmatrix}. \tag{21}$$

Performing the following coordinate transformation:[4,5]

$$r'=r,\quad \theta'=\theta,\quad \mathrm{d}\phi'=\mathrm{d}\phi-\frac{a}{\Delta^*},\quad \mathrm{d}u'=\mathrm{d}t-\frac{r^2+a^2}{\Delta^*}\mathrm{d}r. \tag{22}$$

Then from Eqs. (21) and (22) we get

$$\begin{aligned}
\mathrm{d}s^2 = g'_{\mu\nu}\mathrm{d}x'^\mu\mathrm{d}x'^\nu &= \left(\frac{\Delta^*}{\rho^2}-\frac{a^2\sin^2\theta}{\rho^2}\right)\mathrm{d}u'^2 + 2\mathrm{d}u'\mathrm{d}r + \frac{2a\sin^2\theta}{\rho^2}(r^2+a^2-\Delta^*)\mathrm{d}u'\mathrm{d}\phi' \\
&\quad - 2a\sin^2\theta\,\mathrm{d}\phi'\mathrm{d}r + \frac{\sin^2\theta}{\rho^2}[\Delta^*a^2\sin^2\theta-(r^2+a^2)^2]\mathrm{d}\phi'^2 - \rho^2\mathrm{d}\theta^2 \\
&= \left(\frac{\Delta^*}{\rho^2}-\frac{a^2\sin^2\theta}{\rho^2}\right)\mathrm{d}t^2 - \frac{\rho^2}{\Delta^*}\mathrm{d}r^2 + \frac{\sin^2\theta}{\rho^2}[\Delta^*a^2\sin^2\theta-(r^2+a^2)^2]\mathrm{d}\phi^2 - \rho^2\mathrm{d}\theta^2 \\
&\quad + \frac{2a\sin^2\theta}{\rho^2}(r^2+a^2-\Delta^*)\mathrm{d}t\mathrm{d}\phi.
\end{aligned} \tag{23}$$

Substituting Eq. (20) into Eq. (23) we obtain

$$\begin{aligned}
\mathrm{d}s^2 &= \frac{(r-m)^2+q^2+a^2\cos^2\theta}{r^2+a^2\cos^2\theta}\mathrm{d}t^2 - \frac{r^2+a^2\cos^2\theta}{(r-m)^2+q^2+a^2}\mathrm{d}r^2 - \left[(r^2+a^2)\sin^2\theta+\frac{(2mr-m^2-q^2)a^2\sin^4\theta}{r^2+a^2\cos^2\theta}\right]\mathrm{d}\phi^2 \\
&\quad + \frac{2(2mr-m^2-q^2)a\sin^2\theta}{r^2+a^2\cos^2\theta}\mathrm{d}t\mathrm{d}\phi - (r^2+a^2\cos^2\theta)\mathrm{d}\theta^2.
\end{aligned} \tag{24}$$

Equation (24) can be written as

$$\mathrm{d}s^2 = \frac{\Delta^*}{\rho^2}(\mathrm{d}t-a\sin^2\theta\mathrm{d}\phi)^2 - \frac{\sin^2\theta}{\rho^2}[(r^2+a^2)\mathrm{d}\phi-a\mathrm{d}t]^2 - \frac{\rho^2}{\Delta^*}\mathrm{d}r^2 - \rho^2\mathrm{d}\theta^2. \tag{25}$$

It is verified by rather tedious calculations that equation (25) is the correct solution of the field equations in VGM: $R_{\mu\nu}=-8\pi G(T_{\mu\nu}+T_{\mu\nu(gr)})$. Hence, using the method of analytic continuation of the coordinate r and the coordinate transformation (22), we successfully obtain the required metric of the rotating charged spherical mass in the vacuum. Apparently equation (25) is of the same form as the Kerr–Newmann geometry written in the t, r, θ, ϕ coordinates of Boyer and Linquist.[4] The only difference is that the value of Δ^* is different from Δ. Once more we notice the interesting fact that Δ^* can be simply obtained from Δ by the replacement of $G \to G^* = G(1-GM/2r)$, i.e. $(r^2-2mr) \to (r-m)^2$.

References

[1] S.W. Qian, Commun. Theor. Phys. (Beijing, China) **41** (2004) 377.

[2] S.W. Qian, Commun. Theor. Phys. (Beijing, China) **43** (2005) 1045.

[3] Hans Stephani, *General Relativity*, English ed., Cambridge University Press, Cambridge (1982).

[4] C.W. Misner, K.S. Thorne, and J.A. Wheeler, *Gravitation*, W.H. Freeman & Company, San Fransisco (1973).

[5] Eric Poisson, *A Relativist's Toolkit — The Mathematics of Black-Hole Mechanics*, Cambridge University Press, Cambridge, UK (2004).

Some Interesting Features for External Region of Spherical Symmetric Mass in New Theory of Gravitation VGM

Some Interesting Features for External Region of Spherical Symmetric Mass in New Theory of Gravitation VGM

QIAN Shang-Wu

Physics Department, Peking University, Beijing 100871, China

(Received January 16, 2006)

Abstract *This paper briefly discusses some interesting features for the external region of the spherical symmetric mass in the new theory of gravitation VGM, i.e. the theory of gravitation by considering the vector graviton field and the metric field, such as pseudo-singularity, curvature tensor, static limit, event horizon, and the radial motion of a particle. All these features are different from the corresponding features obtained from general relativity.*

PACS numbers: 04.20.-q, 04.20.Cv, 04.50.+h

Key words: pseudo-singularity, curvature tensor, static limit, event horizon, radial motion of the particle

1 Introduction

From the new theory of gravitation,[1−3] i.e. the theory of gravitation by considering the vector graviton field and the metric field (VGM), we have obtained external solution of a rotating charged spherical body,

$$ds^2 = \frac{(r-m)^2 + q^2 + a^2\cos^2\theta}{r^2 + a^2\cos^2\theta}dt^2$$
$$- \frac{r^2 + a^2\cos^2\theta}{(r-m)^2 + q^2 + a^2}dr^2$$
$$- \left[(r^2+a^2)\sin^2\theta + \frac{(2mr - m^2 - q^2)a^2\sin^4\theta}{r^2 + a^2\cos^2\theta}\right]d\phi^2$$
$$+ \frac{2(2mr - m^2 - q^2)a\sin^2\theta}{r^2 + a^2\cos^2\theta}dtd\phi$$
$$- (r^2 + a^2\cos^2\theta)d\theta^2 , \tag{1}$$

where $q = \sqrt{G}Q$ and Q is the charge of the spherical body, $m = GM$ and M is the total mass of the spherical body, a is the angular momentum per unit mass. In this paper we shall further discuss the properties of external spacetime of such a rotating charged spherical mass.

2 Pseudo-singularity

When $q = a = 0$, i.e. for an uncharged static spherical body, equation (1) reduces to

$$ds^2 = \left(1 - \frac{GM}{r}\right)^2 dt^2 - \frac{dr^2}{[1 - (GM/r)]^2} - r^2 d\theta^2$$
$$- r^2\sin^2\theta d\phi^2 . \tag{2}$$

Only in this particular case, there exists pseudo-singularity $r = r_c = GM$. Using the method of introducing new coordinates which is similar to Kruscal coordinates,[4] we can remove this singularity when the Schwartzschild coordinates (t, r, θ, ϕ) are transformed to Kruskal-like coordinates (u, v, θ, ϕ) such that

$$ds^2 = f^2(u, v)(dv^2 - du^2) - r^2(d\theta^2 + \sin^2\theta d\phi^2) , \tag{3}$$

where $f(u, v)$ is a function of the variables u and v. In order to compare with the corresponding results in general relativity (GR) we introduce the coordinate transformation $\rho = r - GM$, then equation (2) can be transformed to confomal metric

$$ds^2 = \left(1 + \frac{GM}{\rho}\right)^{-2}dt^2 - \left(1 + \frac{GM}{\rho}\right)^2(d\rho^2 + \rho^2 d\theta^2$$
$$+ \rho^2\sin^2\theta d\phi^2) . \tag{4}$$

Using Eddington and Robertson's expansion formula,[5]

$$ds^2 = \left[1 - 2\alpha\frac{GM}{\rho} + 2\beta\left(\frac{GM}{\rho}\right)^2 + \cdots\right]dt^2$$
$$- \left(1 + 2\gamma\frac{GM}{\rho} + \cdots\right)$$
$$\times (d\rho^2 + \rho^2 d\theta^2 + \rho^2\sin^2\theta d\phi^2) , \tag{5}$$

we readily see that for GR, $\alpha = \beta = \gamma = 1$, whereas for VGM $\alpha = \gamma = 1$ and $\beta = 3/2$. This result explains why to the precision of the second order of GM/r, the results obtained by using VGM are different from those by using GR. Pavelle[6] pointed out that the necessary and sufficient condition for the tenability of Birkhoff's theorem is that the energy-momentum tensor is static and diagonal. Since $T^{\mu\nu}_{(\text{gr})}$ in the external space is static and diagonal, hence in VGM the Birkhoff's theorem is also tenable.

3 Curvature Tensor

The existence of vector graviton field influences the value of the metric tensor, and hence the value of the curvature tensor, which is different from that in GR. For example, we consider the static uncharged spherical mass, i.e. Eq. (2). From this equation we readily calculate the following nonzero components of the curvature tensor. For convenience of comparison with those by using GR, we list the corresponding values by using GR in the square bracket,

$$R^0_{101} = \frac{2GM[r - (3GM/2)]}{(r^2 - GMr)^2} , \quad \left[\frac{2GM}{r^2(r - 2GMr)}\right],$$

$$R^0_{202} = -\frac{GM}{r}\left(1 - \frac{GM}{r}\right) , \quad \left[-\frac{GM}{r}\right],$$

$$R^0_{303} = -\frac{GM}{r}\left(1 - \frac{GM}{r}\right)\sin^2\theta , \quad \left[-\frac{GM}{r}\sin^2\theta\right],$$

$$R^1_{212} = -\frac{GM}{r}\left(1 - \frac{GM}{r}\right) , \quad \left[-\frac{GM}{r}\right],$$

$$R^1_{313} = -\frac{GM}{r}\left(1 - \frac{GM}{r}\right)\sin^2\theta , \quad \left[-\frac{GM}{r}\sin^2\theta\right],$$

$$R^2_{323} = \frac{2GM}{r}\left(1 - \frac{GM}{r}\right)\sin^2\theta , \quad \left[\frac{2GM}{r}\sin^2\theta\right]. \tag{6}$$

Since curvature tensor in VGM is different from that in GR, hence the geodesic deviation[4] will be different for VGM and GR. Theoretically we can use the experimental results of tidal forces to testify the above two theories.

4 Static Limit

The static limit (surface of infinite redshift) can be determined by the condition $g_{00} = 0$. From Eq. (1) we know that for the rotating charged spherical body,

$$g_{00} = \frac{(r - m)^2 + q^2 + a^2 \cos^2 \theta}{r^2 + a^2 \cos^2 \theta}. \tag{7}$$

We readily see that when $q = a = 0$, $r = m = GM$ the condition $g_{00} = 0$ is satisfied, i.e. for static uncharged spherical body there exists static limit, which is the surface $r = GM$. Otherwise only for $q = 0$ and $\theta = \pi/2$, i.e. only for uncharged spherical body there exists static limit $r = GM$, which is a circle in the equatorial plane.

In our previous paper,[1] we have already mentioned that in the case of uncharged static mass g_{00} and g_{11} do not change their sign when they cross the surface of infinite redshift, i.e. when they cross the static limit. Therefore, for VGM in this case, t is always a timelike coordinate, and r is always a spacelike coordinate, they cannot change their role when they cross the static limit. This unchangeability of the nature of timelike coordinate and spacelike coordinate still exists in the case of rotating charged spherical body. From Eq. (1) we know

$$g_{00} = \frac{(r - m)^2 + q^2 + a^2 \cos^2 \theta}{r^2 + a^2 \cos^2 \theta} \geq 0, \tag{8}$$

which cannot change its sign, i.e. from positive to negative, no matter how the value of r changes. Similarly

$$g_{11} = -\frac{r^2 + a^2 \cos^2 \theta}{(r - m)^2 + q^2 + a^2} \leq 0, \tag{9}$$

which cannot change its sign, i.e. from negative to positive, no matter how the value of r changes.

5 Event Horizon

Equation of event horizon $f(x^\mu) = 0$ satisfies the null supersurface condition[7]

$$g^{\mu\nu} \frac{\partial f}{\partial x^\mu} \frac{\partial f}{\partial x^\nu} = 0. \tag{10}$$

Because of the spherical symmetry in this case, the equation of horizon can be written as $f(r, \theta) = 0$. Let $f(r, \theta) = R(r)\Theta(\theta)$, then from Eqs. (1) and (10) we get

$$[(r - m)^2 + q^2 + a^2]\Big(\frac{1}{R}\frac{dR}{dr}\Big)^2 = 0. \tag{11}$$

Since for rotating spherical mass $a^2 > 0$, equation (11) cannot be satisfied, hence event horizon cannot exist in this case. Similarly for charged sphere $q^2 > 0$, event horizon cannot exist. Only when $q = a = 0$, $r = m = GM$ can the condition (11) be satisfied, but as we already explained in Ref. [1], since the unchangeability of the nature

of timelike coordinate and spacelike coordinate, such surface is not a one-way membrane, hence it is only a static limit, not an event horizon.

6 Radial Motion of Particle

The equation of motion of a particle in gravitational field is

$$\frac{d^2 x^\mu}{d\tau^2} + \Gamma^\mu_{\nu\lambda} \frac{dx^\nu}{d\tau} \frac{dx^\lambda}{d\tau} = 0. \tag{12}$$

For the radial motion of a particle in the gravitational field of the spherical symmetric static mass, equation (12) reduces to

$$\ddot{r} + \tfrac{1}{2} N' e^{N-L} \dot{t}^2 + \tfrac{1}{2} L' \dot{r}^2 = 0, \quad \ddot{t} + N' \dot{r} \dot{t} = 0, \tag{13}$$

where

$$e^N = e^{-L} = A = \big(1 - GM/r\big)^2, \quad N' = dN/dr,$$
$$\dot{r} = dr/d\tau, \quad \dot{t} = dt/d\tau. \tag{14}$$

From the second equation of Eq. (13) we have

$$\dot{t} = B e^{-N} = B/A = B/[1 - (GM/r)]^2, \tag{15}$$

where B is the integration constant. Substituting Eq. (15) into the first equation of Eq. (13), after some tedious calculations, finally we obtain

$$\frac{dV}{dt} = \frac{A'}{2A}(3V^2 - A^2), \tag{16}$$

where $A' = dA/dt$, and $V = dr/dt$ is the radial velocity of the particle. From Eqs. (14) and (16) we readily know that in VGM the equation of radial motion of a particle in the gravitational field of the spherical symmetric static mass is

$$\frac{dV}{dt} = -\frac{GM}{r^2}\Big[\Big(1 - \frac{GM}{r}\Big)^4 - 3V^2\Big]\Big(1 - \frac{GM}{r}\Big)^{-1}, \tag{17}$$

where as in GR, $A = 1 - (2GM/r)$, the corresponding equation of motion is

$$\frac{dV}{dt} = -\frac{GM}{r^2}\Big[\Big(1 - \frac{2GM}{r}\Big)^2 - 3V^2\Big]\Big(1 - \frac{2GM}{r}\Big)^{-1}. \tag{18}$$

In the case of weak field and low velocity, both equations (17) and (18) reduce to Newton's equation of motion

$$dV/dt = -GM/r^2. \tag{19}$$

Of course, equation (19) is the desired result.

We have briefly discussed some interesting features in the external region for the spherical symmetric mass in the new theory of gravitation VGM, such as pseudo singularity, curvature tensor, static limit, event horizon, and the radial motion of a particle. All these features are different from the corresponding features obtained from general relativity, in principle these differences can be testified by observations or experimental results.

References

[1] S.W. Qian, Commun. Theor. Phys. (Beijing, China) **41** (2004) 377.

[2] S.W. Qian, Commun. Theor. Phys. (Beijing, China) **43** (2005) 1045.

[3] S.W. Qian, Z.Z. Zhong, and Z.Y. Gu, Commun. Theor. Phys. (Beijing, China) **44** (2005) 855.

[4] H.C. Ohanion, *Gravitation and Spacetime*, W.W. Norton & Company, New York (1976).

[5] H.P. Robertson, *Space Age Astronomy*, Academic Press, New York (1962) p. 228.

[6] R. Pavelle, Phys. Rev. D **19** (1979) 2876.

[7] R.M. Wald, *General Relativity*, University of Chicago Press, Chicago and London (1984).

Bibliography

[1] S.-W. Qian and Z.-Y. Gu, On the semiclassical propagator for any one-dimensional admissible potential, *Ann. Phys.* **250**, pp. 420–432 (1996).

[2] S.-W. Qian, G.-Q. Xie and Z.-Y. Gu, Separation of variables treatment of the invariant quadratic in momentum for any admissible potential, *Ann. Phys.* **266**, pp. 497–502 (1998).

[3] D. W. MacArthur, Special relativity: Understanding experimental tests and formulations, *Phys. Rev. A* **33**, p. 1 (1986).

[4] C. M. Will, General relativity at 75: How right was Einstein? *Science* **250**, pp. 770–776 (1990).

Index